SpringerWienNewYork

CISM COURSES AND LECTURES

The series presents lecture notes, monographs, edited works and proceedings in the field of Mechanics, Engineering, Computer Science and Applied Mathematics.
Purpose of the series is to make known in the international scientific and technical community results obtained in some of the activities organized by CISM, the International Centre for Mechanical Sciences.

INTERNATIONAL CENTRE FOR MECHANICAL SCIENCES

COURSES AND LECTURES - No. 506

SPORT AERODYNAMICS

EDITED BY

HELGE NØRSTRUD

NORWEGIAN UNIVERSITY OF SCIENCE AND TECHNOLOGY
TRONDHEIM, NORWAY

SpringerWienNewYork

This volume contains 195 illustrations

Printed in Italy

ISBN 978-3-211-99945-5
e-ISBN 978-3-211-89297-8

PREFACE

Sport aerodynamics constitutes the science of aerodynamics coupled to the human activity of sports, i.e. the biomechanics of the human body under the influence of aerodynamic forces. It also encompasses the use of various equipments (or aids) in performing the individual sport activity. The aerodynamic interaction often implies the task to minimize a drag force which must be overcome by the human power output or a gravity force acting on the body. In the areas of, say, soccer or ski jumping the lift force also plays an important role. Bicycling is a further example where sport equipment is essential in performing the art of exposing the human body to the air environment. Hence, three fundamental areas of subjects will be covered in the book such as

A. *Basic aerodynamics (lift, drag, friction etc.)*
B. *Basic biomechanics (sport medicine, performance analysis etc.)*
C. *Sport equipment design (suits, helmets, shoes etc.)*

In order to pay attention to the environmental influence of the individual sport activities, the book will be divided in the following subdivisions:

- *Track running (human power, heat balance)*
- *Ice skating (suits, medical issues)*
- *Cross-country skiing (flat terrain, uphill, downhill)*
- *Ball aerodynamics (tennis, soccer, golf, cricket and baseballs)*
- *Ski jumping (suits, ski equipment)*
- *Downhill and speed skiing (terminal velocity)*
- *Bicycling (equipment, group cycling)*

The underlying physical phenomena of the sport activities listed above will be throughly discussed together with the basic equations and the physical quantities involved. Since the theme "Sport Aerodynamic" spans a wide variety of fluid mechanical and biomechanical disciplines, extensive theoretical exposition will be limited. However, many examples will be given with numerical solutions. The reader will also be guided into the literature which exists for the various topics discussed, so she or he can go into a deeper study of the particular sport activity at wish.

Helge Nørstrud

CONTENTS

CONTENTS

Basic Aerodynamics

Helge Nørstrud
Department of Energy and Process Engineering,
Norwegian University of Science and Technology,
Trondheim, Norway

1 Introduction

Aerodynamics is basicly the pressure interaction between a body and the surrounding air, see Figure 1. The body can be stationary in a flowfield, i.e. in a windtunnel (Figure 2) or the body moves in still or unsteady air (Figure 3). In sport aerodynamics the athlete will encounter various forces and a graphical overview is given in Figure 4.

Figure 1. The megaliner Airbus A 380 (Photo: Airbus).

To analyse an athlete and/or his equipment in performing the sport activity under consideration we have three methods at hand, i.e. either to make an experimental test or to perform a theory and finally to adopt the computer for solving the underlying equations, see Figure 5.

Figure 2. A speed-skier in a windtunnel at sea level measures a drag area of 0.13 m^2 and will experience a wind force of 6.26 kg at a windspeed of 100 km/h (gale/storm on the Beaufort scale). If the skier stands in an upright position the drag area will be 5 times larger and, hence, the wind force will be 31.3 kg (Photo: NTNU).

Figure 3. Ingard Strand (NOR) jumps from Strandkolvet in Norway (1300 m altitude) and reaches a velocity of 200 km/h (Photo: Magnus Knutsen Bjørke).

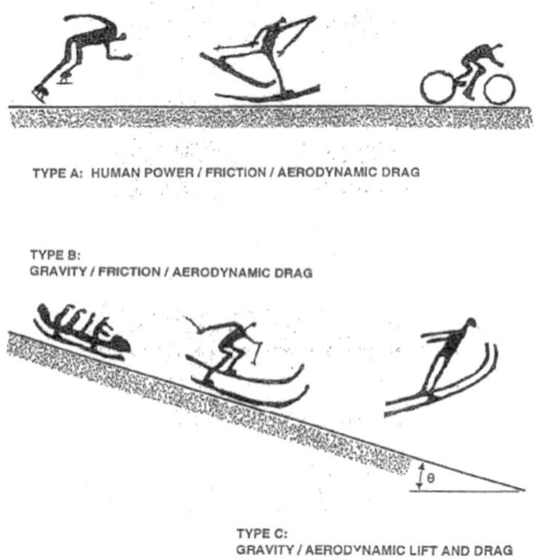

Figure 4. Sport aerodynamics and the forces involved.

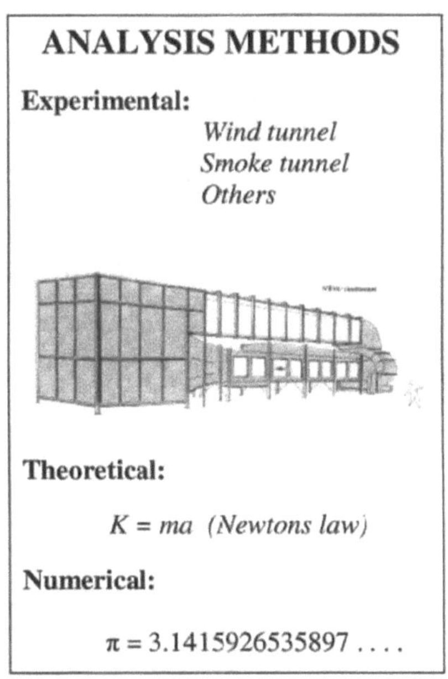

Figure 5. The three basic methods of analysis in aerodynamics.

2 Fundamental definitions

In integrating (or summing up) the steady and static pressure field p [Pa] $= p(x, y, z)$ over the body under investigation we will obtain the resulting aerodynamic force acting on the body. This force will be divided into two components, i.e. firstly the drag force D [N] acting parallell to the free stream velocity (e.g. the wind tunnel velocity) or the velocity of the body. The other component is the lift force L [N] acting normal to the drag force and these components are expressed as

$$D = \frac{1}{2}\rho_\infty C_D A U^2 \tag{1}$$

and

$$L = \frac{1}{2}\rho_\infty C_L A U^2 \tag{2}$$

Here ρ_∞ [kg/m^3] denotes the air density in the free stream, A [m^2] is a reference area and U [m/s] is the velocity in the wind tunnel or of the body. The two dimensional coefficients $C_D[-]$ and $C_L[-]$ are respectively the drag coefficient and the lift coefficient. The reference area A is the frontal area of the body in the case of drag evaluation, but for lifting bodies (like an airplane) the reference area is the lifting area (or the wing area for an airplane). The product $C_D A$ [m^2] is referred to as the drag area and is an indirect measure of the drag D, see also Figure 2. Since A is a fixed geometric value the coefficient C_D plays an important role in reducing the drag. Some values for C_D with reference to the frontal area are given in Figure 6.

Another dimensionless coefficient often used is the pressure coefficient $C_p[-]$ defined as

$$C_p = (p - p_\infty)/q_\infty \tag{3}$$

Here the lower index ∞ refers to the free stream and $q_\infty = 1/2\rho_\infty U_\infty^2$ is the dynamic pressure. The static pressure p acts normal to the body surface and gives rise to the socalled pressure drag on the body. The shear stress tangential to the body surface is responsible for the viscous drag.

Induced drag denotes the third part of the total drag and is related to the lift produced by a body, see Figure 7.

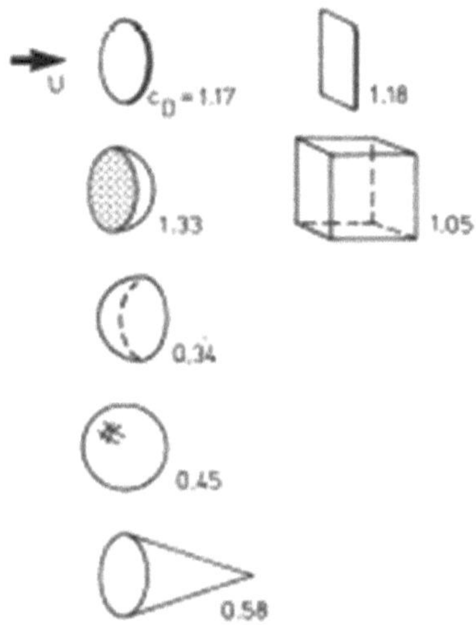

Figure 6. Drag coefficient for various bodies.

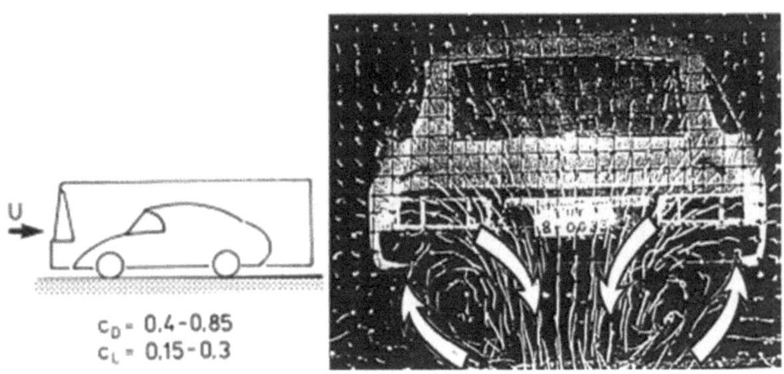

Figure 7. Vortical flow behind a lifting car (Photo: Unknown).

A body (like a car, an airplane or a ski jumper) generates lift due to the pressure differences above and under the body and trailing vortices will appear in the wake. This is visualized with a flow screen in a wind tunnel (Figure 7).

3 Similarity parameters

The most important dimensionless parameter in aerodynamics is the Reynolds number $Re[-]$ defined as

$$Re_\ell = U\ell/\nu \tag{4}$$

and describes the ratio between the inertia force and the viscous force. Here U is the flow velocity, ℓ [m] is a characteristic length and ν [m^2/s] is the kinematic viscosity which for air is equal to the value 0.000015. The critical Reynolds number designates a value where e.g. the drag of a body changes very rapidly.

Another similarity parameter is the Mach number $M[-]$ which is the ratio between the local flow velocity u [m/s] and the speed of sound c [m/s], i.e.

$$M = u/c \tag{5}$$

For $M < 0.3$ the flow can be regarded as incompressible and at a velocity of $u = 70$ m/s ($= 252$ km/h) at sea level, where $c = 343$ m/s; the Mach number will be $M = 0.2$.

4 The earth atmosphere

Since air is the key medium in sport aerodynamics we will introduce the following picture of Mount Everest (see Figure 8).

The atmospheric data below shows the air pressure and density from sea level up to the altitude of $z = 9000$ m.

It is convenient to adopt the approximation for an isothermal atmosphere, i.e.

$$\rho(z) = \rho_0 \exp(-\beta z) \tag{6}$$

where the air density $\rho(z)$ is defined through the value at sea level $\rho_0 = 1.225$ kg/m^3, the altitude z [m] and the scaling factor $\beta = 0.0001064$ m^{-1}.

Figure 8. Mount Everest (Photo: Internet).

Table 1. Meteorological data taken from the U.S. Standard Atmosphere, 1962.

GEOMETRIC ALTITUDE, z [m]	AIR PRESSURE, p[mb]	PRESSURE RATIO, $p/p_0[-]$	**AIR DENSITY,** ρ **[kg/m^3]**	DENSITY RATIO, $\rho/\rho_0[-]$
0	1013.25	1.0	**1.2250**	1.0
200	989.45	0.977	**1.2017**	0.981
400	966.11	0.953	**1.1786**	0.962
600	943.22	0.931	**1.1560**	0.944
800	920.78	0.909	**1.1337**	0.925
1000	898.76	0.887	**1.1117**	0.907
3000	701.21	0.692	**0.9093**	0.742
5000	540.48	0.533	**0.7364**	0.601
7000	411.05	0.406	**0.5900**	0.482
9000	308.01	0.304	**0.4671**	0.381

5 Acknowledgement

The author wants to thank very much Mr. Ørjan Sakariassen for his valuable and friendly help in converting the lecture notes (written in Word) to the required LaTeX system.

Factors influencing on running

Lars Sætran
Norwegian University of Science and Technology N-7491 Trondheim, Norway

1 VO2max

Running economy is important, but can it be altered? It seems that persons beginning exercise definitely become more efficient with training, as do persons who are already trained but who continue heavy training. Conley et al. (1981) followed a single runner during 6 months of interval training and found that the subject's running efficiency improved by between 9 and 16% at three different speeds. However, his body weight also fell about 6%, which could have been the more important factor explaining the improved running efficiency. Subsequently these authors showed that the running efficiency of Steve Scott, Americas premier 1500 m runner of the early 1980s, improved with interval running. Svedenhaug and Sjodin (1985) showed that the running efficiencies of a group of elite Swedish distance runners improved between 1 and 4% during the course of one year, changes that where in the range of those measured in the adolescent runners is studied by Daniels and Oldridge (1971). Svedhaug and Sjodin speculated that the continual improvements in the running performances of these Swedish athletes were due to slowly progressive improvements in their running efficiencies rather than to increase in the VO2max values, which were relatively fixed, increasing only during that phase of the season when the athletes were performing high-intensity interval-type training.

Athletes appear to choose stride lengths at which they are most efficient, that is, at which oxygen uptake is the least (Cavanagh and Williams 1982). When forced to take either shorter or longer strides but to maintain the same running pace, athletes become less efficient and require an increased oxygen uptake. With training, runners increase the length of their strides and reduce their stride frequency (Nelson and Gregor 1976). Some researchers believe that this optimises running efficiency because increasing stride length is more economical than increasing stride frequency.

Although VO2max values differ between the sexes, gender has no effects on running efficiency, trained men and women are equally efficient

(Maughan and Leiper 1983). Race may influence running efficiency, researchers have found that Asians and Africans utilise 17% less energy than Europeans when lying, sitting, or standing, but no studies have compared energy uses of these groups during exercise. In a study of elite runners of different racial groups, researchers found no race-related differences in running economies (Noakes 1991).

2 Extra weight

Clothing weight is another factor that can influence an athletes efficiency. Stevens (1983) calculated the effect of the weight of clothing on marathon racing performance. He found that the typical nylon vest and shorts worn by marathon runners weighted 150 g, 100% cotton shorts and vest weighted 234 g, and the heavy tracksuits weighted 985 g. Stevens calculated that changing from nylon to cotton clothing would increase a world-class runners marathon time by about 13 seconds and an average 3:40 marathoners time by about 23 seconds. Running in a full tracksuit should increase the average runners marathon time by about four minutes.

However, laboratory experiments do not necessarily substantiate these calculations. Cureton et al (1978) found that the addition of up to 5% of body weight to the torso increased the oxygen cost of running by only about 2.5%. Extrapolation these data suggest that the addition of even 1 kg of extra weight to the torso in the form of clothing would increase the oxygen cost of running by less than 0.5

Extra weights added to their legs or feet appear to have a far greater effect on the running economy.Martin (1985) found that the addition of 0.5 kg to each thigh or to each foot increased the oxygen cost of running by 3.5 and 7.2%, respectively, values considerably higher than those found by Cureton et al (1978). A number of other studies show that the addition of 1 kg to the feet increases the oxygen cost of running by between 6 and 10%, or about 1% per 100 g increase in the weight of footwear. The increase is the same in men and women.

Clearly, a 1% savings in energy expenditure during a standard marathon race, for example, is not inconsiderable, if translated directly into a 1% improvement in performance it would mean a savings of 77 seconds at world-record marathon pace, equivalent to a sub 2:07 standard marathon. But we have yet to prove that these energy savings will cause an equivalent improvement in running performance.

In-shoe orthotics used in the treatment of a number of running injuries will increase shoe weight and therefore might influence running economy adversely. In the study of Burkett et al (1985), the addition to an 80-g orthotic device to each running shoe increased the oxygen cost of running by about 1.4%, smaller increases, 0.4 to 1.1%, were reported by Berg and Sady (1985). These studies indicate that the added weight of the orthotic device decreases running economy in direct proportion to its weight.

Work at the Nike Sport Research Laboratory has shown that the air pockets used in the midsole of different Nike air running shoes reduces the oxygen cost of running by 1.6 to 2.8% at a running speed of 16 km/hr. If these savings directly translate into equivalent improvements in racing performance, then they are significant, at least for the top athletes. Further research is needed to study this possibility.

3 Running surface, gradient and wind speed

Obviously, prevailing conditions such as running surface, gradient, and wind speed and direction will have considerable effects on the runners economy. The influence of the running surface on the oxygen cost of running was first noted by Passmore and Durnin (1955), who reported that the oxygen cost of walking across a plowed field was 35% greater than the cost of walking at the same speed on a smooth, firm surface. Running on sand has a similar effect (Wyngand et al 1985). McMahon and Green (1979) suggested that optimising the spring constant on a running track will likely improve running performance and running economy (and reduce injury risk).

One of the first scientists to study the influence of wind speeds on running performance was G. Pugh, whose work on the effects of altitude on athletic performance is among the classic contributions on the topic. Pugh performed four different studies designed to measure how wind speed and the gradient of the running surface influence the oxygen cost of running (Pugh 1970). His studies showed that the extra cost of running into a facing wind increased as the square of the wind speed. Thus the oxygen cost of running into a 66-km/h headwind increases by 30 ml/kg/min. Similarly, running up an 8% incline increases the oxygen cost of running by about 20 ml/kg/min.

The figure indicates that for each 1 km/hr increase in running speed, the oxygen cost increases about 4 ml/kg/min. Thus, the increased oxygen cost of 30 ml/kg/min caused by running into a 66 km/hr wind would cause a 7.5 km/hr were reduction in running speed. Similarly, an 8% gradient would slow the runner by 5 km/hr.

Pugh also showed that at the speeds at which middle-distance track events are run (6 m/s or about 67 seconds per 400 m), about 8% of the runner's energy is used in overcoming air resistance. But by running directly behind a leading runner (or drafting) at the distance of about 1 m, their

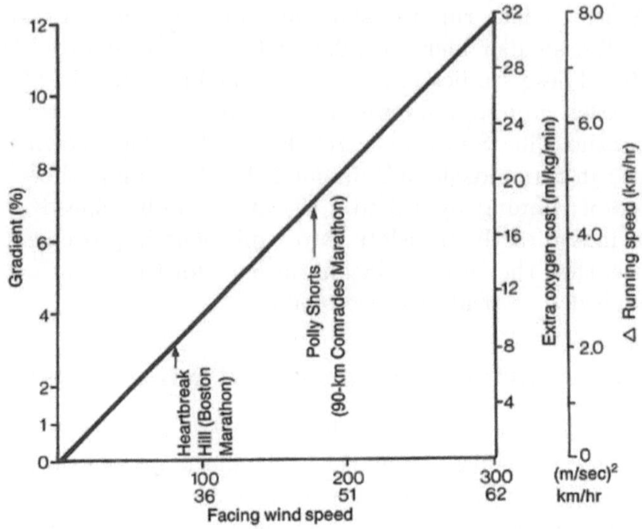

Figure 1. Extra cost of oxygen for running uphill and in a facing wind

athlete can save 80% of that energy. In a middle-distance race this would be equivalent to a savings of about 4 seconds per lap. However, Pugh considers it unlikely that in practice the following athletes would ever be able to run as close to the lead runner to benefits to this extent. By running slightly to the side of the lead runner, the following runner would probably benefit by about 1 second per lap (Pugh 1970).

Another researcher to study the benefits of drafting was C. Kyle (1979). His calculations suggest that at word-record mile pace a runner running 2 m behind the lead runner would save about 1.66 seconds per lap, which generally confirmed Pughs estimations. Kyle calculated that the benefits of drafting in cycling are much greater than in the running, some 30% or more. In addition, the larger the group and the farther from the front the cyclist rides, the more the cyclist benefits.

In contrast, the aerodynamic drag is increased when runners are positioned abreast because the larger frontal area results in a larger sheared drag (Bassett et al 1987).

These findings explain why track athletes find pacers to be such essential ingredients in aiming for world track records. In addition, these findings explain why the world records in the sprints are set at altitude. During sprinting, the energy cost of overcoming air resistance rises to between 13 and 16% of the total cost of running. Thus, the sprinter benefits greatly by running at an altitude where air resistance is considerably reduced. It is interesting that when the runner is racing on a circular track and optimum strategy is to accelerate into the wind and to decelerate when the wind is from behind, the opposite of what one would expect (Hastell 1974).

M. Davies (1980) extended Pughs findings. Davies used essentially the same techniques as Pugh but included observations on the effects of downhill running and of following winds of different speeds.

Davis found that when the runner was measured on the treadmill, facing winds of up to 18 km/hr had no effect of the oxygen cost of running, but that the same conditions on the rode will have a very marked effect. On the treadmill, the athlete does not move forward and thus does not expend energy overcoming air resistance. However, an athlete who runs on the rode into a wind of 18 km/hr faces an extra wind speed equal to that of his or her running speed plus that of the prevailing wind.

The practical relevance of this is that on a calm day, anyone running slower than 18 km/hr will not benefit by drafting in the wake of other runners. However, runners stand to gain considerably by drafting when running at faster speeds or when running into winds that, when added to the running speeds, would make the actual wind speed greater than 18 km per hour.

Of course, the world marathon record is run at a faster pace than 18 km/hr. This means that athletes intend on setting world records would be well advised to draft for as much of the race as possible. Front running in the marathon is always as wasteful of energy as is front running on the track. One can only assume that as the runners begin to realise this fact, we shall see pacesetters in marathon races just as we have them in track races.

The only way besides drafting to reduce wind resistance is to run with a following wind, the speed of which is at least equal to that of the runner. Davies calculated that under these circumstances, the removal of energy required to overcome wind resistance at world marathon pace (19.91 km/hr) would increase their running speed by about 0.82 km/hr, equivalent to a reduction in the racing time of about 5 minutes. Similarly, drafting in a tightly knit bunch for the entire race would reduce air resistance by about 80%, allowing the runner to run about 4 minutes faster.

Davies found that the effects of a tail wind on the oxygen cost of running was about half that of a facing wind. Thus, a following wind of 19.8 km/hr is of little assistance to runners running slower than 18 km/hr, but a following wind of 19.8 km/hr would assist in a world marathon record attempt to the extent of a 0.5 km/hr increase in speed. Higher following wind speeds of 35 to 66 km/hr would improve running speeds by 1.5 to 4 km/hr.

At higher facing wind speeds, the oxygen cost of running increases enormously. Wind speeds of 35 km/hr would reduce running speeds by about 2.5 km/hr, speeds of 60 km/hr by about 8 km/hr.

Finally, Davies calculated the additional oxygen cost of running uphill and the energy savings by running downhill. He found that the energy savings during downhill running equaled only half of the energy that would be lost when running on an equivalent uphill gradient. Uphill running increased the energy costs by about 2.6 ml/kg/min for each 1% increase in gradient. This is roughly equivalent to a reduction in running speed of about 0.65 km/hr. Downhill running was associated with a reduction in the oxygen cost of running by about 1.5 ml/kg/min for each 1% gradients, equivalent to an increase in speed of about 0.35 km/hr.

The practical value of this information is twofold. First, it indicates that time lost going uphill can never be fully regained by running an identical downhill gradient. Second, the data shown in Figure A can be used to estimate how much time you can expect to lose or gain on the particular section of a race (if you know the gradient of that section).

4 Aerodynamics

Kyle (1986) studied the aerodynamic drag effects of athletic clothing and showed that the following factors increased the aerodynamic drag experienced by the runner: shoes with exposed laces 0.5%hair on limbs 0.6%, long socks 0.9%, short hair 4%, loosely fitting clothing 4.2% and long hair 6%. He also calculated that by reducing aerodynamic drag as little as 2%, equivalent to a haircut, a runner would reduce his or her running time over 100 m by 0.01 seconds and in the standard marathon by 5.7 seconds. Even better results could be achieved by running in a custom-fitted speed suit with a tight-fitting hood to cover their hair and ears. Such a suit made of polyurethane-coated, stretchable nylon reduces aerodynamic drag by smoothing the airflow around the streamlined areas of the chin, ears, and hair, and by eliminating the flapping of loose clothing. Calculations suggest that wearing such clothing would reduce running time in the 100 m race by 0.284 seconds (3%) and by 1:34.50 (1%) in the standard marathon (Bassett et al 1987). Unfortunately, this clothing is impracticable for marathon runners because it's streamlining prevents heat loss. The first attempt to use the streamlined hood in Olympic relay competition had a disastrous results - the 1988 United States Olympic Games 100 m relay team was disqualified when one runner received the baton outside the legal zone because he was unable to hear the approach of the other runner!

5 Controlling body temperature

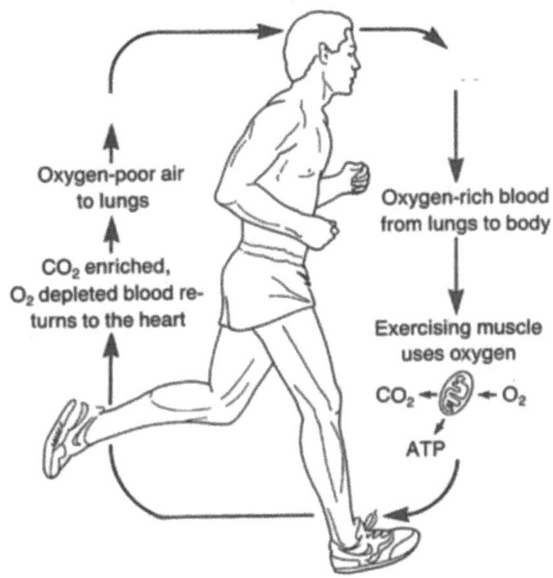

Figure 2. A runner's metabolism

A runner faces a major problem: the excess heat produced by muscle contraction. Humans are homeotherms, and to live they must keep their body temperature within a narrow range (35 to 42°C) despite wide variations in environmental temperatures and differences in levels of physical activity.

During exercise the conversion of chemical energy stored in ATP (adenosine triphosphate) into mechanical energy is extremely inefficient, so that as much as 70% of the total chemical energy used during muscle contraction is released as heat rather than as athletic endeavor.

Thus, when Don Ritchie wins an ultramarathon at an average pace of 16.3 km/hr he utilises about 56 kJ of energy every minute or 18,482 kJ in the 5 1/2 hours that he runs. Of this, only 5942 kJ help transport the athlete from the start to the finish of the race, the remaining 12,540 kJ are

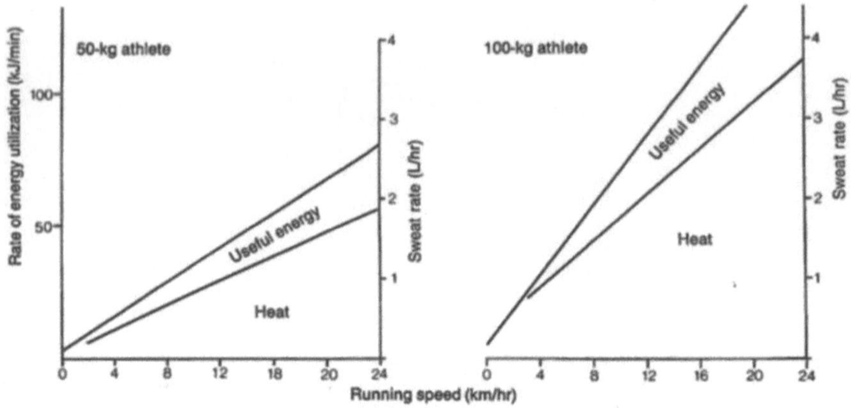

Figure 3. Energy

nothing more than a hindrance, as they serve only to heat the athlete. Were the athlete unable to lose that heat, his body temperature would rise above 43°C, causing heatstroke.

To prevent disastrous overheating and heatstroke and to control the increased heat associated with exercise, the body must be able to call upon and number of very effective heat-losing mechanisms.

6 Mechanisms for heat loss during exercise

As exercise begin, the blood flow to the muscles increases. Not only does the heart pump more blood, but blood is preferentially diverted away from nonessential organs and toward the working muscles and skin. As blood passes through the muscles, it is heated, and it distributes the added heat throughout the body, particularly to the skin. In this manner, as well as by direct transfer from muscles lying close to the skin, heat is conducted to the skin surface. Here, circulating air currents convect this heat away, and any nearby objects whose surface temperature is lower than the skin temperature attract this heat, which travels by electromagnetic waves in the form of energy transfer known as a radiation.

In another method of heat loss, surface heat evaporates the sweat produced by the sweat glands in the skin. Sweating itself does not lose heat,

heat is lost only when the sweat actually evaporates. The efficiencies of all these mechanisms depend on a variety of factors, most of which are open to modification by the athlete.

7 The exercise intensity

As the intensity of exercise increases, the body must decide whether to pump more blood to the muscles to maintain their increased energy requirements or to assist heat dissipation by increasing the skin blood flow. Faced with conflicting demands, the body always favors an increased blood flow to the muscles. The result is that while body heat production is increased, the ability to lose that heat is decreased.

It appears that athletes running at word-record speeds, at least in races up to 16 km, develop marked limitations to skin blood flow and therefore have limited abilities to lose heat. They thus run in micro-environments in which their abilities to maintain heat equilibrium depend entirely on the prevailing environmental conditions (Pugh 1972). If these conditions are unfavourable, the athletes will continually accumulate heat until their body temperatures reach the critical level at which heatstroke occurs. Two famous athletes running at word-record pace who developed heatstroke when forced to race in unfavourable environmental conditions were Jim Peters in the 1964 Empire Games Marathon and Alberto Salazar in the 1980 Falmouth 12 km Road Race.

8 Environmental factors affecting heat loss

The air temperature and wind speed determine the amount of heat that can be lost from the skin by convection, which is the heating of the surrounding air by the skin.

High facing-wind speeds cause a large volume of unwarmed air to cross the skin in unit time and therefore allow for greater heat loss by convection. Running itself produces an effective wind speed that aids convective heat loss but may not be sufficient to increase heat loss adequately in severe environmental conditions. Obviously, a wind coming from behind the runner at the same speed that the runner is moving forward will cause him or her to run into a totally windless environment, which will prevent convective heat loss. In contrast, the wind speed developed by cyclists appear to be sufficient to compensate even for severe conditions, which explains why heat is not as great a problem for endurance cyclists as it is for runners.

At rest the body skin temperature is about 33°C. If you exercise in environmental temperatures greater than 33°C, heat cannot be lost by convection, because the air temperature is higher than that of the body surface. In this case the direction of heat transfer is reversed, and the superficial tissue gain heat from the environment. In these conditions, the only avenue for heat loss is by sweating. Sweating removes 1092 to 2520 kJ of heat per litre of sweat evaporated depending on whether the sweat evaporates or, as usually happens, a large percentage drips from the body without evaporating. As the air humidity increases, the bodys ability to lose heat by this mechanism decreases.

The body can absorb additional heat from the environment, in particular from the sun. The body temperature is cooler than that of the sun, thus the body will absorb radiant energy from the sun. Obviously, the amount of radiant energy to which the athlete is exposed is greater when there is no cloud cover and least when cloud cover is absolute.

These three environmental factors that determine the athlete's ability to lose heat - wind speed, the humidity and temperature of the air, and the radiant energy load - are measured in the wet bulb globe temperature (WBGT) index.

The WBGT index integrates the measurement of radiant energy (as the temperature of a black globe - the globe temperature) with the wet bulb temperature (measured by the thermometer covered by a wick permeated by water). The difference between the wet bulb temperature and the prevailing air (dry bulb) temperature is a measure of the humidity of the air and therefore is also a measure of the ease with which sweat will evaporate from the athlete. Furthermore, wind blowing over the wet wick of the wet bulb thermometer will increase the rate of evaporate cooling and the therefore lower the wet bulb temperature. In this way, the prevailing wind speed also influences the WBGT.

9 Dehydration

With sweating, fluid is removed from the body causing dehydration, which may be compounded by vomiting and diarrhea. Whyndham and Strydom (1969) drew attention to what they believed to be the dangers of dehydration in predisposing the athletes to heatstroke. They studied runners in a 32 km road race and showed that the body temperature of athletes who became dehydrated by more than 3% of their body weight approached values previously recorded only in victims of heatstroke. In addition, they found that athletes weighing 70 kg or more who had not drunk during the race incurred 5% water deficits and had markedly elevated body temperatures after the race.

The conclusion that these authors drew is now believed to be incorrect

(Noakes 1991). In particular, it seems that dehydration is not the most important factor determining the body temperature during exercise. Nevertheless, the findings of Whyndham and Strydom (1969) drew attention to the potential dangers of the International Amateur Athletic Federation rule number 165:5, which stipulated that marathon runners could not drink fluids before the 11th kilometre mark of the standard marathon and thereafter could drink only every 5 km. This ruling discouraged marathon runners from drinking during races and promoted the idea that drinking during running was unnecessary and a sign of weakness. This rule was eventually repealed.

That early marathon runners were not used to drinking fluids regularly during races is shown by the trivial amounts drunk by Athur Newton during his races. Jim Peters (1957) described the conventional wisdom in the following statement about marathon racing:

There is no need to take any solid food at all and every effort should also be made to do without liquid, as the moment food or drink is taken, the body has to start dealing with the digestion and in so doing some discomfort will almost invariably be felt.

Although dehydration is not the critical factor predisposing athletes to heatstroke during exercise, marked dehydration does have detrimental effects. Skin blood flow is reduced, and body heat storage (and therefore body temperature) is increased by dehydration. However, the effects are somewhat less dramatic than generally believed (Noakes 1991).

10 Clothing

Apart from aesthetic reasons, the reason for wearing clothing is to trap a thin layer of air next to the body. Because air is a poor conductor of heat, this thin layer rapidly heats to body temperature and act as an insulator preventing heat loss. Clearly, any clothing that is worn during exercise in the heat must be designed for the opposite effect - to promote heat loss.

Marathon runners have learned that the light, porous clothing such as fish net vests best achieve this heat loss. In contrast, T-shirts or heavy sweatsuits, particularly when soggy with sweat, become very good insulators, preventing adequate heat loss.

Novice runners, particularly those who might consider themselves overweight, often train in full tracksuits in the heat. Many neophyte athletes probably believe that the more they sweat, the harder they must be exercising and therefore the greater the weight they stand to lose. The unfortunate truth is that in running, the energy cost is related only to the distance run. Thus, to lose more weight, one needs to run a greater distance.

Excessive sweating will effect a sudden loss of weight by dehydrating the body, this is the procedure used by boxers, jockeys, and wrestlers in making the weight. By exercising in the heat for as little as half an hour, one can lose as much as 1 kg, but this is a fluid loss that will be rapidly replaced if the athlete rehydrates by drinking. In contrast, to lose a real kilogram of body weight one must expend about 37500 kJ of energy, equivalent to running about 160 km.

A runner who lives in a moderate climate seldom (if ever) needs to train for any period in a track suit. By doing so, the runner merely increases discomfort and promotes conditions favourable for heatstroke.

11 Heat acclimatization

When athletes who have trained exclusively in cool weather are suddenly confronted with hot, humid conditions, they suffer immediate and dramatic impairments in performance. However, with perseverance and continued training in the heat, performance soon improves, returning to normal within a short period.

The process underlying this adaptation is termed Heat Acclimatization. It begins after the first exposure to exercise in the heat, progress is rapid, and is fully developed after 7 to 10 days. Only by exercising in the heat can one become acclimatized to the heat. The optimum method for achieving this is to train daily in the heat for periods of 2 to 4 hours for 10 days. Once established, heat acclimatization is fully retained for about two weeks. Thereafter, it is lost at rates that vary among individuals. It is best retained by those who stay in good physical condition and who re-expose themselves to exercise in the heat at least every two weeks.

Important changes occur with heat acclimatization: Heart rate, body temperature, and sweat salt (sodium chloride) content during exercise decreases, whereas sweating rate increases due to increased secretory capacity of the sweat glands. In addition, metabolic rates and the rate of muscle and blood lactate accumulation are decreased by heat acclimatization (Young et al 1985), as is the rate of muscle glycogen utilisation.

Heat acclimatization confers considerable protection from heat injury. Equally important, in competition in the heat, the heat-acclimatized athlete will always have the edge over an equally fit, but unacclimatized opponent.

12 Sponging

As the skin temperature rises, it causes blood to pool in the veins of the arms and legs. This is because the elevated skin temperature paralyses the veins, which dilate and soon fill with a large volume of blood.

This blood is effectively lost from the circulation and can only be returned to the circulation if the skin temperature is again lowered. This can be achieved by Sponging (this is literally wetting the skin with a sponge).

A recent study confirmed that wetting the skin did indeed lower the skin temperature during exercise but did not aid heat loss (Bassett et al 1987). Thus the benefits of sponging during exercise probably relate to its effect on the central circulation.

13 Factors explaining impaired running performance in the heat

Anyone who has to run a marathon or longer race in the heat knows that such races are much more difficult than are races of the same distance run in cool conditions. The most likely explanation for this comes from recent studies showing that pre-cooling of the body or of the active muscles either prior to or during exercise prolongs endurance time to exhaustion in both dogs and humans (Kozlowski et al 1985). This cooling keeps the muscle temperature lower during subsequent exercise and alters the metabolic response by decreasing the rate of muscle glycogen utilisation, muscle lactate accumulation, and the fall in muscle high-energy phosphate contents, thereby allowing the cooler muscle to exercise for longer. By inference we may conclude that the sustained elevation of muscle temperature that occurs during prolonged exercise may be one of the most important factors limiting endurance performance.

14 Calculating sweat rate during exercise

To calculate your sweat rate or to figure how much your current drinking pattern during races falls short of replacing your sweat losses, you could try the following experiment.

Weigh yourself (naked) on the scale reading in kilograms, immediately before (WB) and immediately after (WA) a run in conditions and at a race pace to which you are accustomed. Measure carefully the total amount of fluid (F) in litres that you ingest while running. You can then calculate your sweat rate fairly accurately.

$$SweatRate\,(L/hr) = \frac{(WB - WA) + F}{RunningTime(hours)} \tag{1}$$

Your fluid replacement will have been adequate if, after races longer than 30 km, you have lost less than 2 to 3 kg and are not dehydrated by more than 3%, calculated by this equation:

$$Dehydration\,(\%) = \frac{WB - WA}{WB} x100 \qquad (2)$$

Bibliography

Conley, Krabenbuhl, Burkett. 1981. *The Physician and Sportsmedicine* 9, 107-115

Svedenhaug, Sjodin. 1985. *Canadian J of Applied Sport Sciences* 10, 127-133

Daniels, Oldridge. 1971. *Medicine and Science in Sports* 3, 161-165

Cavenagh, Williams. 1982. *Medicine and Science in Sports and Exercise* 14, 30-35

Nelson, Gregor. 1976. *Research Quarterly* 47, 417-428

Maughan, Leiper. 1983. *European J of Applied Physiology* 52, 80-87

Noakes. 1991. *The Lore of Running.* Leisure Press

Stevens. 1983. *J of Sports Medicine and Physical Fitness* 23, 185-190

Cureton, Sparling, Evans, Johnson, Kong, Purvis. 1978. *Medicine and Science in Sports* 10, 194-199

Martin. 1985. *Medicine and Science in Sports and Exercise* 17, 427-433

Burkett, Kohrt, Buchbinder. 1985. *Medicine and Science in Sports and Exercise* 17, 158-163

Berg, Sady. 1985. *Research Quarterly* 56, 86-89

Passmore, Durnin. 1955. *Physiological Reviews* 35, 801-836

Wyngand, Otto, Smith, Perez. 1985. *Medicine and Science in Sports and Exercise* 17, 237

McMahon, Greene. 1979. J of Biomechanics 12, 893-904

Pugh. 1970. *J of Physiology* 207, 823-835

Kyle. 1979. *Ergonomics* 22, 387-397

Brownlie, Mekjavic, Gartshore, Mutch, Banister. 1987. *Annals of Physiology and Antrolopogy* 6. 133-143

Hastell. 1974. *Institute of Electrical and Electronic Engineers Transactions of Biomedical Engineering* 22, 428-429

Davies. 1980. *J of Applied Physiology* 48, 702-709

Kyle. 1986. *Scientific American* 254, 92-98

Pugh. 1972.*British J of Physical Education* 3, IX-XII

Wyndham, Strydom. 1969. South African Medical J 43, 893-896

Peters, Johnston, Edmunson. 1957. *Modern Middle and Long Distance Running*, Nicholas Kaye, London

Young, Sawka, Levine, Cadarette, Pandolf. 1985. *J of Applied Physiology* 59, 1929-1935

Bassett, Nagle, Mookerjee, Darr, Ng, Voss, Napp. 1987. *Medicine and Science in Sports and Exercise* 19, 28-32

Kozlowski, Brzezienska, Kruk, Kaciuba-Uscilko. 1985. *J Applied Physiology* 59, 766-773

Cycling Aerodynamics

Giuseppe Gibertini and Donato Grassi
Dipartimento di Ingegneria Aerospaziale, Politecnico di Milano, Milano, Italy

Abstract A general introduction presents the main concepts about biker and bicycle aerodynamics. A decription of the drag reduction problem is presented and the athlete position effects as well as the main bicycle components effects are examinated. Advices are proposed to improve performances (taking the international regulations as a constant reference).

1 Introduction

In the present treatise, among the severall possible items related to cycling and wind, the resistance to the bicycle and biker progression due to the relative wind (produced by the motion itself) will be the focal point.

Other possible items related to aerodynamics could be the effects of natural wind (cross-wind, favourable or contrary wind) as well as ventilation problems (see, for example, Bruühwiler et al. (2004). Among these, only cross-wind effects will be briefly described.

Aerodynamic resistance is a non negligible topic for many kind of bicycle competitions but it is the fundamental problem when the velocity is particularly high as in time trial competitions. For this reason the present treatise is focalized on time trial competitions.

2 The UCI regulations

The problem of bicycle aerodynamics is very complex as a great number of variables should be considered. In order to limit the degrees of freedom of the problem, only the solutions that comply the regulations of the Union Cyclist Interanationale (UCI) will be taken in account.

In the following some articles of the UCI Cycling Regulations (UCI, 2007) particularly related to aerodynamics argument will be cited and commented.

The article 1.3.006 says: 'The bicycle is a vehicle with two wheels of equal diameter. [...]'. This definitively excludes same exotic aero-bike (see for example Kyle (1991c) with the front wheel smaller than the rear one.

The article 1.3.013 says: ‘The peak of the saddle shall be a minimum of 5 *cm* to the rear of a vertical plane passing through the bottom bracket spindle. [...]’.

The article 1.3.022 says (the reference to specific diagrams are skipped in the following citation) : ‘In competitions other than those covered by article 1.3.023, only the traditional type of handlebars may be used. The point of support for the hand must be positioned in the area defined as follows: above, by the horizontal plane of the point of support of the saddle [...]’. While the article 1.3.023 says: ‘For road time trial competitions [...] an extension may be added to the steering system. The distance between the vertical line passing through the bottom bracket axle and the extremity of the handlebar may not exceed 75 *cm* [...]’. It’s important to outline, as done by UCI in a recent official dispatch, that the general indications of article 1.3.022 apply for time trial competitions too, except for what explicetely modified by the article 1.3.023, thus the upper limit for the handlebars position is the same for all the competitions. As it will be explained in the following, all together these three articles (1.3.013, 1.3.022 and 1.3.023) prohibit extreme positions, like Obree’s or ‘superman’ position, but allow for a quite aerodynamic arrangement (the so called ‘time trial position’) that, if accurately adjusted for each specific biker, leads to very good results in term of aerodynamic resistance (very close to the values obtainable with extreme positions).

3 The air resistance

The bicycle belongs to the group of vehicles that live the athlete body exposed to the wind. Furthermore the bike surface is rather small with respect to the biker surface and therefore the main part of the aerodynamic force acts on the athlete body whose position is, nevertheless, strongly related to the shape and the dimensions of the bicycle itself.

On the aerodynamic point of view the biker can be regarded as ‘bluff body’ and, generally speacking, the bike too can be considered bluff. The bluffness leads to the fact that the aerodynamic resistance is mainly pressure drag (instead of friction drag) and thus, on a very general point of view, it’s more important to reduce the frontal area than to reduce the wet area. An other general consideration is: as lift (positive or negative) is not required, it’s better to keep it as small as possible in order to avoid the production of induced drag.

Aerodynamic drag is essentially proportional to the square of the speed.

Usually it is expressed as:

$$D = \frac{1}{2}\rho V^2 S C_D \tag{1}$$

where S is a reference surface (usually, for bluff bodies, it's the projected frontal area) and C_D is the dimensionless drag coefficient. In aerodynamics drag is defined as the projection of the aerodynamic force along the direction of the relative wind. This means that if the relative wind is aligned with the bike (no matter if it's due to bike motion only or to natural wind too) the drag coincides with the aerodynamic force opposite to the bike motion (let's call it F_x) but in case of lateral wind the two concepts are not the same.

As the present treatise is essentially focalized on what produces resistance to the motion and adsorbs power from the biker, the F_x forced will usually be considered in the following (and therefore the C_x coeffcient) although sometime the term drag will be used for brevity when no risk of ambiguity is present.

For a certain bike and biker (in a specific position) the C_x coefficient is essentially a constant as it varies slightly with the velocity (with the Reynolds number to be more correct), thus aerodynamic resistance is essentially proportional to the square of the velocity and its importance grows more and more as the velocity increases. For this reason the drag reduction is very important in time trial races were the velocity is in the order of 14 m/s (about 50 km/h) and aerodynamic resistance is more than 90% of the total resistance. Following Kyle (1989) this can easily be estimated by the following equation:

$$R = gm(C_{rr_1} + C_{rr_2}V) + \frac{1}{2}\rho V^2 S C_x \tag{2}$$

where R is the total resistance, gm is the weight of rider and bike, C_{rr_1} is the static rolling resistance coefficient, C_{rr_2} is the dynamic rolling resistance (including wheel bearing losses and dynamic tire losses).

Typical values for the rolling resistance coefficients are $C_{rr_1} = 0.0023$ and $C_{rr_2} = 0.115 \times 10^{-4}$ s/m, while a mass of 75 kg and a drag area $SC_x = 0.24$ m^2 can be representative for a time trial biker.

With this values and a velocity of 50 km/h, corresponding to 13.89 m/s, a rolling resistance of 1.8 N is obtained and an aerodynamic resistance of 26.6 N that's the 94% of the total.

As already said, on the point of view of its aerodynamic drag the biker-bicycle system can be considered as a bluff body: the rider is rather obviously a bluff body but the bicycle too is essentially a non streamlined object except for some detail. In any case the athlete drag is more important of the bike drag (in the order of two thirds of the total amount (Kyle, 1989)), thus the biker position is the focus point for performances improving. Nevertheless, for extreme competitions, also a few percents of drag reduction can make the difference thus a good design of bike components can be important as well as helmet and dress choose.

Following Kyle (1989) it's possible to have an estimation of the time reduction due to the drag reduction in a time trial competition. The basic idea is to evaluate a typical value for the biker power simply multiplying a typical resistance for a typical mean velocity and keep the same power value to estimate a new mean velocity (and therefore a new time) with a different aerodynamic resistance. In the Table 1 the time reduction for some values of SC_x reduction are listed for three different race lenghts. The reference times and the biker powers for the three race lenghts (1 km, 4 km and 40 km) are computed on the base of resonable velocities: 57 km/h, 50 km/h and 48 km/h respectively. The table shows that also a small drag reduction can produce appreciable results. Of course this is a very rough model of the realty but neverless it gives an idea of the order of magnitude of the time gain.

Table 1. Time trial race time reduction due to drag reduction.

SC_x reduction $[m^2]$	Time reduction for 1 km race $[s]$	Time reduction for 4 km race $[s]$	Time reduction for 40 km race $[s]$
0.001	0.09	0.39	4.07
0.005	0.43	1.97	20.4
0.01	0.87	3.96	41.2
0.02	1.77	8.04	83.6

Aerodynamics is governed by non linear equation thus effects summation is not rigorously applicable. Nevertheless there is a need of separate the different effects in order to understand the main phenomena and in order to guide the optimization. For this reasons, althought it's not completely correct, the different part effects will be presented separately in the following.

4 Wind tunnel testing

The main way to study the cycling aerodynamic is the experimental testing in wind tunnel. A possible alternative to wind tunnel tests could be mea-

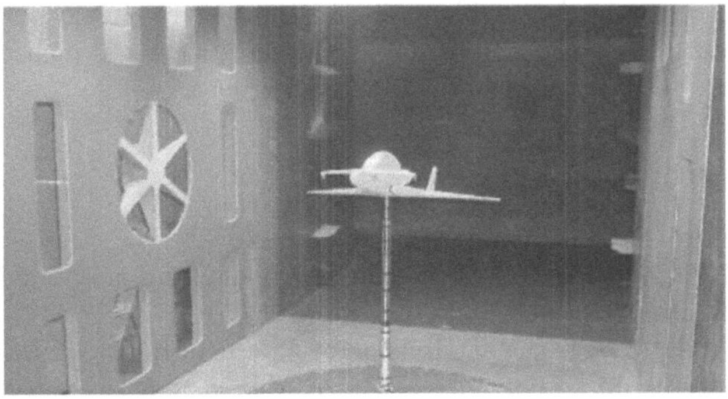

Figure 1. The GVPM aeronautical test section

surements in field conditions (as explained, for example, in Martin et al. (2006) or in the work of Grappe et al. (1997)) but, generally, the wind tunnel tests are more repetible and documentable.

In order to perform full-scale wind tunnel tests (allowing to involve the real athletes) large facilities are necessary. In fact it is well known (see for example Barlow et al. (1999)) that, in order to have realistic test conditions, the solid blockage, i.e. the ratio between the projected frontal area of the bicycle and biker combination and the sectional area of the test room should be lower than 10% (lower than 5% it'd be much better).

4.1 The large wind tunnel of Politecnico di Milano

The wind tunnel of Politecnico di Milano (GVPM) is a low speed large facility. Due to its size and its adjustable flow velocity , the aeronautical test section (see Figure 1 and Table 2) is well suitable for full scale testing in the field of sport research (for cycling, sled, etc.).

Table 2. GVPM aeronautical test section main data

Width	Height	Min velocity	Max velocity	Turbulence
4.02 m	3.84 m	10 km/h	200 km/h	$< 0.1\%$

The facility is equipped with a special roller system allows to make both the wheels spinning. A turning table on the floor allows for yawed conditions and therefore for cross flow cycling conditions. The tests are usually carried out with the real athletes inside the test room although this leads to a problem of test repeatibility and long and repeated measurements are necessary to get reliable results (Flanagan, 1996).

5 The rider

As said before, the biker body drag is dominant to the other effects (it was found by Kyle and Burke (1984) that the biker drag is about from 60% to 70% of the total amount).

As a matter of fact, the attribution of a percent quota of drag to the bike and the biker could be considered contestable, as the two effects cannot be really decoupled, and it's here considered nothing more than an indication.

The fact that the biker drag is the most part of the total drag is rather easily inferred if the body surface is compared with the bike one, both in terms of projected frontal area and in terms of wet area. Of course, as the human position is the focus, the bicycle dimensions (frame and handler dimensions) have to be considered as they strongly influence the biker body attitude: as a matter of fact, the bicycle is a very rigid constrain to the rider position and several strange bicycles have been developed in the past in order to keep the rider in a very aerodynamic position. Of course these new solutions have to take in account the roles specific of the considered competition.

5.1 The biker positions

Generally speaking the main way to reduce the drag is the reduction of the projected frontal area and this is essentially obtained keeping the body aligned with the wind as more as possible. That's the reason why, in the recent past, very good performances have been obtained with rather extreme positions with the biker body highly stretched forward with horizontal torso. Depending on the arms attitude these extreme biker positions can be essentially divided in two typologies: the first is the Obree's position with the arms drawn up below the chest; the second is the superman position with the arms stretched straight forward (Kyle and Weaver, 2004) as sketched in Figure 2.

Both these positions produce a large drag reduction, up to 30%, respect to the standard (drops) position (Bassett et al., 1999; Lukes et al., 2005).

But UCI decided to stop this rush to exotic solutions (see, for example, Broker et al. (1997)) fixing strict constraints to the bike shape in order to go back to more traditional cycling (as told before, it's the bike that mainly determines the biker position) and so both Obree's and superman position

The Obree's position. The superman position.

Figure 2. The extreme aero-positions.

are now banned. In fact, the Obree's position needs to raise the handlebars more than permitted by article 1.3.022 (see Section 2 on UCI regulations) and the superman position requires to protrude forward the hands more than permitted by article 1.3.023 (see again Section 2).

Two series of systematic tests have been carried out in GVPM facility, in order to evaluate the biker position effect. The first series of tests was carried out on January 2007 (Gibertini et al. (2008)) with a 1.75m tall man (in the following he's called "biker A") while the second tests series was carried out one year later (january 2008) with a taller 1.8m athlete (Biker B)

For this activities, a raised ground plane has been mounted in the test section slightly reducing its height from 3.84 m to 3.6 m. Nevertheless, as the biker and bike projected frontal area is about 0.3 m^2 the blokage is quite small (in the order of 2%).

The bicycle was fixed to a special strut with a six components balance below it. The strut is composed by a beam with a fork to fix the rear wheel axle and a couple of rollers (one under each wheel) connected by a belt in order to drive the front wheel too (Figure 3). The system provides an adjustable resistance torque to the rear roller producing a realistic rider effort and thus a realistic body attitude.

As the aim of the tests was the study of biker position a rather standard bike configuration has been used in both the two tests series: the frame was a rather traditional one (each athlete used it's own one) as well as the helmet. The clothes too were not extremely optimized (they were rather common ones) while a typical time trial wheels choose has been adopted (a lens disk

The bicycle fixed to the support

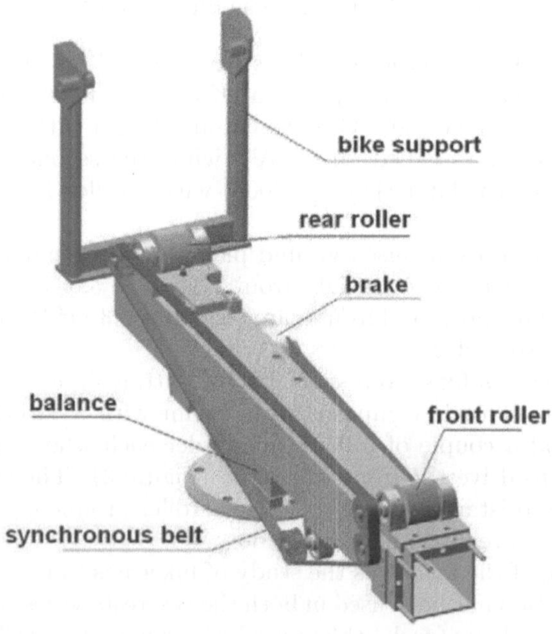

The support with the rollers and the balance

Figure 3. The system for cycling tests in GVPM wind tunnel

rear wheel and a 18 flat spokes aero-rim front wheel). The handlebars were traditional for the traditional positions while aero-bars have been used when time trial positions were tested.

Three typical traditional cycling positions have been tested (only two of them with biker A) while severall time trial positions (with different adjustement of handlebars) have been investigated with the biker A in order to identify the best position (i.e. the position with the smaller drag) and only a few slight variations around this optimal position have been tested with biker B. The three traditional positions were (following the nomenclature of Heil (2002)) the 'stem position' (upright torso position with the hands placed near the stem of the handlebars), the 'brake hoods position' (with the hands placed on the brake hoods as in the second sketch of Figure 4) and the 'drops position' (with the hands on the drops portion of the handlebar as in the third sketch of Figure 4). In Figure 5 the time trial position is sketched so that it can be compared with the other ones.

Figure 6, 7 and 8 shows the bikers inside the wind tunnel in the three traditional positions, while Figure 9 shows the optimal time trial position that produced a quite low drag area ($SC_x = 0.223\ m^2$ for biker A and $SC_x = 0.235\ m^2$ for biker B): the torso is horizontal to be aligned with the wind, the forearms are horizontal as well and the head is turned down (just maintaining a bit of forward visibility). In this attitude all the athlete body is pointed to reduce projected frontal area and to avoid to produce unnecessary lift. In Figure 10 the view from above shows the forearms kept attached each to the other.

A similar comparison between the fundamental biker positions has been presented by Grappe et al. (1997), based on measurements in field conditions. The athlete was 1.75m tall (as biker A) and the general bike configuration was the same as in the present GVPM tests (but they used less aerodynamic wheels). Of course there are a lot of details that can be different and the bikers antropometry was different (also for the 1.75m biker), but a comparison is nevertheless interesting (particolarly in terms of percent differences).

As can be seen in the Table 3 the values found for the drops position (that is assumed as reference position in the table) are quite similar and almost equal for the two less tall men (the larger drag of athlete B with respect to the other two ones is reasonably scaled with their dimensions). Considering all the tested positions they look in rather good agreement (in terms of percent difference respect to the reference drops position) with the exception of the time trial position for which the GVPM results show an higher drag reduction, in very good agreement between biker A and biker B. It has to be pointed out that the GVPM value is the result of a

The stem position. The brake hoods position.

The drops position.

Figure 4. The traditional positions.

careful adjustment that may be was not carried out by Grappe et al., as they were more focalized on Obree's position. Therefore this comparison is very interesting as it suggests that results not so far from the ones that can be obtained with Obree's position can be achieved with regular time trial position.

It can be interesting to examine how muche these drag differences depends on the different projected area and how much they depends on the differences in the drag coefficients. In Table 4 the estimated frontal projected areas and the related drag coefficients C_x are presented and percent difference are reported again with comparison to the drops position.

The tabulated C_x values, so close to 1, demonstrated that biker is really a bluff body as espected. Nevertheless some non-negligible differences between the drag coefficient are present and it's rather interesting (and a bit surprising) that the smallest value is associated to the brake position (that,

Figure 5. The time trial position.

Figure 6. The stem positions in the wind tunnel tests (biker B).

Biker A

Biker B

Figure 7. The brake hoods positions in the wind tunnel tests.

Biker A

Biker B

Figure 8. The drops positions in the wind tunnel tests.

Biker A

Biker B

Figure 9. The optimal time trial position.

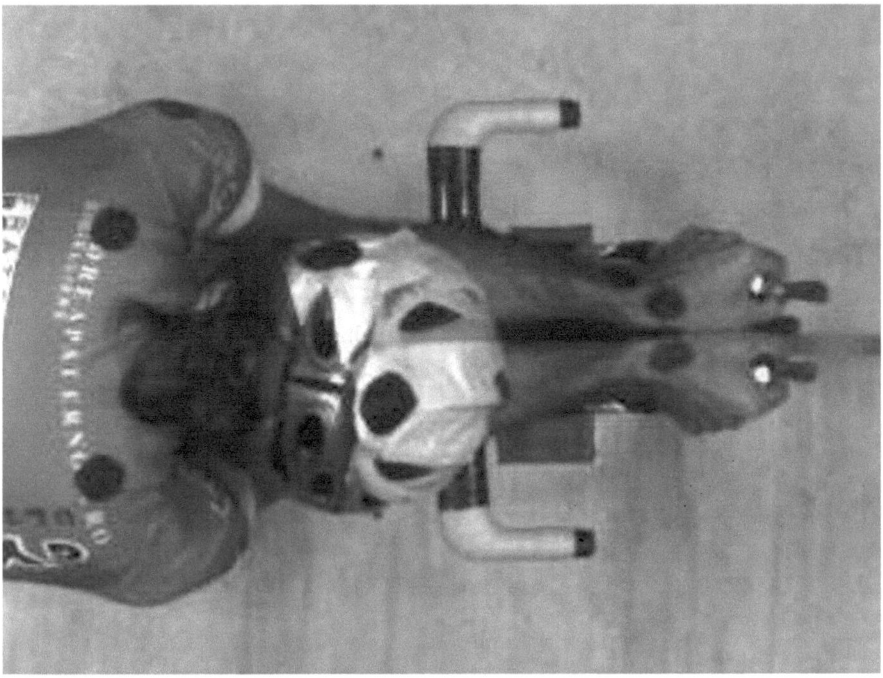

Figure 10. Time trial position seen from above: the detail of forearms

unfortunately, has the lergest area).

Of course, as already mentioned, these SC_x absolute values are specifically related to the specific athlete involved in the test: in the experience of the authors at GVPM, the best SC_x that can be obtained in time trial position varies from 0.20 m^2 to 0.29 m^2 depending on athlete size.

The head, the torso and, for a certain extent, the arms too, dont interact too much with the bicycle aerodynamics as they are rather above the bike structure but the same assertion is clearly not valid for the legs. Parker et al. (1996) have carried out a series of wind tunnel testing studying the interference between the legs and the bike frame. Two kind of frames has been used in the tests: a standard open frame and a particular (not UCI rules complying) closed aero-frame. They found that, for the standard frame, the best legs spacing is the standard spacing.

Table 3. Comparison between GVPM tests results and measurements in field conditions by Grappe et al. (1997).

Position	Grappe et al. (1997)		GVPM biker A		GVPM biker B	
	SC_x $[m^2]$	respect drops p.	SC_x $[m^2]$	respect drops p.	SC_x $[m^2]$	respect drops p.
Stem position	0.299	+8%			0.318	+10%
Brake hoods position			0.282	+3%	0.304	+5%
Drops position	0.276	0%	0.275	0%	0.289	0%
Time trial position	0.262	−5%	0.223	−19%	0.235	−19%
Obree's position	0.216	−22%				

Table 4. Projected frontal areas and drag coefficients for the different positions of biker B

Position	Projected frontal areas		Drag coefficient	
	S $[m^2]$	respect drops p.	C_x	respect drops p.
Stem position	0.386	+9%	0.824	+1%
Brake hoods position	0.400	+13%	0.760	−7%
Drops position	0.355	0%	0.814	0%
Time trial position	0.297	−16%	0.792	−3%

5.2 Ergonomics considerations

The present treatise doesn't undertake the complex problem of cycling ergonomics but some simple considerations can be made on the presented results. The proposed optimal time trial position is surely rather unconfortable and can jeopardize a good respiration so is reasonable to say that the remarcable advantage of this position would be really effective for short time-trial competition while a more confortable position should be considered in the case of a long race. As an example, a time-trial position essentially identical to the proposed optimal one but with the forearms slightly more separated can allow for a better respiration without to jeopardize too much the aerodynamic efficiency. In Figure 11 these two positions are showed from the front: the more "large" position produces an SC_x increase of 0.007 m^2 that corresponds to a 3%.

5.3 The head and the helmet

Beside the fact that the first aim of an helmet should be the head safety, it can be considered a sort of biker head fairing too. Actually, this was more true in the past (a streamlined helmet could reduce the SC_x, with respect to a traditional road helmet up to 0.02 m^2) but, nowadays, the more strict safety rules prescribes thicker helemets than, of course, have an higher SC_x (often not so different from the traditional ventilated helmet).

An interesting point is the head position. It's well known that some extreme aero-helmet that are designed for a specific head position work quite bad when the posture is not the required one, but this is not the only case where the head position is important. For example it can be interesting to observe the simple case of a biker wearing a traditional helmet: the two positions presented in Figure 12 have difference in the SC_x of about 0.002 m^2 (of course the less aerodynamic posture is the second one with the head turned up).

The “narrow” forearms posture. The “large” forearms posture.

Figure 11. The frontal view of two time-trial positions

6 The bicycle

6.1 The frame

The traditional frame is composed by round (or almost round) tube that are typically bluff bodies. Several attempts have been made to reduce the frame drag and actually from a traditional round tube frame to a very streamlined one (with airfoil shaped tube sections) the measured drag area (without anything but the frame itself) can decrease from a value around 0.05 m^2 to a value as small as 0.03 m^2. Unfortunately in the authors experience this rather valuable improvement is not replicated by the tests with the biker (that probably disturbs too much the flow around the frame, strongly reducing the effect of its sophisticated design).

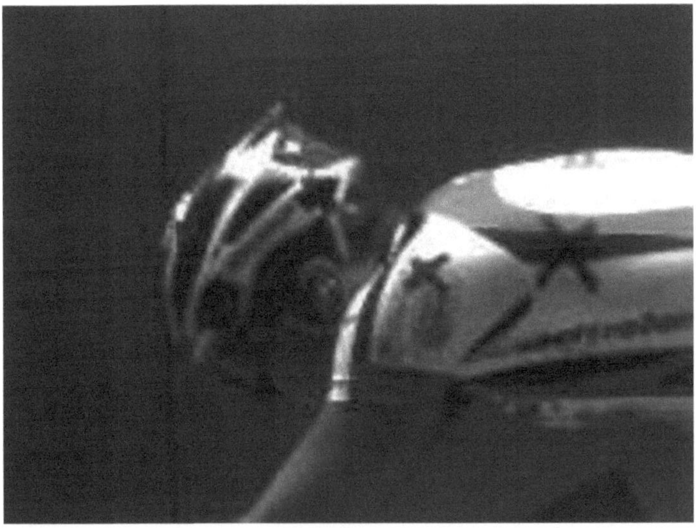

Head down

Head up

Figure 12. Comparison between two head postures.

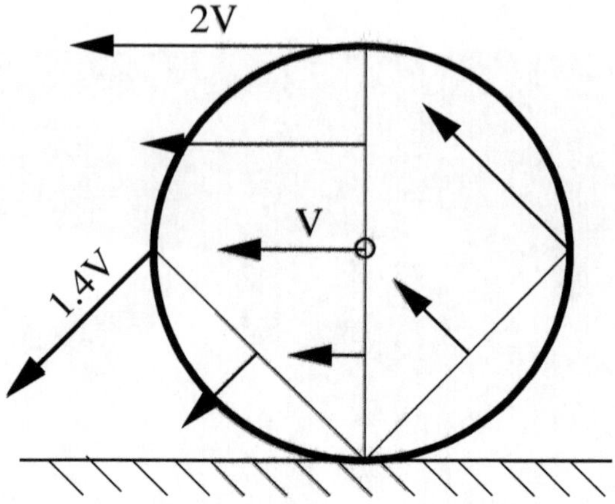

Figure 13. Rolling wheel velocities distribution

6.2 The wheels

The wheel aerodynamics has been widely study. As a matter of fact, although bicycle wheels have a very small frontal area, nevertheless they produce a non-negligible drag due to the fact that they have larger velocity above the axle (up to twice the bike velocity at wheel top as shown in Figure 13) and drag is proportional to the square of the speed (Sayers and Stanley, 1994).

Traditional wheels have a very poor aerodynamic as spokes, tire and rim section are all bluff bodies. Thus many solutions have been proposed in order to reduce the wake of tire and spokes. To reduce the wake of the tire, the rim is shaped in such a way the tire-rim section results to be more streamlined. On the other hand, to reduce the spoke energy dissipation their number is reduced and their section is aerodynamically shaped. The extreme results of this design philosophy are the three-spoke aero-wheels and the lens disk wheel. It's rather difficult to give definitive and general data about wheels drag as it can depends on small details of any commercial product. Generally speacking, tests carried out on isolated single spinning 700mm wheels (Kyle, 1990, 1991a) demostrades a drag area of about 0.024 m^2 for the traditional wheel (with 36 round spoke and standard rim) while much lower values have been found with the aerodynamic wheels, down to about 0.01 m^2 (with some lens-wheels or three-spoke wheels). In the cited tests, not only the air resistance force has been measured but also the aerodynamic resistant torque on the axle (this term too dissipates power, thus it has been reduced to an equivalent drag force and added to the real force in order to obtain a value that represents the total air resistance).

Actually a lot of model are proposed by manufacturers and is impossible to generalize and to say a priori which one is better. Furthermore the test of the wheel alone risks to be rather misleading as the important point is the whole bicycle behaviour. An important point is the effect of the wake produced by the biker legs and by the biker itself on the rear wheel (generally speacking, the wake is characterized buy a lot of turbulence and by a lower velocity).

A serious problem of disk wheels is their behaviour in cross-flow conditions (Tew and Sayers, 1999), particularly for the front wheel: the side aerodynamic force and the moment respect to the steering axis can produce severe driving difficulties (Suryanarayanan et al., 2002).

On the other hand, a problem with the three-spokes wheel is related to safety and up to now they are forbidden for massed start races (UCI, 2007).

In the experience of the authors, the lens disk is a very good choice for the rear wheel: due to its size with respect to the wake vortices size, the lens wheel has a stabilizing effect, while low solidity wheels (not disk) generally contribuites to turbulence increase and therefore produce a larger drag. On the contrary, the advantage of disk wheels is not so sensible in the front position when compared with other good aero-wheels (beside the fact that the front disk wheel is usually not used because it's very dangerouse in case of cross-wind).

Differently, comparing traditional rim wheels with low solidity aero-rim wheels, the larger difference is produced in the front position while the difference in the rear position is very lower. Two reasons can explain this fact: first of all, as already said, the velocity in the wake is lower and thus, in general, produces lower forces, then it has to be considered that the flow in the wake is quite irregular and probably the flow local velocities produce some separations also on the aero-rim.

Some tests were carried out at GVPM to measure the side-force at different crosswind velocities. In these tests only crossflow natural wind has been simulated as the longitudinal wind component was considered as due to the bike speed only and all the force cefficient were computed on the base of longitudinal dynamic pressure (i.e. computed on the base of the bike speed).

As an example, at 50 km/h with a 8.8 km/h crosswind (witch means a yaw angle of 10°) a side-force area SC_y from 0.31 m^2 to 0.45 m^2 was measured (depending on bike size and athlete size and position) when two lens wheels were used, while rather lower values (from 0.22 m^2 to 0.36 m^2) were

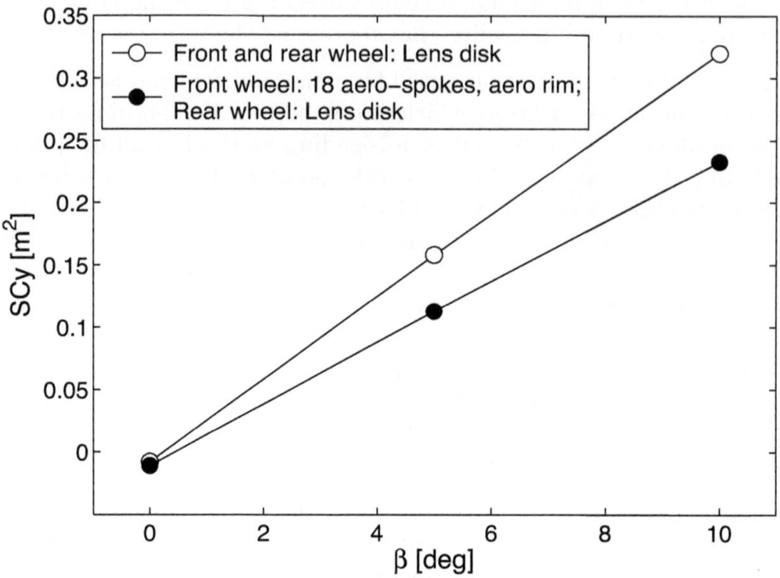

Figure 14. Measured SC_y for two different wheels combinations at different yaw angles.

obtained when the front lens wheel was substituted with a 18 flat spokes aero-rim wheel. Of course not all the side force is due to wheels but, in any case, the rather valuable difference between the two cited configurations (about 0.09 m^2) depends on the different front wheels.

A particular behaviour was measured by Tew and Sayers (1999) that registered a 'jump' in the lens disk wheel side force versus yaw angle: starting from zero yaw angle the side force grows linearly with the angle up to 5° then from 5° to 8° it decreases rapidly and then, starting from the 8° angle, it grows again with the initial slope. This rather surprising result was never repeated in GVPM measurements as can be seen, for example, in the graph of Figure 14 where a very regular and linear dependency of side force from yaw angle is clearly visible. These results refer to test carried out with the biker A in the traditional drops position (as can be seen in Figure 15). It's also interesting to observe that, in presence of lateral wind the front lens wheel produces a rather relevant "sail effect" (it's lift has a non-null component in the bike motion direction) that reduces the resistance. This is not, of course, the case of the spoked wheel (an holed sail is not a good

Test with lens disk front wheel.

Test with 18 spokes aero-rim wheel.

Figure 15. Crosswind tests at GVPM.

sail!). This effect was already observed by other authors (see, for example, Kyle (1991b))

7 Acknowledgements

The authors want to thank the bikers Enrico De Angeli (the biker A) and Simone Salet (the biker B) that produced the present results with their high quality performances. Thanks also to Luca Guercilena and Luca Minesso that supported the present activity with their advice and thanks to Gabriele Campanardi, Carlo Macchi and Alex Zanotelli for their technical collaboration.

Bibliography

J.B. Barlow, W.H Rae, and A.Pope. *Low-speed wind tunnel testing*. John Wiley & Sons, 1999.

D.R. Bassett, C.R. Kyle, L. Passfield, J.P. Broker, and E.R. Burke. Comparing cycling world hour records, 1967-1996: modeling with the empirical data. *Med. and Sci. in Sports and Exercise*, 31:1665–1676, 1999.

J.P. Broker, E. Knapp, and R. Shafer. Control of rider positions on bicycles: Proposed solutions at the 1997 uci technical meeting. *Cycling Science*, 8(1):7–13 and 24, 1997.

P.A. Bruühwiler, C. Ducas, R. Huber, and P.A. Bishop. Bicycle helmet ventilation and confort angle dependence. *Eur. J. Appl. Physiol.*, 92: 698–701, 2004.

M.J. Flanagan. Considerations for data quality and uncertainty in testing of bicycle aerodynamics. *Cycling Science*, 8(1):7–10 and 22–23, 1996.

G. Gibertini, G. Campanardi, D. Grassi, and C. Macchi. Aerodynamics of biker position. VI International Colloquium on: Bluff Bodies Aerodynamics and Applications, 2008.

F. Grappe, R. Candau, A. Belli, and J.D. Rouillon. Aerodynamic drag in field cycling with special reference to the obree's position. *Ergonomics*, 40:1299–1311, 1997.

D.P. Heil. Body mass scaling of projected frontal area in competitive cyclists not using aero-handlebars. *Eur. J. Appl. Physiol.*, 87:520–528, 2002.

C.R. Kyle. Wind tunnel tests of bicycle wheels & helmets. *Cycling Science*, 2(1):27–30, 1990.

C.R. Kyle. New aero wheel tests. *Cycling Science*, 3(1):27–30, 1991a.

C.R. Kyle. The effect of crosswind upon time trials. *Cycling Science*, 3(3 and 4):51–56, 1991b.

C.R. Kyle. Wind tunnel tests of aero bicycles. *Cycling Science*, 3(3 and 4): 57–61, 1991c.

C.R. Kyle. The aerodynamics of handlebars & helmets. *Cycling Science*, 1 (1):22–25, 1989.

C.R. Kyle and E.R. Burke. Improving the racing bicycle. *Mechanical Engineering*, 106:34–35, 1984.

C.R. Kyle and M.D. Weaver. Aerodynamics of human-powered vehicles. *Proc. Instn Mech. Engrs Part A: J. Power and Energy*, 218:141–154, 2004.

R.A. Lukes, S.B. Chin, and S.J. Haake. The understanding and development of cycling aerodynamics. *Sports Engineering*, 8:59–74, 2005.

J.C. Martin, A.S. Gardner, M. Barras, and D.T. Martin. Aerodynamic drag area of cyclists determined with field-based measures. *Sportscience*, 10: 68–69, 2006.

B.A. Parker, M.E. Franke, and A.W. Elledge, editors. *Bicycle Aerodynamics and Recent Testing*, AIAA-96-0557, 1996.

A.T. Sayers and P. Stanley. Drag force on rotating racing cycle wheels. *J. Wind Eng. Ind. Aerodyn.*, 53:431–440, 1994.

S. Suryanarayanan, M. Tomizuka, and M.Weaver. System dynamics and control of bicycles at high speeds. Proceedings of the American Control Conference, pages 845–850, 2002.

G.S. Tew and A.T. Sayers. Aerodynamics of yawed racing cycle wheels. *J. Wind Eng. Ind. Aerodyn.*, 82:209–222, 1999.

UCI. *UCI Cycling Regulations: Part 1 General Organization of Cycling as a Sport*, 2007.

Performance factors in bicycling: Human power, drag, and rolling resistance

Wolfram Müller* and Peter Hofmann*†

* Human Performance Research^{Graz}, Karl-Franzens University and Medical University of Graz, Austria

† Institute of Sport Science, Karl-Franzens University of Graz, Austria

Abstract The maximum power that an athlete can produce is a question of the time domain of interest which can be hours or just fractions of a second in a given discipline (human power spectrum). The human body functions as a "biological machine". Like any other motor the "muscle force machine" has a limited efficiency. In every sportive movement another combination of the loading types influences the performance. Consequently, there is a diversity of human performance factors. A muscle cell's work is the conversion of useful chemical energy to mechanical work. There are basically three different types of energy supply modes available to the human organism. Two of the body's three energy systems do not require oxygen for the production of adenosine triphosphate (*ATP*). The production, the transport and usage of lactate, which results from glycolysis and glycogenolysis, represents an important means to link glycolytic and oxidative metabolism (*lactate shuttle*). Various physiological variables are influenced by the working status of the body and can be used for performance diagnosis and training control.

The equations describing the necessary mechanical power in cycling for overcoming the rolling resistance, the air resistance, and the power necessary for climbing a slope are described as well as the measurement of power-output by means of mechanical cycle ergometers. Depending on the discipline it can be absolute power or power with respect to the body weight which is of dominating relevance. Power values obtained by cycling athletes of various levels and cycling disciplines are discussed on the basis of results published in the literature.

Individual optimization of drag by means of wind tunnel measurements is essential for high performance. Measurements of drag in elite cyclists should be done during effort because static measurements underestimate drag in real competition or training situations. The position of the athlete, his body dimensions, and the aerodynamic features of the bicycle and the equipment have a pronounced impact on the performance.

To move things is all that mankind can do, for such the sole executant is muscle, whether in whispering a syllable or in felling a forest.

C. Sherrington, Linacre Lecture, 1924

1 The diverse manifestations of physical performance

The accomplishment of physical performance in sports is much more than just power production by the athlete's muscles and does not directly equate to the mechanical power in the sense of the definition of power P as work W divided by time t. Seeing the human body just as a "muscle force machine" would cover just one out of a manifold of aspects which determine human performance. Solving a complex movement task necessitates synchronous and successive coordination of the activity of many muscle groups, and qualities such as flexibility, precision of motor activity, balancing ability, rhythm, and anticipation can be of essential importance. In sports a series of terms characterizing "strength" are in use, e.g. maximal strength, explosive strength, or strength endurance. The muscle contractions necessary for performing various movement tasks may vary in intensity, duration, and frequency.

In physics, work is the scalar product of the force vector $\vec{F}$ and the displacement vector $\vec{s}$: $W = \vec{F} \cdot \vec{s}$. The power $P = W/t$, is measured in Watt [W]. The insertion of the equation for work results in $P = \vec{F} \cdot \vec{s}/t$. The ratio $\vec{s}/t$ is the velocity $\vec{v}$. Thus, power can also be written as the scalar product of force and velocity $P = \vec{F} \cdot \vec{v}$. All static elements in sports (isometric load tasks), as for instance a handstand or a cross on the rings, result in a mechanical output power of zero. This can happen despite of high forces because there is no movement of the object ($\vec{s} = 0$, hence $\vec{v} = 0$): $P = \vec{F} \cdot \vec{s}/t = \vec{F} \cdot \vec{v} = 0$, in this static case.

In general, the measurement of output power during a movement is not simple due to the complexity of many human movement tasks. Ergometers (greek: *ergon* means work, *metron* means measure) are instruments for measuring the physical power of the athlete. For instance, the mechanical cycle ergometer: The power necessary for accelerating the flywheel and for overcoming the set friction on the flywheel (and for overcoming the chain- and bearing friction) equals the mechanical output power generated by the muscles of the athlete. At a constant velocity the set frictional force F_f acting at the flywheel multiplied by the circumferential velocity of the flywheel v_c is equal to the power absorbed: $P_f = F_f \cdot v_c$. Highly trained male athletes in cycling are able to produce a mechanical power of 400 W or even more over longer periods of time (1; 2; 3; 4); female cyclists perform around 300 - 350 W and about 250 - 250 W mean power during competition (5). Short peak power values of sprinters may be as high as 1800 W (6) or even above. Output power during a jump can be measured by means of a force plate which enables the measurement of the force time curve with a temporal resolution of some ms. During jumping movements of trained athletes the peak power can exceed 3000 W for fractions of a second.

The maximum power that an athlete can produce is a question of the time domain of interest which can be hours or just fractions of a second in a given discipline. The *Human Power Spectrum* is the result of a series of maximal power measurements in all time domains of relevance in sports (7; 8; 9). Fig. 1 shows examples of typical maximum power values obtained by a sprinter (a) and (b) or by an endurance athlete (c) and (d) in all time domains ranging from fractions of a second to several thousand seconds.

The physiological background of 'endurance power' is described in section 3 of this chapter (*Oxidative endurance*), the Sprint Power (*SP*) and the Transition Power (*TP*) as well as the tests for jumping power (*JP*) and peak jumping power (*PJP*) determination is discussed in section 4 (*The human power spectrum*).

The human body functions as a „biological machine" ("a muscle force machine"). Like any other motor the „muscle force machine" has a limited efficiency η. The total efficiency is the generated mechanical energy (work) divided by the supplied chemical energy. For static loads the efficiency η is 0, in trained cyclists η can gain values up to approximately 25% to 30% (10; 13). The pedal rate (cadence) influences the efficiency (14) and muscular efficiency can also improve substantially during the maturation of an athlete (13).

Due to the different manifestations of the concept strength in sports, *maximal strength*, *speed strength*, *explosive strength*, *strength endurance*, and other definitions have been introduced which mirror different force-time courses occurring in dependence on the movement task to be solved. Quite often confusion occurs because these definitions appear to be sloppy from the point of view of physics and therefore cannot easily be translated into the language of physics, although they all are just specific cases of force-time functions. *Physics only needs one concept for force and power; all*

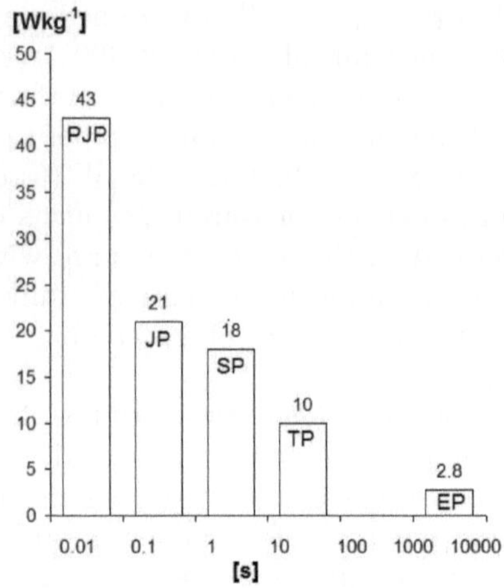

(a) Sprinter: relative power

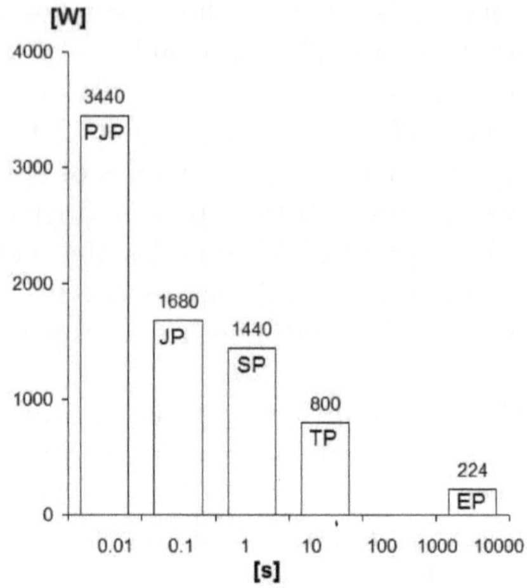

(b) Sprinter: absolute power

Figure 1: Maximum power values. The typical power measurement results shown in the figures give an outline of the athletes' maximum power abilities obtained on a cycle ergometer in all time domains of relevance in sport. Results typical for a sprinter are shown in subfigures (a) and (b), and for an endurance athlete in subfigures (c) and (d). For data evaluation the software *Human Power Spectrum* (www.bewotech.com) was used. Peak and mean jumping powers during a squat jump, termed *PJP* and *JP*, respectively, were measured with a *Kistler Quattro Jump* force plate. All other power values were obtained on *Monark* mechanical cycle ergometers (www.rotosport.com) equipped with a *Power Analyzer* (www.bewotech.com).

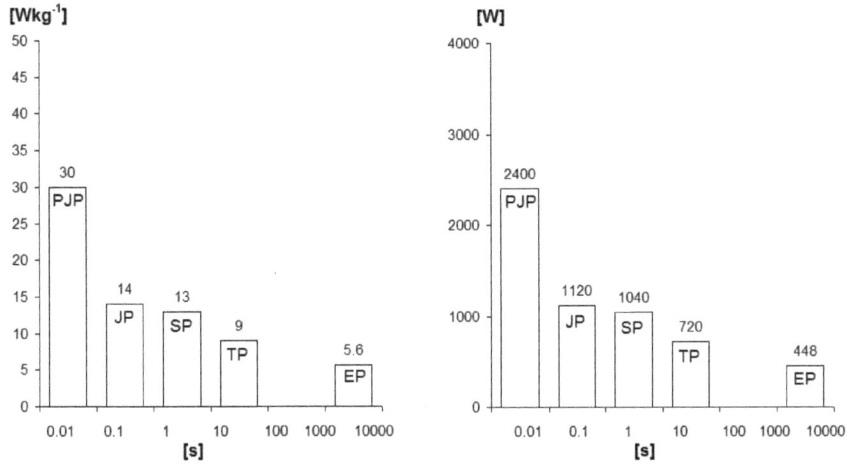

(c) Endurance athlete: relative power

(d) Endurance athlete: absolute power

Figure 1: (continued) *SP* is the maximum power obtained with the *Sprint Power Test* in the time domain of a few seconds, and *TP*, the transition power in these plots is the power the athlete is able to produce during 30 s; the term *TP* is used here because the immediate, the glycolytic, and the oxidative energy modes are involved in this time domain (10) p. 32. For the transition power test the Wingate test was used. Endurance power was estimated by means of the second lactate turn point (*LTP2*) (11; 12).

strength concepts in sports are contained in the force concept of physics. Obviously, the athlete's ability to adapt her/his[1] force-time courses to the goal of the movement task is fundamental for optimising muscular power out-put, which is essential for many athletic and sporting activities. Biomechanical considerations obviously show that a high power level is an essential performance factor in many sports. However, it is being discussed by training practitioners controversially today at what training loads power output can be optimally increased (15). In many cases, high power output may be needed at very high contraction velocities of the muscles – a situation in which the obtainable muscle force is far from its maximum.

In every sportive movement another combination of the loading types and

[1]The masculin form of the personal noun *his* stands for both sexes. For the sense of grammatical convenience only one form has been chosen.

their quantitative specifications influences the performance. Consequently, there is a diversity of human performance factors. The evaluation of physical performance becomes even more complex through the influence of cognitive and psychological processes. For instance, a football player's performance depends on his ability to perform at a high level in a short period of time (sprint power; explosive strength), on his endurance, ball technique, cognitive skills, his ability to anticipate, as well as on psychological factors such as stimulation by the crowd. In game sports "game intelligence" is used as a term to refer to the player's ability to react to stimuli and to anticipate the opponent's actions in an efficient way. Similarly, a skier's performance is also a very complex processes: With the aid of his physical abilities and intricate cognitive processes, with which he chooses the optimal path, he strives to reduce his race time. For such complex sportive tasks, fascinating planning abilities are available to the athlete which include the synchronous "real time" interpretation of the whole stream of information going to the brain from inside and outside the body. The sensory-motor performance of a skier, a mountain biker, or a football player, as examples for many other sports, by far exceeds what we can nowadays contemplate as being possible for intricate and elaborate robotics; not to mention what is meant by "performing" in aesthetic sports. Complex movement tasks necessitate a profound "implicit" knowledge of the athlete on his extremely complicated body's mechanics and how to control it.

The load of the body leads to the alteration of physiological parameters. For instance, a well-trained endurance athlete will have a lower heart rate at the same load than an untrained person, who reaches his peak performance already while accomplishing this task. The basis of sports medical performance diagnosis is the measurement of physiological quantities during a defined muscle load. There is no universal performance factor which maps the entire physical performance spectrum; however, there are test batteries which can provide information about different energy production modes and motor abilities of the athlete.

The quality factors objectivity, reproducibility, and validity of performance tests:

- Objectivity is the quality of being free from personal bias. The results of a test should not vary from examiner to examiner. A test has high objectivity if similar results are achieved with different examiners. Strict objectivity, however, is an unattainable goal, and therefore subdisciplines of sport science have developed their working criteria for objectivity. Objectivity refers to the testing procedure as well as the evaluation of the test results.

- Reproducibility or precision is a characteristic of a measurement or an experimental procedure which produces consistent results on two or more separate occasions. Reproducibility of tests of human performance is never perfect because of biological variance due to factors such as mood, daytime, or previous experience. Reproducibility (reliability, precision) can be maximized by standardizing as many factors as possible of those which would otherwise influence the investigated effect under study. It is the main challenge of biometry, which is the application of methods of statistical mathematics to measurements in biological contexts, to gain information despite the limited reproducibility due to stochastic effects and biological variances.

- Validity: Even if a feature has been treated with a high degree of objectivity and reproducibility, this does not necessarily mean that the achieved data are meaningful for the interpretation of the sporting performance. The *maximum oxygen consumption*, for instance, is a valid measure for oxidative endurance, but it has no significance for the short-term peak power of an athlete because in the latter case the required energy is produced without the usage of oxygen. Also the *resting pulse rate* relates to the endurance performance. However, due to the considerable differences between comparably performing endurance athletes it is a less valid measure for the prediction of the endurance performance (e.g. performance in a bicycle race or running time in a marathon) than the maximal oxygen consumption.

Note on *accuracy* and *precision* of a measurement or experimental procedure: A test should measure a physiological quantity of interest, e.g., heart rate as a function of the mechanical output power, with sufficient accuracy and precision. These two terms, accuracy and precision, have distinctly different meanings: For instance, if the dispersion around a target point in a firing task with a pistol is large, the mean of the deviations can be small and can even coincide with the inner annulus of the target disc - such a pistol would be accurate but not precise. On the other hand, a pistol which always hits the same small area, but far from the centre of the target: This pistol is inaccurate although it is precise.
The term *systematic error* (determining the accuracy) refers to a consistent bias in the reading, for instance, of power displayed by an ergometer which is not correctly calibrated, whereas *random error* refers to fluctuations in readings from measurement to measurement (determining precision).

2 Loading types and energy supply modes

A cell's work is the conversion of useful chemical energy to other forms of energy, e.g. to mechanical work. This needs a substance which can act both as an energy receiver and as an energy donor. In most cells of most organisms this substance is the adenosine triphosphate (*ATP*). The cell gains the useful energy stored in the *ATP* by respiration, i.e., the conversion of the chemical energy of foodstuffs into a useful chemical form. There are basically three different types of energy supply available to the human organism. Varying in type across different kinds of sport the body has to accomplish mechanical power, which:

(A) Has to be upheld over a longer period: Endurance performance such as long distance cycling and running using oxidative energy sources.

(B) Has to be performed over short periods (seconds) at a peak value: Such as in sudden accelerations in ball games. This is done by means of the immediately available energy source which is composed of three components: firstly, the *ATP* itself; secondly, the *ATP* generation by means of creatin phosphate (*CP*) reacting with adenosine diphoshate (*ADP*) with the help of the enzyme creatine kinase; and thirdly, the production of *ATP* molecules by means of two *ADP* molecules with the enzyme myokinase (which is also referred to as adenylate kinase).

(C) Additionally, it may be necessary to perform a considerably higher power than the power which can be generated in a steady – state between energy turn over and supply of nutrients and oxygen in the muscle cell. A substantial part of the power in this case is generated in nonoxidative way by breaking down glucose (a simple sugar) and glycogen (stored carbohydrate composed of many glucose subunits) in the muscle cell. These nonoxidative processes are termed glycolysis and glycogenolysis, respectively. Skeletal muscle (white and red glycolytic fibers) can break down glucose rapidly and can produce significant quantities of *ATP* for short periods. The efficiency of glycolysis ($\eta = 46.8\%$) compares favorably with that of oxidative enzymatic processes (10), p.75.

This occurs in addition to the availability of energy stored in *ATP*, in *CP*, and in *ADP* (immediate energy sources) which only lasts for the first few seconds. The breakdown of glucose or glycogen (the latter stores the major part of the energy since the concentration of free glucose is very low in skeletal muscle) results in the production of *ATP* molecules and lactate. This form of nonoxidative energy supply dominates in intensive work loads in a range between 20 and 60 seconds.

Two of the body's three energy systems (the immediate and the nonoxidative) do not require oxygen for the *ATP* production. By convention, these systems are referred to as anaerobic. The third system, the oxidative system, is referred to as the aerobic system because it only works by using O_2. Because much of glycolysis has little to do with the presence of O_2 (10), p. 67, and because there is a direct link between glycolytic and oxidative processes, due to lactate being used as an energy source itself again which is metabolized at the mitochondria in the presence of O_2, the term anaerobic should be used with caution. The production, the transport and usage of lactate represents an important means to link glycolytic and oxidative metabolism (10), pp. 31-42.

3 Oxidative endurance

The endurance performance strongly depends on the maximal energy which is supplied with oxygen consumption. Through the oxidation of food only a certain maximal energy per time interval can be gained in every organism, which, however, can be strongly influenced by training (10), p. 231.

3.1 The oxidative conversion of the energy available through the nutrition into mechanical work

For the contraction of the muscle cell and other forms of cell work the energy set free by the common chemical intermediate adenosin triphosphate (*ATP*) is used. Cells, tissues, and organs are designed to maintain almost constant cellular *ATP* concentrations over wide ranges of turnover rates (homeostasis). Performance in sports often depends largely on the energy available and thus on the biochemical mechanisms supporting *ATP* homeostasis. The processes in the human body which lead to the forming of the *ATP* – molecules need energy. Only a certain percentage of the energy originally set free during the oxidation of food is available in ATP. Almost all processes in the body which require energy use *ATP*. Apart from muscle contraction, the energy stored in *ATP* is also required for biosynthesis, active transport processes in the body, and for the information processing in the nervous system. The conversion of the energy stored in *ATP* into mechanical muscle work (contraction work of the muscles) takes place with an efficiency of up to approximately 50% (10), p.41. This is remarkable considering the fact that state-of-the-art internal combustion machines attain not more than 30%. Approximately 50% of the potential chemical energy released from foodstuffs is stored in the common chemical intermediate, *ATP*. *ATP* and its storage form, creatine phosphate (*CP*), serve as the immediate cellular energy source for endergonic processes (10), p. 41. For the entire chain of the energy conversion from the oxidation of the nutrients to the generation of mechanical work on a cycle ergometer total degrees of efficiency η of uo to 25% to 30% have been reported (10), p. 55. Depending on the kind of physical work the total efficiency lies within the interval between 0% (isometric muscle contraction) and a maximum of around 25%. The total efficiency in swimming lies only between 3 and 6%, the highest values can be found in cycling (25 – 30%). In rowing about 15% and in weight lifting about 10% have been measured (16) p. 380 and (10) p. 54–58. The efficiency also depends on the training level: If a person is trained to perform a certain movement, usually the efficiency is higher than in untrained individuals. Cyclists achieve better values in cycle ergometer tests than endurance athletes of other sports because they work more economically. Vice versa, runners perform better on treadmills when compared to bicyclists. One reason for this is that with an increase of the technical level, the specialisation for a certain movement leads to muscle stimulation which is predominantly focused on the most efficiently working muscle groups, so that little energy is wasted to irradiation (13).

3.2 Heart rate and blood lactate concentration in dependence on physical power

The loading of large muscle groups as in running (treadmill) or cycling (cycle ergometer) is necessary for the examination of the oxidative endurance performance. The cycle ergometry has the advantage that the mechanically achieved power can be accurately measured. Instead of power the running velocity is measured in treadmill tests because determination of the mechanical power necessary to obtain a certain running velocity cannot be done without a computer modelling approach which would have to include details of the individual's running style.

We will at first focus on the heart rate (*HR*) and the blood lactate concentration in connection with physical performance. The blood lactate concentration is measured in mmol·l^{-1}.

Figure 2 shows the heart rates as functions of power on the bicycle ergometer in young test persons of different endurance levels. Table 1 indicates characteristic values obtained from these 6 test persons. In order to obtain these graphs the power was increased in 7 W, 14 W , or 21 W steps in 1 minute intervals up to the maximal value obtainable by the individual. If the last power step cannot be completed over the entire minute, only the adequate percentage of the final power step is taken: For instance, for

30 seconds at the last power step (with 14 Watt steps) only 7 Watts will be added to the previously achieved power. As a weight related measure, $P_{\text{max,rel}}$, the achieved peak power through the body mass in W·kg^{-1} is used.

Table 1: Tabular listing of data according to Figure 2: The peak power P_{max} depends on the training level and the weight of the test person: As a weight related measure the achieved peak power through the body mass $P_{\text{max,rel}}$ in W·kg^{-1} is used. The maximum heart rate HR_{max} obtained at the end of this cycle ergometer test is usually about 5 to 10 beats per minute lower than the HR_{max} achievable in running tests.

	Age	Sex	Endurance level	Mass [kg]	P_{max} [W]	$P_{\text{max,rel}}$ [W·kg^{-1}]	HR_{max} [min^{-1}]
(a)	12	f	good	36.0	126.0	3.50	188
(b)	15	m	good	71.5	249.0	3.48	194
(c)	16	m	junior athlete	53.0	236.5	4.46	191
(d)	17	m	hobby athlete	57.0	231.0	4.05	197
(e)	18	m	moderate	75.3	229.8	3.05	182
(f)	18	m	elite athlete	63.0	385.0	6.11	188

Up until 1982 the relationship between heart rate and power was usually described to be linear in exercise physiology text books. Just then Conconi et al. (17) discovered a deflection point in the heart rate performance curve. Several authors have described that this turn point correlates with the maximum lactate steady state (18; 11). However, this turn point where the heart rate curve deflects to the right at a specific power value can be found only in around 85% of young healthy subjects (12). Provided an appropriate test protocol is used, the power at the heart rate turn point (*HRTP*) coincides with a progressive incline of blood lactate concentration, the so-called second lactate turn point (*LTP2*). Both, the *HRTP* and the *LTP2* can be used to predict the maximal lactate steady state (*MLSS*) (11; 19) representing the critical lactate clearance rate described by Brooks et al. (10), p. 504. Both, the heart rate turn point *HRTP* as well as the *LTP2* can be used to estimate this critical lactate clearance rate which determines the maximum physical power at which the individual can keep its blood lactate concentration constant (10), p. 504 (compare to Figs. 3 and 5). As described in detail later, on average 71% of the peak power coincides with the power at the *LTP2* and with the power at the *HRTP*. Therefore, $p = 71\%$ of peak power can also be used instead of the *LTP2* or *HRTP* determination. This was applied in Fig. 2 for indicating the estimated maximum endurance power (broken vertical line). This may be of substantial interest for train-

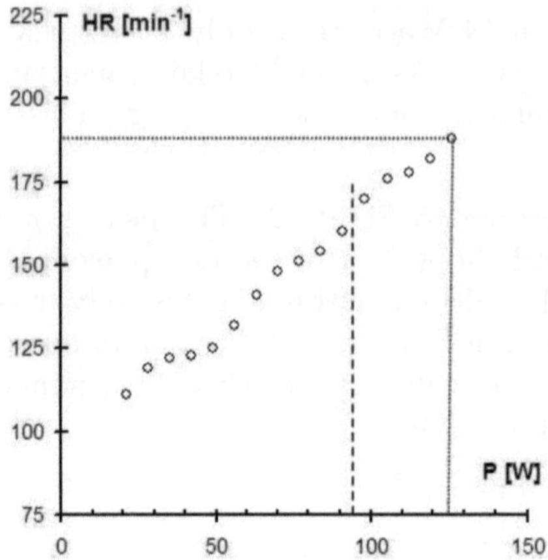

(a) Female, Age 12

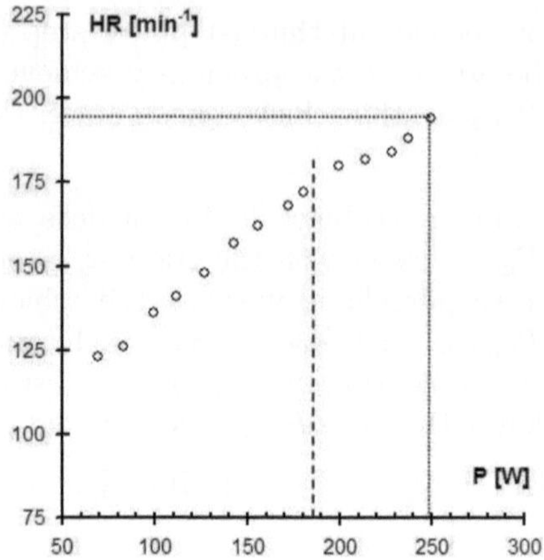

(b) Male, Age 15

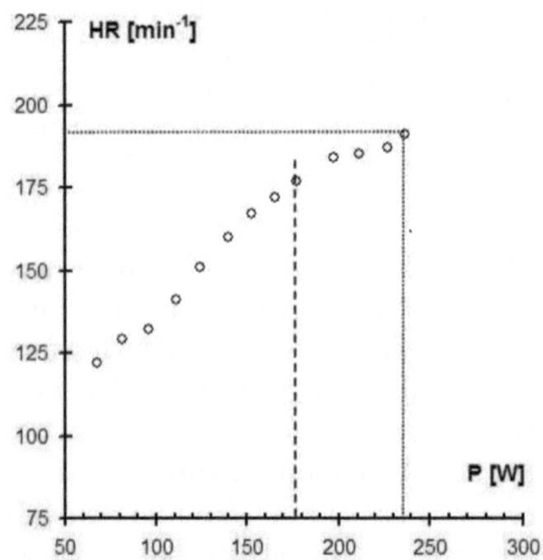

(c) Male, Age 16

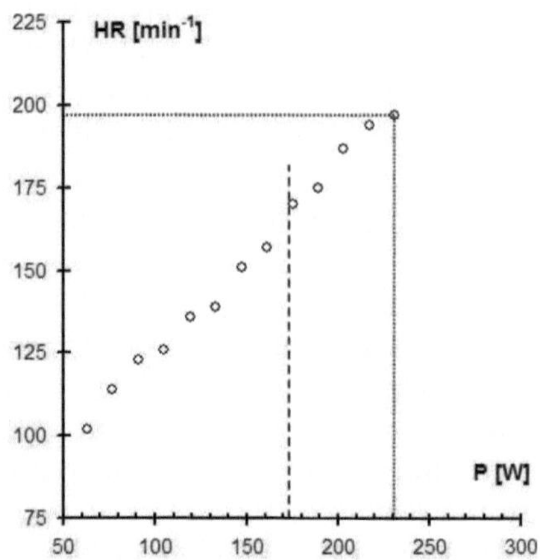

(d) Male, Age 17

Figure 2: The Figure shows the increase of the heart rate in young test persons of different endurance levels with increasing power (compare to Table 1).

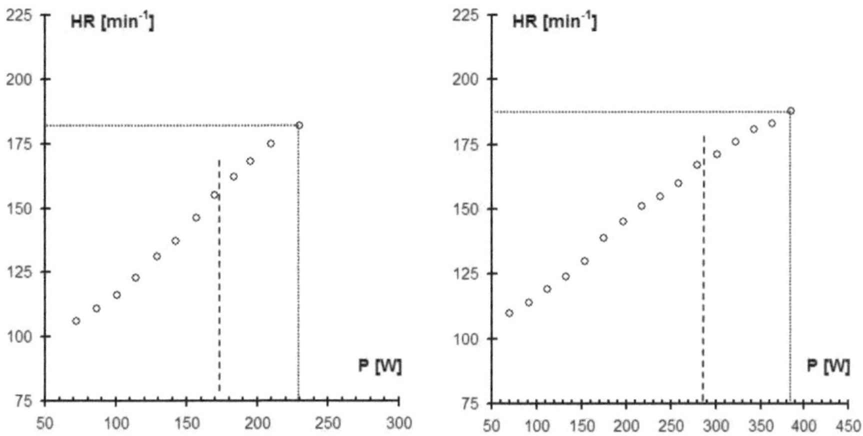

(e) Male, Age 18 (f) Male, Age 18

Figure 2: (continued) The power was increased in 7 W or 14 W or 21 W steps (depending on the weight and the endurance level of the athlete) in 1 minute intervals up to the maximal value obtainable by the individual. 71% of the peak power value was used here for the estimation of the *maximum endurance power* as defined by the *MLSS* concept (broken vertical lines). Figures: W. Müller.

ing practitioners as the *HRTP* determination described by Conconi et al. (17) needs a lot of experience, and is sometimes difficult and in some cases impossible to determine (not all persons show a heart rate deflection) (12). The usage of a percentage p of $P_{\max}$ can easily be applied as a non-invasive tool to determine oxidative endurance performance without any sophisticated analysis method necessary.Table 1 shows the maximum power values $P_{\max}$ obtained by the test persons of Figure 2. 71% of these values were used as an approximation of the *LTP2* (and thus of the *MLSS* power level) in these subjects.

However, this practical approach does not substitute the necessity to describe the causative mechanisms behind the observed phenomenon, such as the transport mechanisms of lactate. They are described in more detail under the topic *lactate shuttle* in the literature (20; 21; 22; 23; 10).

Physiological characteristics of female road cyclists and differences in power between top 20 and non-top 20 female World Cup cyclists were discussed by David T. Martin et al. (5): Mean power during a race was found to be (3.6 ± 0.4) $W \cdot kg^{-1}$ in top 20 and (3.1 ± 0.1) $W \cdot kg^{-1}$ in the non-top female athletes and top 20 produced more than 7.5 $W \cdot kg^{-1}$ during (11 ± 2)% of race time compared to (7 ± 2)% of non-top athletes.

Physiological and performance characteristics of male professional road cyclists were analyzed by Mujika and Padilla (2): Highest mean power values were found in uphill specialists (6.5 ± 0.3) $W \cdot kg^{-1}$ and the associated mean of oxygen uptake was (80.9 ± 3.9) $ml \cdot kg^{-1} \cdot min^{-1}$. These values were closely matched by those of time trial specialists: (6.4 ± 0.1) $W \cdot kg^{-1}$ and (79.2 ± 1.1) $ml \cdot kg^{-1} \cdot min^{-1}$. It was concluded that time trial specialists had an overall performance advantage over the other groups in all cycling terrains (24). Time trials in the World Championships are in the time domain of one hour. Characteristics of track cyclists are reviewed in a publication by Craig and Norton (6); during a 1000 m time trial which lasts approximately one minute a peak power of 1800 W and an average power of 760 W is reported there. For a world record performance peak values of above 1900 W can be assumed.

Often the reference point of "anaerobic performance" is termed *anaerobic threshold* which, however, should not be used any more as it is misleading (10): Confusingly, it is sometimes also used for the point in the lactate curve where blood lactate starts to exceed resting value (often termed lactate threshold) occurring at much lower power values and lactate concentrations (typically around 45% of P_{max} and below 1.5 $mmol \cdot l^{-1}$ blood lactate concentration); this is markedly below the *LTP2* lactate values; these - as well as the *MLSS* lactate values (compare to Fig. 5) - are typically between 2 to 6 $mmol \cdot l^{-1}$ (mean is about 4 $mmol \cdot l^{-1}$). The *LTP2* is the point where lactate starts to accumulate exponentially in the blood because the overall lactate production rate exceeds the whole body lactate elimination rate from this power level on (10). Still, sports medical investigators use a fixed 4 $mmol \cdot l^{-1}$ blood lactate concentration as a criterion for determining the *"anaerobic threshold"* aiming to estimate the *MLSS* this way. The application of such a fixed lactate concentration does not only depend on the applied test protocol and on the nutrition but also results in large errors when the endurance power of an individual is to be determined. This will obviously lead to errors in the training program recommendations. Therefore, diagnosis protocols using fixed lactate concentration values for individual performance diagnosis should be replaced by ***individual turn point*** protocols whatever variable, lactate, heart rate, or ventilation is used.

The objective of the determination of the *LTP2* or the *HRTP* is to estimate the *MLSS* by means of a brief test with sufficient accuracy for individual endurance performance diagnosis. The direct determination of the *MLSS* by performing a series of constant load exercise tests in the vicinity of the person's *MLSS* power level lasting more than half an hour each (time domain of "endurance" power) and being separated by pauses of 2 or more days (for recreation) would - in many cases - cost too much time. When performing such direct tests it is important to perform them in such a way that they include a preceding warm-up phase (on-phase) with a stepwise increase of work loads until the desired power in the *MLSS* vicinity is reached; otherwise, the initial lactate increase would exceed the person's *MLSS* value and this would distort the test result.

A higher power than the one achieved at the *MLSS* cannot be upheld over a longer period (no longer than a couple of minutes) because the continuously increasing lactate level leads to acidaemia which in turn has a negative influence on the metabolic processes that generate muscle force (Fig. 5).

The lactate concentration and the heart rate increase with increasing power during a step test on the bicycle ergometer is shown in Fig. 3(a). The test was started at $P = 40$ W and power was increased by 20 W steps every minute until the maximum power was reached at 360 W. The heart rate turn point *HRTP* and the lactate turn points *LTP1* and *LTP2* are indicated by broken vertical lines. In Fig. 3(b) the lactate concentration and the increase of oxygen consumption are shown. The maximum oxygen consumption is an important marker for the endurance performance of a person. Extremely trained endurance athletes can reach maximum values as high as 80 to 90 ml oxygen per kg body mass and per minute. The volume of air inspired or expired out of the lungs per minute (minute ventilation) also deflects to the right (respiration turn point 2, *RTP2*) when reaching *LTP2* power but oxygen uptake deflection occurs at a higher power level (2 steps above in the example shown). In the mean, the *MLSS* power occurs at 75% of maximum oxygen uptake; however, individual deviations from this percentage are large and therefore training recommendations should not be based on this percentage method. Fig. 4(a) shows that blood lactate concentrations vary in the range from about 1 $\text{mmol}\cdot\text{l}^{-1}$ (level of resting values) up to more than 9 $\text{mmol}\cdot\text{l}^{-1}$.

At the lactate turn point *LTP1* the lactate concentration in the blood starts to increase with a larger slope than in the initial phase of low loads due to

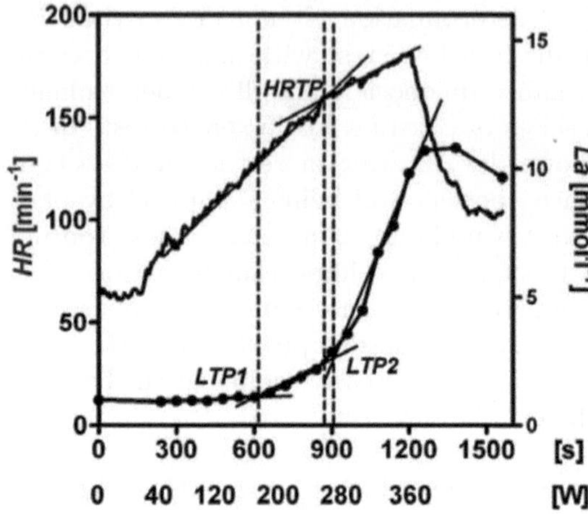

(a) Heart rate *HR* and blood lactate *La* of an endurace athlete

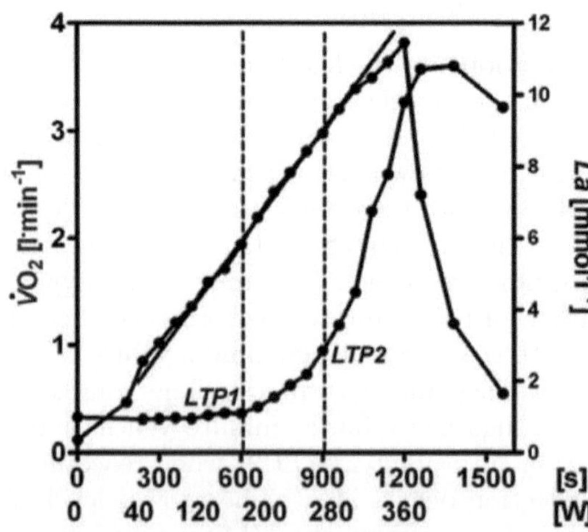

(b) Oxygen uptake $\dot{V}_{O_2}$ in liters per minute and blood lactate

Figure 3: Results from a maximal incremental cycle ergometer step test according to Hofmann (11; 12). (a) Lactate and heart rate curves. *LTP1*, *LTP2*, and *HRTP* are indicated (broken lines).Heart rate *HR* and blood lactate concentration [*La*] of an endurance athlete as functions of the ergometer power. The lactate concentration reaches its maximum delayed, after the end of the step test. (b) Oxygen uptake curve corresponding to (a). Oxygen uptake $\dot{V}_{O_2}$ in liters per minute of the same athlete in the same test.

the muscle having reached its critical lactate clearance rate. Resting lactate values typically ranges from 0.5 to 1.5 mmol·l^{-1}. At the *LTP2* the slope increases again because at this power level the body as a whole system reaches his critical lactate clearance rate (10), p. 216.

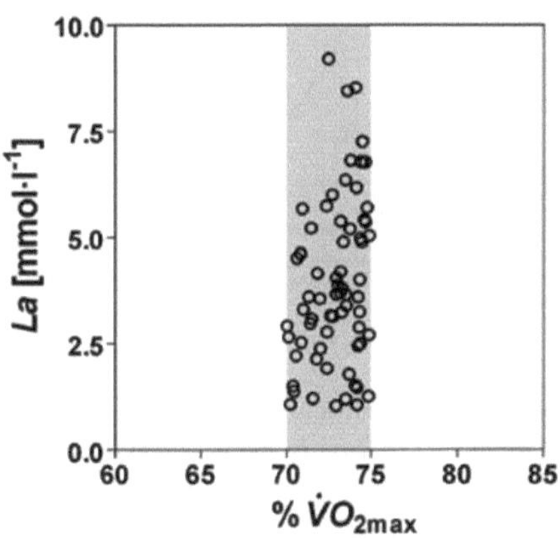

(a) Spread of lactate values

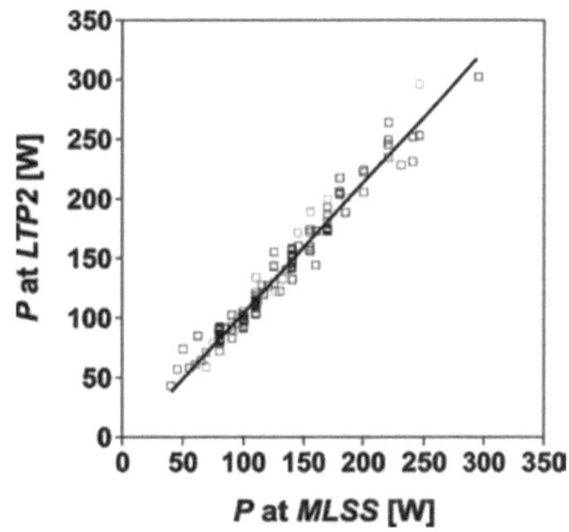

(b) *MLSS* - *LTP2* correlation

Figure 4: (a) Shows the enormous spread of lactate values of a group of test persons ($n = 60$) during a 40 minutes constant load exercise at 70 to 75% of maximum oxygen uptake. The blood lactate concentration varies in the range from resting values up to approximately 9 mmol·l^{-1} indicating that such a training recommendation is misleading. Figure: P. Hofmann, with permission. (b) Correlation of the *MLSS* power and the power at the *LTP2*. The correlation coefficient of 0.98 ($n = 134$) indicates the very high accuracy of this step test protocol and the *LTP2*-evaluation for the *MLSS* estimation. Pokan et al. (25; 26), with permission.

For the 1 minute incremental test described here approximately 15 increments until reaching maximal load are used; the test starts out from about 20% of maximum load. The *LTP2* corresponds excellently to the power value of the maximum lactate steady state (*MLSS*) which can be observed in Fig. 4(b). The slight deviation of the regression line from the diagonal is due to the discrete steps which have to be used for constant load *MLSS* exercise tests: Due to biological variations it would not make sense to strive for a power resolution in the *MLSS* test below 10 W. The correlation coefficient of 0.98 indicates the very high accuracy of this test protocol associated with the lactate turn point 2 (*LTP2*) evaluation.

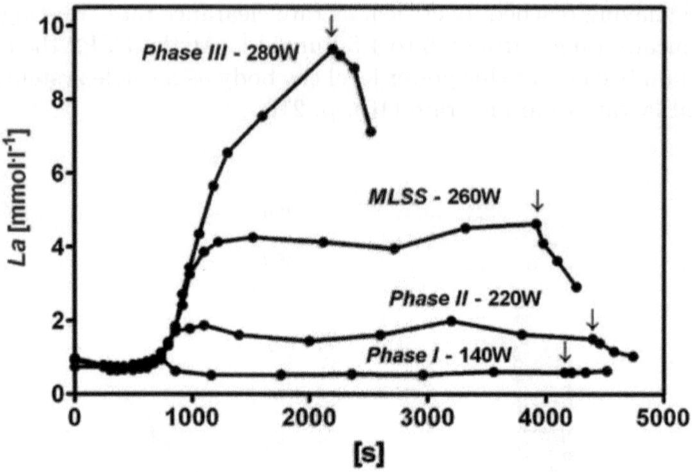

Figure 5: Blood lactate concentrations during various constant load exercise tests with various loads. The term *phase 1* refers to loads lower than the power at *LTP1*, *phase 2* to power values set between the *LTP1* and the *LTP2*, and *phase 3* indicates loads beyond the power level of the *MLSS*. In the later case the person had to interrupt the test due to the accumulated high lactate. Vertical arrows indicate the end of exercise. Figure: P. Hofmann.

A comparison of the maximum power $P_{\max}$ obtained by the athlete at the final power step with the power achieved at *LTP2* shows that the latter is close to 71% of $P_{\max}$. Hofmann et al. (27) have found that the mean of a group of test persons studied ($n = 75$) was (71.4 ± 3.8)%. The means for women ($n = 30$) and for men ($n = 45$) were (72.3 ± 3.2)% and (71.0 ± 4)%, respectively. The individual's small deviation from the mean of 71% allows us to use this percentage of $P_{\max}$ as a useful indicator of the *MLSS* power level[2]. In the group of female test persons the maximum power

[2]A further improvement of the *MLSS* prediction by means of a percentage p of the maximum power $P_{\max}$ obtained in the step test can be obtained by considering that p slightly depends on the maximum power per kg body mass ($P_{\max,\text{rel}}$) obtained by the test person. For the male group we obtain:

$$p = 100 \cdot \left[0.71 + k \cdot \left(P_{\max,\text{rel}} - 4.6\right)\right] \qquad \text{with} \qquad k = \frac{3.5\%}{1\ \text{W} \cdot \text{kg}^{-1}} \tag{1}$$

values obtained ranged from 2.4 W·kg^{-1} to 5.4 W·kg^{-1} (mean 3.5 W·kg^{-1}), and in the group of males from 3.0 W·kg^{-1} to 6.1 W·kg^{-1} (mean 4.6 W·kg^{-1}).

In test protocols with shorter intervals, the heart rate would develop under the shown curve and the lactate concentration curve would be shifted to the right (it takes some time for the lactate to get from the muscle cells, where it has its origin, into the blood circulation) and vice versa for longer time intervals.

Recent scientific papers (28; 29; 30) and reviews (31) show that it is important for endurance athletes to work out with blood lactate values or heart rates corresponding to the vicinity of the *MLSS* and - for limited time spans - even above it; however, this demand has to be balanced with the demand for long training durations, e.g. 3 to 6 hours a day (amounting to 20 to 30 hours a week) in top level endurance sports, which can only be performed in low or medium pace training regimes. A training typically composed of 70% low pace, 20% medium and fast pace, 6 to 7% extensive interval (with pauses of 30 to 60 s), and 3 to 4% intensive interval training (with pauses of 3 to 5 minutes) is called *polarization training* (28), p. 947. Table 2 indicates typical training regimes with respect to the power at the *MLSS* which is 100% per definition in this table.

Table 2: Endurance training regimes: Values with respect to the power at the *MLSS* being 100% per definition. The *MLSS* can be found by direct determination of the *MLSS*, or estimated by means of the *LTP2*, or the *HRTP*, or the second ventilatory turn point *VT2*, or by means of the *p*-value evaluation described above.

	percentage of the *MLSS*-power [%]	approximate work-out time [min]	typical pause [min]
Regeneration	50 - 60	20 - 40	
Ultra long training	60 - 70	180 - 360	
Low pace	70 - 80	60 - 180	
Medium pace	80 - 90	60 - 90	
Fast pace	90 - 97	15 - 60	
Extensive intervals	97 - 103	3 - 5	0.5 - 1.0
Intensive intervals	> 103	0.2 - 1.0	3 - 5

Ignoring the necessary training pauses or too high intensities over too long periods may lead to the overtraining syndrome; long pauses have to be made in such cases: overall, a long lasting build-up of performance level can break down immediately this way. Endurance performance development needs patience.

3.3 Indirect measurement of the energy turnover by means of oxygen consumption and carbon dioxide exchange

Energy turnover takes place even if the organism is at rest so that the vital functions are upheld: The heart accomplishes mechanical work, the respiratory air has to be accelerated, bioelectrical processes in the body require energy, just like protein synthesis and many other vital processes. All energy forms occurring within the body: mechanical, electrical, and chemical, are finally transformed into thermal energy: In the blood vessels kinetic energy is transformed into heat through internal friction (viscosity), the bio-electrical currents in the body generate heat just like any current in a resistor. The chemical reactions in the body are exothermic, which means that thermal energy is produced. The whole energy which is generated in the body has to be externally diverted because body temperature is basically held constant (through intricate control mechanisms), usually within the range of a degree. With the aid of hermetic chambers the diverted heat energy can be determined. This direct way of determining the energy turnover through measurement of the warming of the environment is called *calorimetry.* A better term would be *heat energy measurement.* The relationship between the obsolete unit calorie for thermal energy and the SI energy unit Joule is: 1 cal = 4.18 J. In exercise physiology and in nutrition science the obsolete unit [kcal] is still used: 1 kcal = 1000 cal = 4180 J. Here, for all forms of energy or work the SI-unit Joule [J] and for power Watt [W] is used.

The body gains energy through oxidation of food ("fuel") with carbon dioxide as the chemical end product which is emitted during the exhalation. In the oxidation of nutrients a stoichiometrical relationship between the O_2 consumption and the production of CO_2 persists. The energy turnover can be calculated from the amount of consumed O_2. The energy which is released in the chemical reaction ("internal oxidation") of 1 litre O_2 (at 1 atm air pressure) with the specific substance (nutrient) is called *energy equivalent* (obsolete term: caloric equivalent). For glucose ($C_6H_{12}O_6$) an energy of 21.1 kJ per litre of consumed oxygen is released, while lipid oxidation releases 19.6 kJ and oxidation of proteins 18.8 kJ. The average of approximately 20 kJ per litre O_2 can be used for varied nutrition. During a spiro-ergometric test the exhaled CO_2 is measured too. The ratio of CO_2/O_2 can be used to identify the type of nutrient. For pure glucose oxidation this ratio is 1.00 for stoichiometrical reasons.

Of course, the mass, respectively the amount of oxygen, contained in 1 liter depends on the air pressure. At sea level the air density is 1.2923 kg·m^{-3}(for standard atmospheric conditions, the air pressure at sea level is 1 atm = 101325 Pa = 1.01325 bar = 1.033227 at). Air is a mixture of nitrogen and oxygen. Pure oxygen O_2 (mole masse is 32 g), at a pressure of 1013.25 hPa, has a density of 1.429 kg·m^{-3}, therefore in one litre O_2 there are 1.429/32 = 0.04465 moles of oxygen. The energy released through 1 litre O_2 should be rather related to the amount of 0.04465 moles which is independent of the air pressure. One mole of oxygen O_2 (32 g) corresponds to 22.4 liters O_2(at 1 atm) and the released energy in the body equals 22.4 l · 21.1 kJ·l^{-1} = 472 kJ for oxidation of pure glucose. W. Müller suggests to replace the term energy equivalent by the term molar energy equivalent, which is 22.4 times the value of the energy equivalent. The energy released by 1 mole O_2 equals the oxidation of exactly 1/6 mole glucose: Glucose + $6O_2 \rightarrow 6CO_2$ + $6H_2O$. This reaction equation shows that the respiratory exchange ratio for the oxidation of glucose is 1.00 (the amount of emerging moles CO_2 equals the amount of consumed moles O_2).

For the exact determination of the energy turnover the proportion of the quantities of the three nutrient groups has to be known. In a pure oxidation of carbohydrates the ratio between CO_2 and O_2 (the respiratory exchange ratio *RER*) equals 1.0, in pure oxidation of lipids the *RER* is 0.70, and in the oxidation of proteins 0.81. In Central European eating habits the protein percentage in food uptake is relatively constant at approximately 15%. The analysis of ventilation gases O_2 and CO_2 (spirometry) permits the indirect (stoichiometrical) determination of energy turnover. At *RER* is 1.0, 0.9, 0.82 (mean value), 0.8, and 0.7 the energy turnover is 21.1, 20.6, 20.2, 20.1, and 19.7 kJ per litre O_2, respectively.

Bodily work enhances the energy demand which implies a greater need for oxygen uptake and increased ventilation. For men the mean of $\dot{V}_{O_2}$ is around 3.3 litres per minute, in women around 2.2 l·min^{-1}. Highly-trained endurance athletes can reach more than 6 l·min^{-1}(corresponding to more than 80 ml·kg^{-1}·min^{-1}) at their maximum endurance power level of 6 W·kg^{-1} or even above. By using an average value of about 20 kJ per litre oxygen, an oxygen consumption of 3.3 liters would yield a power of 66 kJ per minute, i.e. 1100 W, at 2.2 l this would results in 730 W. The oxidative energy turnover in elite endurance athletes can exceed 2000 W. Again, it has to be stated that only about 25% of the energy can be transformed into mechanical work.

3.4 Overview on endurance performance diagnosis

(1) Lactate is a parameter which is highly sensitive to exercise. It is frequently used for the evaluation of oxidative endurance performance. The measurement of lactate concentration in the blood (lactate testing) is inexpensive and is widely used in professional sport for the purpose of performance diagnosis and control of training.

(2) State-of-the-art lactate performance diagnosis should use the lactate turn points (*LTP1* and *LTP2*) instead of fixed lactate concentration levels for "aerobic" performance diagnosis due to the large individual deviations associated with the latter approach. The *LTP2* enables a reliable estimation of the *MLSS*. Of course, this holds only true for the particular step test protocol using 15 ($\pm$1) 1-minute power steps starting out from approximately 20% of the maximum power the test person can achieve. For the *LTP* analysis lactate measurements should be done each minute during the test.

(3) In Fig. 5 the concept of the *maximum lactate steady state*, *MLSS*, is illustrated. It refers to the maximum power at which a lactate steady state can be maintained over a period of at least 20 minutes (after the initial lactate overshoot at the beginning of the power production). Alternatively, this point can be defined as the *critical lactate clearance point* (10), p. 504. The power at the *MLSS* is the highest power level at which the capacity to clear lactate is sufficient; though lactate clearance may not be maximal at this point and clearance can increase if lactate concentration rises.

(4) The correct and detailed interpretation of the lactate performance curve also requires knowledge about several internal and external influencing factors: In the case of a depleted glycogen store the lactate level in the muscle cell is considerably decreased at a given work load. This can occur, due to an exhausting training preceding the test or a lack of food; in this case the lactate curve is shifted downwards; when using fixed lactate threshold levels (e.g. 4 $mmol \cdot l^{-1}$) for endurance power estimation this would lead to overestimation of endurance abilities.

(5) Lactate is used as a parameter to evaluate the degree of exhaustion in *maximal tests*. In an ergometer step test complete exhaustion is indicated by lactate values of 8 $mmol \cdot l^{-1}$ or above. Extremely well trained endurance athletes are an exception: Marathon runners or cyclists possess a high percentage of type-I-fibers, which build relatively small amounts of lactate, and often they are not able to exceed a value of

6 mmol·l^{-1}. Extreme values of lactate concentration in the blood can go up to 20 mmol·l^{-1} which can be obtained by highly trained athletes in repetitive intensive uphill sprints or similar exhausting tasks.

(6) The blood lactate concentration of the individual athlete can be measured for training regulation purposes; however, repetitive blood lactate tests are unpleasant and necessitate sanitary precautions. Therefore, the heart rate found in a preceding ergometer step test at *LTP1* and at *LTP2* is often used to control endurance training.

(7) The *HRTP* coincides with the *LTP2* within a narrow range and is therefore also a useful tool for the *MLSS* estimation. Additionally to the *LTP2* and the *HRTP*, the respiratory ventilation slope, which also changes at the power of the *MLSS* (respiratory turn point *RTP* (32)), could, in principle, also be used for the *MLSS* estimation. It is most interesting to see that all three, *LTP2*, *HRTP*, and the *RTP* occur at the power level of the *MLSS*; this indicates a concerted change of physiological processes underlying the transition from lactate steady state physical work to an energy mode where lactate production rate exceeds the maximum lactate elimination rate (and thus the lactate concentration in the blood increases exponentially).

(8) The corresponding heart rate at the *LTP2* is on average 90% of $HR_{\max}$; however, in 7 to 8% of all cases a turn point of the heart rate cannot be detected, in other words, the determination of the threshold according to Conconi is not applicable to everyone. In these cases the heart rate corresponding to the *LTP2* is on average 86% of $HR_{\max}$. In another 7 to 8% of cases a turn to the left (increasing slope of the *HR*) can be observed which is associated to a heart rate of 82% of $HR_{\max}$ at the *LTP2*.

(9) There is a very useful relationship (percentage p) between the maximum power reached in the 1-minute ergometer step-test-protocol and the power at the *LTP2*: 0.71 $P_{\max}$ estimates the *MLSS* with good accuracy too. Notice: It necessitates that the person has reached the individual power maximum in the test. For the supervision of training on the ergometer the power value 0.71 $P_{\max}$, which coincides with the power at the *LTP2* and the *HRTP*, can be taken as a benchmark for the selectable training intensities. A further increase in accuracy can be obtained when considering that p increases slightly with $P_{\max}$.

(10) When talking about heart rate in running, it has to be taken into account that the maximum power as well as the *MLSS* are usually at higher heart rates than on a cycle ergometer due to the larger muscle masses involved (typical difference of 5 -10 beats per minute).

(11) Endurance training does not significantly influence maximum heart rate $HR_{\max}$, but aging does. $HR_{\max}$ for work-out on a bicycle can be approximated by the Åstrand formula: $HR_{\max} = 211{,}3 - 0.922 \cdot a$, with a being the age of the person (33). The Åstrand formula results in accurate mean values when large groups of persons are under study, however, individual $HR_{\max}$ values can deviate remarkably, up to 30 beats per minute from the calculated value. When running, 8 beats per minute typically have to be added to the Åstrand formula result. A simple formula for a rough approximation of $HR_{\max}$ expected to be reached with intensive interval running tests is : $HR_{\max} = 220 - a$. Submax-tests are usually based on such estimations of $HR_{\max}$ and therefore the results obtained for individuals can be misleading.

(12) The resting heart rate declines due to endurance training. In extreme endurance athletes the resting pulse rate can be as low as 30 beats per minute. Reference values for untrained women and men are 70 beats per minute and 60 beats per minute, respectively; children have higher values.

(13) The maximal oxygen uptake per minute divided by the person's body mass yields the *relative maximal oxygen consumption*, usually specified in ml O_2 per minute and per kg body weight. This value is a major parameter in performance diagnosis. Means for people between 20 and 30 years of age: Men 45 $\text{ml}\cdot\text{kg}^{-1}\cdot\text{min}^{-1}$, women 38 $\text{ml}\cdot\text{kg}^{-1}\cdot\text{min}^{-1}$. The relative maximal oxygen uptake of endurance athletes is considerably higher and can reach values between 80 and 90 $\text{ml}\cdot\text{kg}^{-1}\cdot\text{min}^{-1}$, corresponding to an approximately doubled energy turnover. The endurance athlete can constantly attain about double the mechanical power of an untrained individual. Practically, the athlete's continuous power output is even higher, as exercise generally leads to an improvement not only of the maximum power but of the efficiency as well.

(14) Training recommendation concepts based on a percentage of maximum oxygen uptake are misleading due to the large individual variations from the value of 75% at which *MLSS* is formed in the mean (compare to Fig. 4(a)).

(15) The energy turnover determination directly by measurement of heat release is difficult and therefore the indirect determination of energy turnover by means of spiro-ergometry is the standard in all laboratories concerned with exercise physiology and sports medicine.

(16) Submax performance tests give only a rough estimate of a person's maximum power output ability because estimated $HR_{\max}$ values are used for this purpose. In a sub-maximal test the participant does not work as far as the achievable peak power value but breaks off earlier. E.g., 90% of the calculated maximum heart rate according to the Åstrand formula can be used as a criterion for stopping the test (PWC_{i}) (33).

4 The human power spectrum: Maximum power is a function of the activity duration

In all sports where the body or other masses are to be accelerated with the goal to obtain a high velocity, the physical power is a major factor. The maximum power production ability of a person is a function of the activity duration (7). Basically three energy conversion modes: immediate energy source (nonoxidative), glycolysis (nonoxidative), and oxidative energy source keep the cellular concentration of the chemical intermediate *ATP* - which powers *muscle* contraction - almost constant over a wide range of turn-over rates (*ATP* homeostasis).

(1) For time domains **below 1 s** the jumping power can be used as an indicator of the athlete's maximum power production. Athletes can produce very high amounts of power during a jump. The analysis of mean jumping power with respect to a vertical jump (e.g. a squat jump) can be made by means a force plate measurement of the ground reaction force according to $P(t) = F(t) \cdot v(t)$, with $v(t) = \frac{1}{m} \cdot \int_{t=0}^{t} F(t)\ dt$, with t being the acceleration time and v the velocity of the center of gravity. Highest power values during jumping occur in the last phase of a jumping movement when both the force and also the velocity are high.

(2) For the time domains **of 1 s to 10 s** the Sprint Power Test procedure (9; 8) can be used for determining maximum power output of an athlete: This test was designed to measure maximum short-time muscle power (time domain of a few seconds) on a mechanical weight ergometer of the *Monark-type*. The velocity of the flywheel can be measured accurately (e.g. by means of a *Power Analyzer*; rotation time accuracy ± 0.001 s; www.bewotech.com) which also allows the determination of the rotational energy of the fly-wheel: $E_{\mathrm{rot}} = \frac{I \cdot \omega^2}{2}$, with I being it's moment of inertia and ω the angular velocity. For the *Sprint Power Test* the athlete performs a short sprint at 4% load to obtain the maximum, followed by a pause of 5 minutes, followed by the next load step, etc. (4, 8, 10, 12, 14, 16%) until maximum power decreases for the first time: the maximum of the sprint power function is referred to as *SP* (7; 9; 8). A similar approach can also be used for testing the arm and upper body power by means of an arm ergometer. Results from a Sprint Power Test performed on a bicycle ergometer are shown

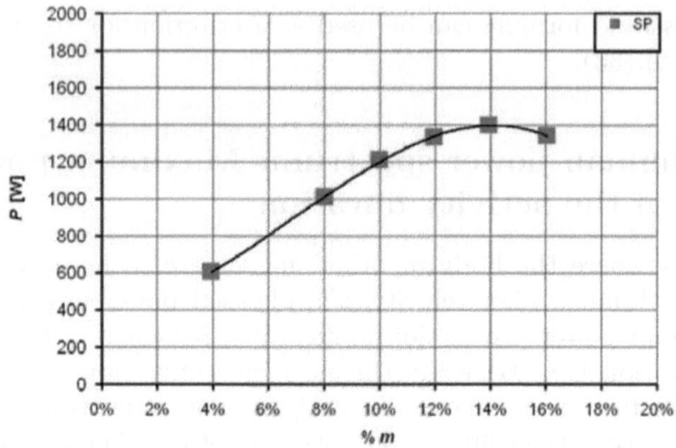

(a) Maximal obtainable power as a function of load

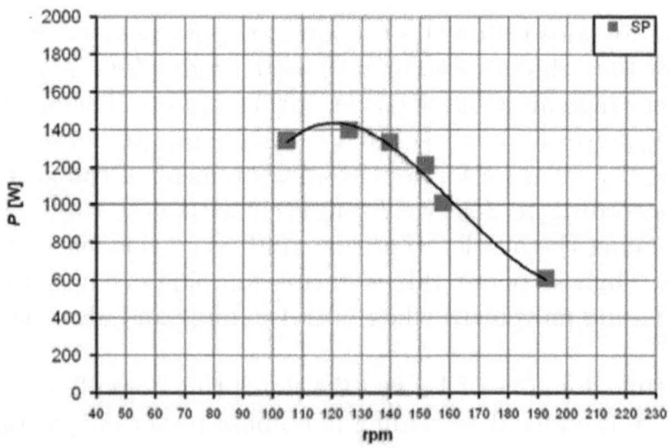

(b) Maximal obtainable power power as function of pedal frequency

Figure 6: Sprint Power Test by W. Müller, 2005 (7; 8; 9). Maximum power (1400 W) was obtained in this example at a load of 14% (of the athlete's body weight) on the mechanical ergometer. The athlete produced the maximum power at 126 pedal rounds per minute.

in Figure 6. In the groups of 28 male (16.2 $\pm$ 0.8) years and 22 female (15.6 $\pm$ 0.9) years finalists of Austrian-wide talent test of the best young athletes from all kinds of sports the following results were found: The mean value of the maximum power obtained in the Sprint Power test was (16.1 $\pm$ 1.6) $W \cdot kg^{-1}$ in the male, and (12.8 $\pm$ 1.5) $W \cdot kg^{-1}$ in the female group. A high correlation ($r = 0.90$) was found between the maximum sprint power and the maximum pedal frequency (8).

(3) For the interval **10 to 100 s**, all-out tests on a mechanical ergometer can be applied. According to (7) all tests of this type are termed Transition Power tests because the dominating power production mode changes from nonoxidative to oxidative in this time domain. Although measurement accuracy and precision problems are associated (W. Müller, personal observation) with the maximum pedal frequency which it starts out from (34; 35), the classical 30 s *Wingate* test in particular with a weight W equal to 7.5% of the person's body weight on the mechanical ergometer has been used for the power spectra shown here; the *Wingate* test was used here because this test is well known (36). The electronic device *Power Analyzer* is capable of measuring the mean power, the lowest power, and the maximum power with high accuracy also in test situations where the angular velocity of the flywheel changes. The test duration as well as the time span for the maximum and minimum power determination can be selected in order to adapt the test to any given situation in a sport. In order to increase the reproducibility and accuracy, tests for this time domain should preferably start out from zero pedal frequency and the rotational energy of the flywheel should be taken into account.

Note: Many of the Wingate test results published in the literature are not corrected for the rotational energy of the flywheel and are thus affected with substantial errors. The *Wingate Anaerobic Test* (36) claims to measure mean power over 30 s and peak power during the first five of the thirty seconds test duration, i.e. from the time on at which the weight (friction force) is put on the flywheel of the ergometer. Since this test starts out from maximum pedal frequency the test person can obtain (before the friction force is applied to the flywheel) the slowing down of the flywheel during the first seconds contributes substantially to the overall power. Ignoring the energy contribution of the flywheel results in high measurement errors; in the extreme case, when the test person does not contribute any power, the flywheel would make 100% of the power. But also including the energy contribution of the flywheel cannot solve the problem of 'peak power' determination in a satisfactory way because firstly, maximum power is usually not obtained with a weight of just 7.5% of the person's body weight on the ergometer (8); an example for this is shown in Fig. 6(a). Secondly, due to muscle physiological reasons maximum muscle power production is not to be expected at the very initial phase of the test when the pedal frequency is still close to the person's maximum velocity.

A *Transition Power Test* which starts from pedal frequency zero would avoid some of the problems associated with the *Wingate* test protocol.

(4) Endurance power (duration of **more than 30 minutes**) can be tested by means of maximum lactate steady state tests (*MLSS*-tests) or by means of ergometer step test protocols. A very effective test protocol (12) starts out from a low power of approximately 20% of the maximum power the athlete can obtain at the end of this step test (typically in the range of 40 W to 80 W) and uses such power increments that maximum power the athlete can obtain is reached in 15 (± 1) steps. The *MLSS*-power can be determined by means of lactate, ventilatory, or heart rate turn point analysis or also by using the relation (percentage p) between the maximum power obtained by the person and the power at *LTP2*, as was described before. Different sports or disciplines necessitate different power spectra. Examples of data typical for a sprinter and for an endurance athlete, both with a body mass of 80 kg, are shown in Fig. 1. The maximum power values obtained in all time domains discussed above and also the relative values in $W{\cdot}kg^{-1}$ are indicated on top of the columns. A comparison of the two athletes shows that the sprinter has substantially higher power values in the short time domains, whereas the endurance power is just about half of the value obtained by the endurance athlete. A comparison of the power values obtained in the various time domains makes clear that maximal power obtainable by an athlete in the sub-second range can be an order of magnitude larger when compared to the time domain of hours (endurance time domain: longer than half an hour). Various power-profiles of different athletes can easily be compared by means of their power spectra. The power spectrum test results in a survey of the maximum mechanical power out-put of a test person in all time domains of relevance in sports.

Final remarks on "human performance": Except for the short discussion in the first part of this chapter we focused here on the mechanical power production of humans and discussed the energy supply mechanisms of the human body as well as the most important physiological responses to physical load. Any human power production is based on principles of bioenergetics. From this point of view the body could be compared to a machine, but it would be simplistic to leave it at this. In order to prevent misunderstandings, it should be pointed out again that the mechanical work a person can produce (conversion of chemical energy into mechanical work) is just one facet of the complex processes going on in a human when performing motions in every day life, in sports, or in the working process. Other major components would be the neuro-mechanical processes of kinesiology involved and the nervous system's abilities for planning and performing the extremely complex regulation tasks associated with human motion. Unlike a machine, the body can adapt to physical stress and improve its functions; conversely, in the absence of appropriate stress the functional potential deteriorates. Performance capabilities change throughout life and can largely be improved by appropriate training. For performance diagnosis in a given sport, the role of mechanical power has to be seen in the context of all other factors determining performance.

5 Power dissipation due to aerodynamic drag and rolling resistance in cycling

When riding a bicycle on a level road the rider must work to overcome the rolling resistance and the drag. The work W is equal to F_t, the tangential force component acting on the tire, multiplied by the path s: $W = F_t \cdot s$. The force acting on the bicycle (ground reaction force) has a component perpendicular to the road surface F_p and the tangential component in the direction of movement F_t (the tread of the tire is bent forward due to this force).

5.1 Rolling resistance

The rolling resistance depends on the consistency of the tire material (rubber type and tread), the tire pressure, the radius of the tire R (inversely proportional), and the composition and structure of the road's surface. The rolling resistance is proportional to the perpendicular force F_p . On a level road the perpendicular force is equivalent to the weight $F_p = m \cdot g$ (on a slope with an angle α, $F_p = m \cdot g \cdot \cos\alpha$), with m being the mass of the bicycle and the rider. The dependence of the rolling resistance F_r on the described factors can be expressed as:

$$F_r = \mu_r \cdot \frac{F_p}{R} \tag{2}$$

The coefficient of rolling friction μ_r has the dimension of a length (contrary to the static and kinetic friction coefficients, which are dimensionless).

The rolling resistance of a bicycle with a rider can be easily measured with a dynamometer, in the simplest case with the aid of a spring balance (Fig. 7). Pulling the bicycle with a constant speed allows determining the coefficient of rolling friction μ_r. The rolling resistance (in first approximation) does not dependent on the velocity. At low velocities (walking speed) the air resistance can be ignored. For racing bicycles with $R = 0.35$ m on smooth asphalt the rolling friction coefficient μ_r is about 0.003 m, for mountain bikes around 0.005 m, depending on the tread; on gravel or in meadows the rolling resistance coefficient can easily become much greater than 0.01 m.

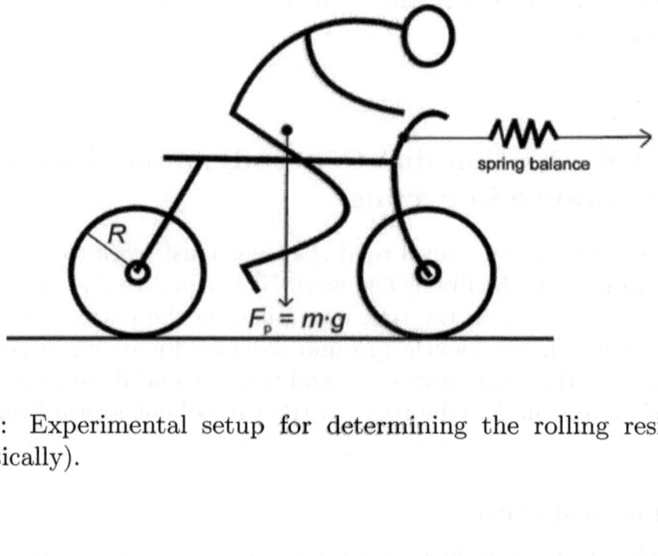

Figure 7: Experimental setup for determining the rolling resistance μ_r (schematically).

$$\mu_r = \frac{F_r}{F_p} \cdot R \tag{3}$$

A very small rolling resistance coefficient is found for steel wheels on a smooth steel surface (10^{-6} m $< \mu_r < 10^{-5}$ m). For car tires on asphalt it lies typically at 7×10^{-3} m. Locomotion with a bicycle makes use of the markedly low rolling resistance; this has fascinated people ever since the invention of it. The wheel is a cultural achievement which has no analogue in nature from which one could have copied it.

5.2 The drag

The drag force F_d increases - conversely to the rolling resistance or to the kinetic resistance – with the square of the velocity:

$$F_d = \frac{\rho}{2} \cdot v^2 \cdot A \cdot c_d \quad (4)$$

with ρ being the air density (1.29 kg·m^{-3} at sea level), A the cross sectional reference area, c_d the drag coefficient, that takes the form of the object into account (c_d can have values between 0.01 and 1.5 depending on the form of an object) and v is the velocity. For two velocities v_1 and v_2 (within the range of Reynold's numbers in which c_d is constant) we get:

$$\frac{F_{d,1}}{F_{d,2}} = \left(\frac{v_1}{v_2}\right)^2 \quad (5)$$

The pronounced role of drag for performance in bicycling makes it very important to minimize it; this has to be done by means of wind tunnel measurements (Fig. 8) because with computational fluid dynamics *CFD* the high accuracy necessary in sports is usually not obtainable when turbulent flow occurs (37).

The Figure 9(a) shows the air resistance F_d of a racing bicycle with a rider. The cross sectional area A is 0.25 m^2 in this case (corresponding to a small rider), c_d is 0.8, and the air density is set to 1.2 kg·m^{-3}. Much effort is spent in top level sports to reduce the drag area D, which is $D = A{\cdot}c_d$.

The rolling resistance F_r is a horizontal line since it is independent from v (Fig. 9: for m = 70 kg, μ_r = 0.003 m and R = 0.35 m); the total frictional resistance F as sum of the drag and rolling resistance is also shown. From the diagram one can also determine the velocity at which the air resistance becomes larger than the rolling resistance (Intersection of the drag curve with the horizontal rolling resistance line). In competitive cycling sophisticated bicycle designs are used in order to minimise drag and to optimise the rider's position.

5.3 The mechanical power during bicycle riding on the level

The required power to overcome the resistance forces at a given velocity $P_1 = F_1{\cdot}v$ is shown in Fig. 9(b). A force must be applied from outside to the

Figure 8: Drag minimisation in the wind tunnel: equipment design and bicyclist's position have a large impact on the aerodynamic forces. Photo: W. Müller.

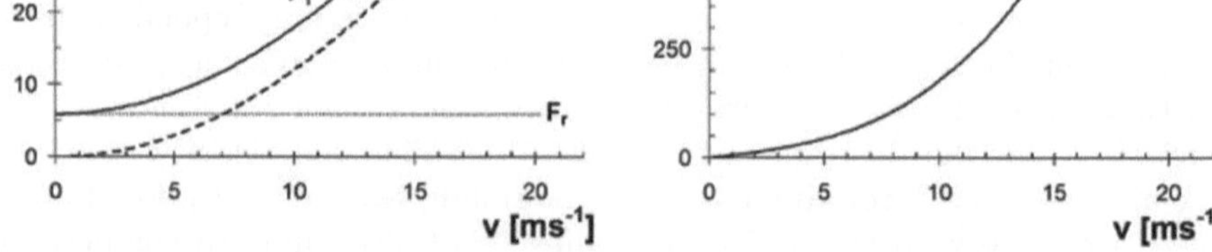

(a) Resistance forces as a function of v (b) Power as a function of v

Figure 9: (a) Resistance forces acting on a rider (including the bicycle): $F_{\text{r}} = 5.9$ N, $F_{\text{d}} = 0.12\ v^2$, $F_1 = F_{\text{r}} + F_{\text{d}}$. (b) Requried power P_1 to overcome the acting force F_1: $P_1 = F_{\text{r}} \cdot v + F_{\text{d}} \cdot v$

observed system (bicycle with rider) to overcome the frictional resistance. In the case of the towing experiment the forwards oriented force applied to the bicycle frame is applied through the spring balance. If the rider was to make use of the pedals the force applied by the rider would result - through the construction of the bicycle - in a backward directed force on the road. According to Newton's third law a force of the same size but opposite direction acts on the system, i.e. the tangential component of the ground reaction force. This ground reaction force acting on the tires is oriented in forward direction which can be directly observed: the tire tread is bent forwards (i.e. in the direction of the movement). Fig. 9 (b) shows the power that the rider has to achieve when he wishes to reach a particular velocity v on a level road. The power required to overcome the rolling resistance increases linearly with the velocity and that to overcome the drag to the third power of the velocity.

5.4 The performance on slopes

On a mountain slope additional work has to be done in order to lift the bicycle with rider (total mass m) in the earth's gravitational field. The potential energy is $E_{\mathrm{pot}} = m \cdot g \cdot H$, with H being the height obtained. The change in potential energy per time is equal to the mechanical lifting power $P_{\mathrm{H}} = \frac{m \cdot g \cdot H}{t}$. The height H depends on the angle of the slope α by $sin\,\alpha = \frac{H}{s}$.

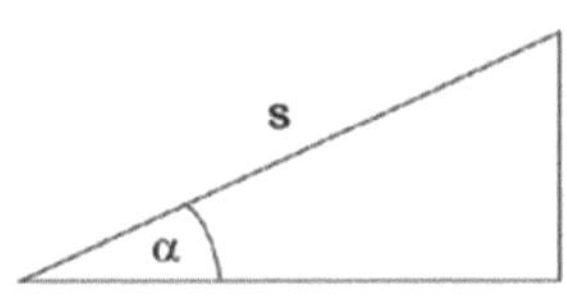

Figure 10: To illustrate the slope angle α of a road.

The performance required for lifting the mass (bicycle with rider) can be given as $P_{\mathrm{H}} = \frac{m \cdot g \cdot s}{t} \cdot sin\,\alpha = m \cdot g \cdot v \cdot sin\,\alpha$. Herewith the total power required can be calculated for riding a bicycle up an incline at a desired speed v:

$$P = P_1 + P_{\mathrm{H}} = F_1 \cdot v + m \cdot g \cdot v \cdot sin\,\alpha \tag{6}$$

The total frictional resistance F_1 includes both the rolling resistance F_{r} and the drag F_{d}. The perpendicular force F_{p} in the rolling resistance equation is to be set as $m \cdot g \cdot cos\,\alpha$ when the road is inclined (slope angle α). Inserting the equations for F_{d} and F_{r} results in:

$$P = \left(\frac{\rho}{2} \cdot v^2 \cdot c_d \cdot A + \mu_r \cdot \frac{m \cdot g \cdot cos\,\alpha}{R}\right) \cdot v + m \cdot g \cdot v \cdot sin\,\alpha \qquad (7)$$

Finally, the consideration of wind $\vec{u}$, additionally to the air stream velocity $(-\vec{v})$ which occurs due to the velocity $\vec{v}$ of the bike, results in the relative wind $\vec{w} = \vec{u} - \vec{v}$; for wind in direction or contrary to the velocity of the bike we get:

$$P = \frac{\rho}{2} \cdot c_d \cdot A \cdot w^3 + \mu_r \cdot \frac{m \cdot g \cdot cos\,\alpha}{R} \cdot v + m \cdot g \cdot v \cdot sin\,\alpha \qquad \text{with } w \equiv |\vec{w}| \quad (8)$$

For calm wind ($\vec{u} = 0$) this equation is identical with the one above because $v \equiv w$ in this case.

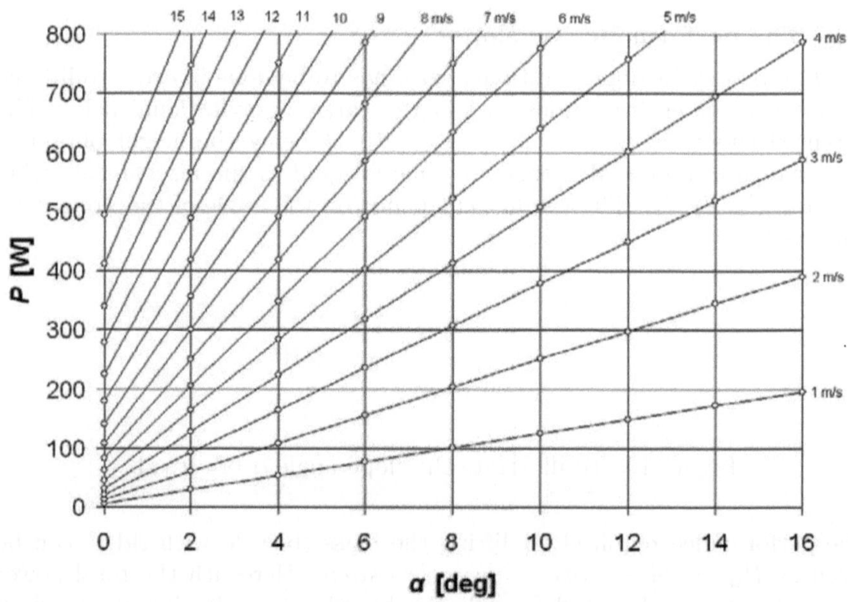

Figure 11: Power, incline and velocity diagram using: $m = 70$ kg, $\mu_r = 0.003$ m, $R = 0.35$ m, $A = 0.25$ m^2; $c_d = 0.8$, $\rho = 1.2$ kg·m^{-3}, $w = 0$, $A \cdot c_d = D = 0.20$ m^2. The D-value chosen for this example of a small rider is at the lower end of the range of D values reported in the literature for aerodynamically optimized positions and bikes (38; 39).

5.5 Interpretation of the equation for total power P

To overcome the air resistance a mechanical power is required that increases with the third power of the velocity; this is true at a wind speed of zero. When a wind $\vec{u}$ occurs, the relative wind speed $\vec{w}$ has to be considered.

With a wind vector $\vec{u}$ blowing with a component from the side too, a torque would act on the bicycle rider and he would have to react to this in order to keep on going straight.

The second term in Eq. 8 shows that the power to overcoming the rolling resistance is proportional to the velocity (the rolling resistance is independent of the velocity $\vec{v}$) and indirectly proportional to the wheel radius R (larger wheels have lower rolling resistance).

The third term in Eq. 8 is the required power for overcoming the incline (angle α). Because the mass occurs in the third term, it can be seen in the equations that light riders on mountain slopes have an advantage. The equation derived allows the determination of the increase of the athlete's power necessary to compensate for increased weight. In reality the best road bicycle competitors on uphill roads are light athletes. This is not necessarily true for mountain bikers as in this sport overcoming difficult passages necessitates high forces. Also for track racing very high forces and absolute power abilities are necessary. Top athletes are able to produce a maximum power of approximately 1.9 kW for time spans of about 10 s. Reaching high speed requires high forces too and that requires a large muscle cross section. Depending on the discipline it can be absolute power or power with respect to the body weight which is of dominating relevance.

In any case the performance is reached through muscle power and this human "power output" can be significantly increased through training ((31). The increase in endurance performance depends largely on the training time when working out below the maximum lactate steady state *MLSS* (low and medium pace training); there is a lower intensity limit above which the training intensity should be: i.e. above the power at the *lactate turn point 1* (*LTP1*). In competitive sport usually 10% and more of the training time is spent in extensive and intensive interval training regimes which are close to or also above the *MLSS* (polarization training, (28)).

The use of an adequate load and with training times of about one hour per day over ten to fourteen weeks can - with an originally untrained person - lead to an increase in the endurance performance of more than 50%. This does apply to healthy people of all ages.

Measurements of power output (using *SRM Trainingssystem*, Jülich, Germany), heart rate, and cadence of 15 professional cyclists during the 2005 Tour de France have recently been published (40). The authors describe results obtained at different terrain categories: flat (*FLT*), semi-mountainous (*SMT*), and mountainous (*MT*) stages. Average power output was: *FLT* (218 ± 21) W or (3.1 ± 0.3) W·kg^{-1}, *SMT* (228 ± 22) W or (3.3 ± 0.3) W·kg^{-1}, and *MT* (234 ± 13) W or (3.3 ± 0.2) W·kg^{-1}. During *MT* stages, the maximum mean power value (*MMP*) for 30 minutes was highest (394 W vs. 342 W) but the maximum mean power value for 15 s was lower (836 W vs. 895 W) compared to *FLT*. Average *HR* was similar between *FLT* (133 ± 10) bpm and *SMT* (134 ± 8) bpm but higher during *MT* (140 ± 3) bpm. Cadence during *MT* was approximately 6–7 rpm lower (81 ± 15) rpm compared to *FLT* or *SMT*.

There is much controversy about the power needed to break the one-hour world record. While the calculated mean power based on wind tunnel data was 510 W for Indurain's record (1) other researchers, using mathematical models, calculated a mean power of only 436 W (38). A comparison of theoretical models for estimating the mechanical power output in cycling has been published by C. Gonzales-Haro et al. (41).

Wind tunnel measurements of aerodynamic drag in elite cyclists during effort (5 W·kg^{-1}) and a comparison of the obtained data with studies on aerodynamic drag in professional cyclists from other research groups using various measurement methods can be found in a publication by J. Garcia-Lopez et al. (39). The authors found that aerodynamic drag increases by more than 30% during effort when compared to static measurements.

6 Measurement of human power on the mechanical cycle ergometer

The measurement of the physical power of bicycle riders is of significant interest: For overcoming the air and rolling resistance and inclines when bicycle riding, a mechanical power P is necessary. At given resistance relationships and slope of the road the maximum deliverable power determines the maximum velocity. A good endurance performance (high power P at the *LTP2*) enables a high velocity over a longer period of time to be reached without tiring. Cycling ergometry is the basis of most performance tests for cyclists. Most of them are stationary devices for the measurement of power while a cyclist pedals against sliding friction, electromagnetic braking, or air resistance. Mobile ergometers allow power measurement during training or competition on the road, in a velodrome or in the laboratory. An analysis of systematic and random errors of various types of ergometers has been published by Paton and Hopkins (42). Here, from all types of cycle ergometers the focus is on the mechanical weight ergometer. It was the first type of cycle ergometer which has already been described in 1954 (43) and it is still being used in many laboratories because of it's major advantage: it is easy to calibrate it correctly.

6.1 The weight ergometer

Using the bicycle ergometer the rider must deliver a power to overcome a variable frictional resistance on a "stationary" cycle. The principle of the weight ergometer of the Monark type can be seen in Figure 12. The test person pedals on the ergometer as on a bicycle, with the crank driving a large cog on the crank axel (52 teeth), and the chain, and a small cog on the axel of the flywheel (front wheel) transmits the energy to the flywheel. The small cog has 14 teeth, whereby this results in that the flywheel rotates 52/14 times faster than the pedal axel (in order to store as much rotational energy so that the apparatus runs smoothly). On the circumference of the flywheel a braking mechanism is used with the help of a rope. Due to the patented construction the value of the frictional force $|\vec{F}_\mathrm{f}|$ acting on the rope is equivalent to the value of the weight $|\vec{W}|$ used. The mechanical power is equivalent to the force W times the circumferential velocity v_c of the flywheel; this power $P_\mathrm{f} = W \cdot v_\mathrm{c}$ is dissipated as thermal energy. The load weight W is equivalent to the load mass m_l put on the ergometer times g, the gravitational acceleration (g = 9,81 m·s^{-1}): $W = m_\mathrm{l} \cdot g$. It has to be taken into account that the mass of the weight basket is 1 kg when using Monark ergometers.

The mechanical cycle ergometer allows determination of the power dissipated at the flywheel. Since it functions without electricity and independently of the actual coefficient of friction between the flywheel and the rope, it can also be used advantageously in the field. Many fundamental exercise physiological investigations from Per-Olof Åstrand (44) in the middle of the last century to current work were carried out on mechanical ergometers with the power being measured at the flywheel. The power at the pedal is noticeably higher (about 5 - 8%) due to the chain friction and the bearings (42).

It is notable that – due to the frictional force clamped to the value of the weight W (by means of the turn round mechanism with the cylinders C and C_1) - the measurement is independent of the frictional coefficient between the flywheel and the rope. Due to the mechanical construction of the weight ergometer the tension in the rope is regulated such that the value of the frictional force is equal to the weight force (torque equilibrium occurs at cylinder C). The automatic adjustment of the rope, when the point of balance of the torques at cylinder C occurs, is reached on the weight ergometer in that both ends of the rope are kept at different distances from the rotational axis of the cylinders C and C_1. If the rope, due to the friction at the flywheel is pulled upon at the start of the rotation, the cylinder C will be turned and due to its larger radius unroll more rope than is rolled up on the smaller concentric cylinder C_1. In other words, the tension in the rope is reduced, which is then more loosely wrapped around the flywheel such that the value of the frictional force F_f equals that of the weight force W. (Equilibrium of forces on the cylinder C means equilibrium of the torques; the end of the rope fixed on cylinder C_1 is in the state of equilibrium force free).

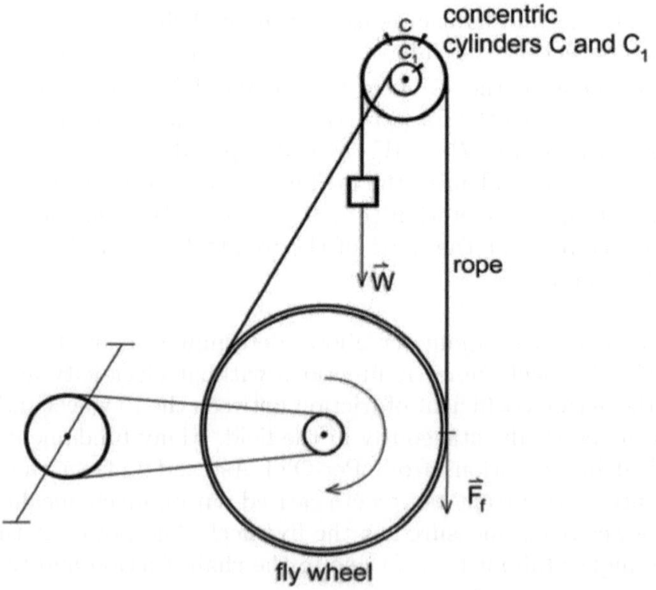

Figure 12: Principle of the weight ergometer (*Monark*).

6.2 Power measurement including the rotational energy of the flywheel

The use of an adequate electronic device for measuring the mechanical power (e.g.: *Power Analyzer*; www.bewotech.com) enables a highly accurate determination of power at the flywheel according to $P_f = W \cdot v_c$, in that the angular velocity of the flywheel $w_f = \frac{v_c}{R}$ is measured with sufficient accuracy directly at the flywheel (and not at the pedal) and the rotational energy is calculated (the fly wheel's moment of inertia of Monark-type ergometers is $I = 0.91$ kg·m^{-2} and its radius $R = 0.2571$ m). The weights can be easily calibrated to an accuracy of 1 g, e.g. by making use of bore holes to adjust for differences in the cast weights. The power at the flywheel can be determined with errors of approximately 1%. The *Power Analyzer* also allows the recording of the maximum and minimum power during a defined time span as well as the average power and also determines the rotational energy of the flywheel. During the acceleration in a sprint a considerable percentage of the mechanical power produced by the subject is needed to accelerate the fly wheel. This kinetic energy has to be considered when human output power is to be determined accurately. Conversely, the flywheel during rotational deceleration sets energy free which the test person does not have to deliver. The rotational energy E_{rot} of the fly wheel depends on its moment of inertia I and the angular velocity ω_f and is given by $E_{rot} = \frac{I \cdot \omega_f^2}{2}$. Any reliable test approach at non-constant rotational velocities of the flywheel must account for the power contribution of the flywheel.

Instead of weights the mechanical bicycle ergometer can also be constructed with a pendulum which makes the handling in training easier. The pendulum can easily be calibrated by means of weights; but highest accuracy for the power measurement on the flywheel is obtained with weight ergometers. It has to be considered that the athlete's power output is noticeably higher (about 5 – 8%) when measured at the pedals due to the chain and bearing friction (42). An additional advantage when compared to many electrically operated cycle ergometers is that weight ergometers are available for power measurements up to 2000 W and several test procedures for maximum power tests necessitate the pedal frequency dependent power characteristic. The mechanical pendulum ergometer can also be operated using a power regulation loop (the position of the pendulum is regulated) in order to perform incremental power tests independently of pedalling frequency.

Acknowledgements I would like to thank C. Cagran for constructive comments and for the layout, A. Fürhapter-Rieger for preparing the manuscript, and T. DeVaney for linguistic improvements.

Bibliography

[1] S. Padilla, I. Mujika, F. Angulo, and J. J. Goiriena. Scientific approach to the 1-h cycling world record: a case study. *Journal of Applied Physiology*, 89(4):1522–1527, 2000.

[2] I. Mujika and S. Padilla. Physiological and Performance Characteristics of Male Professional Road Cyclists. *Sports Medicine*, 31(7):479–487, 2001.

[3] S. Vogt, L. Heinrich, Y. O. Schumacher, A. Blum, K. Roecker, H.-H. Dickhuth, and A. Schmid. Power ouput during stage racing in professional cycling. *Medicine & Science Sports & Exercise*, 38(1):147–151, 2006.

[4] S. Vogt, K. Roecker, Y. O. Schumacher, T. Pottgiesser, H.-H. Dickhuth, A. Schmid, and L. Heinrich. Cadence-Power-Relationship during Decisive Mountain Ascents at the Tour de France. *International Journal of Sports Medicine*, 29(3):244–250, 2008.

[5] D. T. Martin, B. McLean, C. Trewin, H. Lee, J. Victor, and A. G. Hahn. Physiological Characteristics of Nationally Competitive Female Road Cyclists and Demands of Competition. *Sports Medicine*, 31(7):469–477, 2001.

[6] N. P Craig and K. I. Norton. Characteristics of Track Cycling. *Sports Medicine*, 31(7):457–468, 2001.

[7] W. Müller and B. Schmölzer. The human power spectrum: Maximum physical power is a function of activity duration. *Isokinetics and Exercise Science*, 14(2):137–139, 2006.

[8] W. Müller and A. Fürhapter-Rieger. The Wingate Test underestimates maximum power: The Sprint Power Test measures maximum power as a function of load. *Isokinetics and Exercise Science*, 14(2):201–203, 2006.

[9] A. Fürhapter-Rieger and W. Müller. Maximum sprint power on the bicycle ergometer at high load: correlation with maximum pedal frequency at low load. *Journal of Biomechanics*, 39(Supplement 1: 5th World Congress of Biomechanics):S560, 2006.

[10] G. A. Brooks, T. D. Fahey, and K. M. Baldwin. *Exercise Physiology - Human Kinetics and Its Application, 4th edition.* McGraw-Hill, 2005.

[11] P. Hofmann, V. Bunc, H. Leitner, R. Pokan, and G. Gaisl. Heart Rate Threshold Related to Lactate Turn Point and Steady State Exercise on

Cycle Ergometer. *European Journal of Applied Physiology*, 69(2):132–139, 1994.

[12] P. Hofmann, R. Pokan, F.-J. Seibert, R. Zweiker, and P. Schmid. The heart rate performance curve during incremental cycle ergometer exercise in healthy young male subjects. *Medicine & Science Sports & Exercise*, 29(6):762–768, 1997.

[13] E. F. Coyle. Improved muscular efficiency displayed as tour de france champion matures. *Journal of Applied Physiology*, 98(6):2191–2196, 2005.

[14] M. Tokui and K. Hirakoba. Effect of internal power on muscular efficiency during cycling exercise. *European Journal of Applied Physiology*, 101(5):565–570, 2007.

[15] J. Cronin and G. Sleivert. Challenges in understanding the influence of maximal power training on improving athletic performance. *Sports Medicine*, 35(3):213–234, 2005.

[16] H. de Marées. *Sportphysiologie, 9. erweiterte Auflage*. SPORT und BUCH Strauß, 2003.

[17] F. Conconi, M. Ferrari, P. G. Zigli, P. Droghetti, and L. Codea. Determination of the anaerobic thershold by a noninvasive field test in runners. *Journal of Applied Physiology*, 52(4):869–873, 1982.

[18] M. E. Bodner and E. C. Rhodes. A Review of the Concept of the Heart Rate Deflection Point. *Sports Medicine*, 30(1):31–46, 2000.

[19] M. Wonisch, P. Hofmann, F. M. Fruhwald, R. Hödl, G. Schwaberger, R. Pokan, S. P von Duvillard, and W. Klein. Effect of β_1-selective adrenergic blockade on maximal blood lactate steady state in healthy men. *European Journal of Applied Physiology*, 87(1):66–71, 2002.

[20] G. A. Brooks, H. Debouchaud, M. Brown, J. P. Sicurello, and C. E. Butz. Role of mitochondrial lactate dehydrogenase and lactate oxidation in the intracellular lactate shuttle. *Proceedings of the National Academy of Sciences of the Unites States of America*, 96(3):1129–1134, 1999.

[21] G. A. Brooks. The lactate shuttle during exercise and recovery. *Medicine & Science Sports & Exercise*, 18(3):360–368, 1986.

[22] G. A. Brooks. Mammalian fuel utilization during sustained exercise. *Comparative Biochemistry and Physiology - Part B: Biochemistry & Molecular Biology*, 120(1):89–107, 1998.

[23] G. A. Brooks. Current Concepts in Lactate Exchange. *Medicine & Science Sports & Exercise*, 23(8):895–906, 1991.

[24] S. Padilla, I. Mujika, G. Cuesta, and J. J. Goiriena. Level ground and uphill cycling ability in professional road cycling. *Medicine & Science Sports & Exercise*, 31(6):878–885, 1999.

[25] R. Pokan, N. Bachl, W. Benzer, P. Hofmann, K. Mayr, P. Schmid, G. Smekal, and G. Wonisch. Leistungsdiagnostik und Trainingsherzfrequenzbestimmung in der kardiologischen Rehabilitation. *Journal für Kardiologie*, 11(11):446–452, 2004.

[26] P. Hofmann, M. Wonisch, and R. Pokan. Laktatdiagnostik - Durchführung und Interpretation. In *Kompendium der Sportmedizin - Physiologie, Innere Medizin und Pädiatrie*, pages 103–132. Springer Verlag, 2004.

[27] P. Hofmann, K. Dohr, F.-J. Seibert, M. Wonisch, R. Pokan, G. Smekal, and G. Schwaberger. Relationship between Lactate Turn Point and Maximal Performance in Young Healthy Male and Female Subjects of Different Exercise Performance Level. *ECSS 2008 - accepted*, 2008.

[28] J. Esteve-Lanao, C. Foster, S. Seiler, and A. Lucia. Impact of Training Intensity Distribution on Performance in Endurance Athletes. *Journal of Strength and Conditioning Research*, 21(3):943–949, 2007.

[29] A. Philp, A. L. Macdonald, H. Carter, P. W. Watt, and J. S. Pringle. Maximal Lactate Steady State as a Training Stimulus. *International Journal of Sports Medicine*, Feb. 26, 2008(online first), 2008.

[30] K. A. Burgomaster, S. C. Hughes, G. J. F. Heigenhauser, S. N. Bradwell, and M. J. Gibala. Six sessions of sprint interval training increases muscle oxidative potential and cycle endurance capacity in humans. *Journal of Applied Physiology*, 98(5):1985–1990, 2005.

[31] E. F. Coyle. Very intense exercise-training is extremly potent and time efficient: a reminder. *Journal of Applied Physiology*, 98(5):1983–1984, 2005.

[32] T. Meyer, A. Lucía, C. P. Earnest, and W. Kindermann. A Conceptual Framework for Performance Diagnosis and Training Prescription from Submaximal Gas Exchange Parameters - Theory and Application. *International Journal of Sports Medicine*, 26(Supplement 1):S38–S48, 2005.

[33] P. Hofmann, P. Niederkofler, R. Pokan, and V. Bunc. Comparison between heart rate threshold and individual physical working capacity. *Acta Universitas Carolinae - Kinanthropoligia*, 32(1):47–50, 1996.

[34] J. Balmer, S. R. Bird, R. C. Davison, M. Doherty, and P. M. Smith. Mechanically braked Wingate powers: agreement between SRM, corrected and conventional methods of measurement. *Journal of Sports Sciences*, 22(7):661–667, 2004.

[35] D. Micklewright, A. Alkhatib, and R. Beneke. Mechanically versus electro-magnetically braked cycle ergometer: performance and energy cost of the Wingate Anaerobic Test. *European Journal of Applied Physiology*, 96(6):748–751, 2006.

[36] O. Inbar, O. Bar-Or, and J. S. Skinner. *The Wingate Anaerobic Test.* Human Kinetics, 1996.

[37] W. Meile, E. Reisenberger, M. Mayer, B. Schmölzer, W. Müller, and G. Brenn. Aerodynamics of ski jumping: experiments and cfd simulations. *Experiments of Fluids*, 41(6):949–964, 2006.

[38] D. R. Basset Jr., C. R. Kyle, L. Passfield, J. P. Broker, and E. R. Burke. Comparing cycling world hour records, 1967-1996: modeling with empirical data. *Medicine & Science Sports & Exercise*, 31(11):1665–1676, 1999.

[39] J. García-López, J. A. Rodríguez-Marroyo, C.-E. Juneau, J. Peleteiro, A. C. Martínez, and J. G. Villa. Reference values and improvement of aerodynamic drag in professional cycling. *Journal of Sports Sciences*, 26(3):277–286, 2008.

[40] S. Vogt, Y. O. Schumacher, H. Roecker, U. Schoberer, A. Schmid, and L. Heinrich. Power Output during the Tour de France. *International Journal of Sports Medicine*, 28(9):756–761, 2007.

[41] C. González-Haro, P. A. Galilea Ballarini, M. Soria, F. Drobnic, and J. F. Escanero. Comparison of nine theoretical models for estimating the mechanical power output in cycling. *British Journal of Sports Medicine*, 41(8):506–509, 2007.

[42] C. D. Paton and W. G. Hopkins. Tests of Cycling Performance. *Sports Medicine*, 31(7):489–496, 2001.

[43] W. von Döbeln. A Simple Bicycle Ergometer. *Journal of Applied Physiology*, 7(2):222–224, 1954.

[44] P.-O. Åstrand, K. Rodahl, H. A. Dahl, and S. B. Strømme. *Textbook of Work Physiology - Physiological Bases of Exercise, 4th edition.* Human Kinetics, 2003.

Skin Suit Aerodynamics in Speed Skating

Lars Sætran and Luca Oggiano
Norwegian University of Science and Technology, N-7491 Trondheim, Norway

1 Introduction

Performances analysis in sports is complex since many factors linked together contribute to the final result thus a quantification of the effects of different suits on speed skating performances just with field measurements is a difficult task. In fact, field measurements and competition results do not clearly show the effect of using different suits. Industries and Olympic committees have then been pushed to increase the number of laboratory tests on materials, apparels and equipments in order quantify the effects of different textiles on skating suits. In some sports like cycling and speed skating the speed is entirely determined by the equivalence between external power and the power lost both by frictional losses and in order to increase the speed ((Di Prampero et al 1979a) and (Ingen-Schenau 1982)). Forces acting against the athlete and power dissipated are related with the equation:

$$P = FV \tag{1}$$

where F is Force and V is velocity. Drag, in these sports, is the most important among the frictional losses. In cycling and speed skating, when the athlete reach a constant speed, D is about 80% of the total force while the frictional force is only about 20%. Drag acting on the athletes has been often represented as:

$$D = \frac{1}{2} A_p C_d \rho V^2 = KV^2 \tag{2}$$

K has been considered as constant both in cycling (Di Prampero et al. 1979b), and speed skating (Di Prampero et al 1979a). This was mostly due to the fact that the athletes were not able to reach the critical speed which cause the fall in Cd due to the change from laminar to turbulent regime and the consequent reduction of pressure drag. Transition of the flow around the athlete from laminar to turbulent and the consequent drop in terms of drag has been predicted by (Pugh 1970). However he has not been able to reach the critical speed value in his wind tunnel experiments in order to verify his prediction. Pugh prediction is strictly connected to studies carried out in the past on bluff bodies and in particular on spheres and cylinders. The introduction of a surface roughness on cylinders can shift the transition at lower speed. Usually the rougher the surface is, the lower the value of the critical speed is. This behaviour has been shown from Achenbach (1964) experiments for the spheres and from Bearman, Harvey (1983) experiments for cylinders. (Ingen-Schenau 1982) showed that is not constant but dependent on different factors. Among these factors also the speed is mentioned. The effect of different suits has not been taken into account in Van Ingen Schenau experiments but only effects of postures and speed have been analyzed. The technological development of new textiles and materials has improved the aerodynamics of speed skating suits. The use of rough patterns (as it has been done for the golf balls) in order to reach the critical speed at lower speeds reduces the drag of the suits. In the present paper then K will be considered as dependent on the speed and on the different suits.

$$K = K\left(V, suit\right) \tag{3}$$

The effect due to different postures has not been considered since the effect of it has been already discussed by (Ingen-Schenau 1982). In the experiments carried out in the wind tunnel with a full scale doll model has been reached and the fall in happened. These experimental results, associated with field measurements made by Len Brownlie which showed that, on average, skaters who used the Nike swift suits performed almost 1% better than their previous personal record permits to declare that suits have an important effect on speed skating performances. Kuper and Sterken (2003) analyzed the effect of skating suits on skaters performances and they found out that some suits, especially in long-distance events, significantly increase the average skating speed. It has been found that the Nike swift skin suit can increase the speed by up to 0.2-0.3 seconds per lap on a 400m oval. The target of this work is to give an overview on the effects of textiles with different roughness implemented in speed skaters suits with particular attention to legs. A test on 6 different speed skaters suits and on the effect of different textiles on legs athletes has been performed using a doll positioned in the wind tunnel. The position of the doll has been chosen in order to reproduce the position that speed skaters utilize during a 1500m race.

1.1 Forces acting on a skater

The total power due to friction losses that a speed skater needs to overcome during his race is given by the sum of power spent to overcome the drag resistance (P_D) and power needed to overcome the skate-ice friction (P_F).

$$P_T = P_D + P_F \tag{4}$$

Aerodynamic forces Aerodynamic forces are about 80% of the total forces acting on a skater. (Ingen-Schenau 1982) showed that the intensity of the aerodynamic forces acting on a speed skater is related to the position assumed during the race. It has been shown that for both θ_2 and θ_3 (fig.3) there is a linear correlation between the angles and the drag. The greater the θ_2 or θ_3 are, the higher the drag is. From previous works De Koning et al (1992b) the average of the angle θ_3 is about 25degrees and the average of the angle θ_2 is about 115degrees. The total power spent in order to act against the aerodynamic resistance can be calculated as:

$$P_D = \frac{1}{2} A_P C_D \left(V, suit\right) \rho V^2 V \tag{5}$$

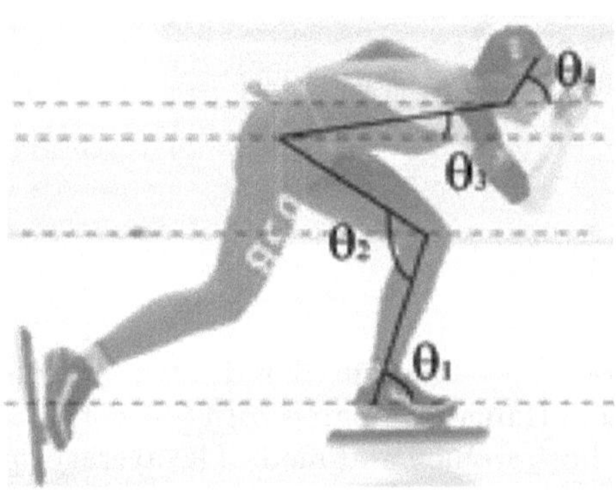

Figure 1. Angles that characterize a speed skater posture

Friction Experiments have been carried out in order to measure the friction coefficient (μ) for a speed skater. The mean friction coefficients for the straights and curves are, respectively, 0.0046 and 0.0059 De Koning et al (1992a). The total power spent to overcome the friction losses can be written as:

$$P_F = \mu mgV \tag{6}$$

1.2 Speed

Data regarding 28 men athletes and 17 women athletes competing in a national Norwegian speed skating competition (Hamar 13-14 October 2007) have been acquired. For the 1500m race, the average speed for each interval has been calculated for each athlete and distance-speed curves regarding the averaged performaces have been plotted for women and men competition. These data have then been compared to the data relatives to the world record in the 1500m. Furthermore, the average speed for the different disciplines (500m, 1000m, 1500m, 3000m, 5000m, 10000m) have been calculated for the world records performances. In short distance disciplines the speed is higher and increases and varies during the race while in long distance disciplines, where a constant speed during the whole race is important, the speed is lower. A difference of about 1.5m/s between ladies and men has been found.

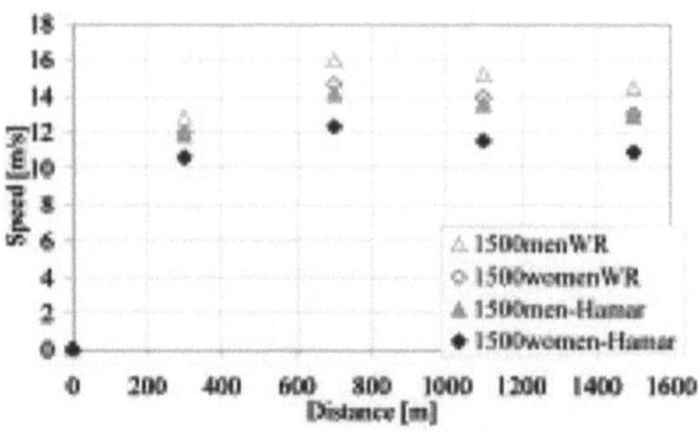

Figure 2. (a)Average speed measured at 4 different time intervals during a 1500m competition in Hamar compared with the world records. Speeds have been measured for both women and men. The average speed for the men in the national competition in Hamar is comparable with the speed measured for the word women record. (b)Average speed during world record events for women and men. Each point on the graph represents the average speed calculated during the whole event. Women have, in average, a lower speed than men.

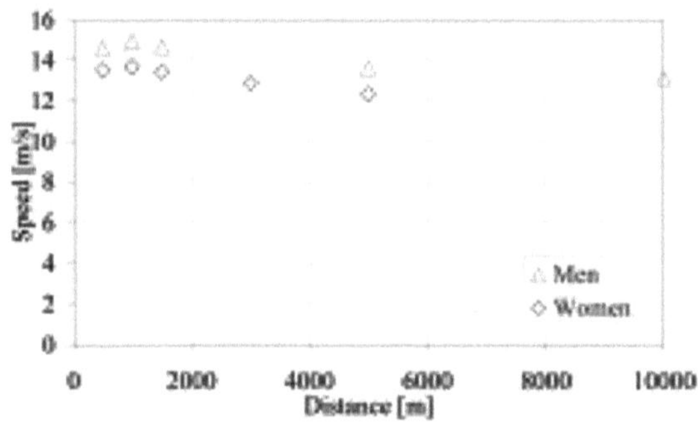

2 The power balance model

In order to evaluate the effect of using different skin-suits, the power balance model developed by Koning De Koning et al (1992b) has been adopted. There are several examples of power equations which can be used to estimate performances in different sports. For speed skating, the equation can be written as De Koning. (1992b, 2005)

$$P_u = P_D + P_F + \frac{dE_{mcb}}{dt} \tag{7}$$

Where P_u is the total power in output generated by the athlete, P_D is the power lost with frictional losses due to drag, P_F is the power lost with frictional losses due to friction between skates and ice and dE_{mcb}/dt is the power output spent to increase the kinetic energy.

The efficiency of the power balance model in speed skating performance estimate has been shown by De Koning. (1992b, 2005); in these articles P_u has been experimentally calculated measuring the oxygen uptake during the skating action De Koning. (1992b, 2005). The power balance equation 7 can be used to estimate a reference power output P_{uR} if friction coefficient (μ), drag coefficient (C_d) and the instantaneous velocity of the skater are known. In this case P_{uR} is the only unknown and it can be calculated as a function of time. The instant speed has been calculated from data acquired during a Norwegian national speed skating competition (Hamar 13-14 October 2007). A constant friction coefficient (μ = 0,005) have been used. A constant C_{dA} (C_{dA} = 0.3) has been estimated using the data acquired by (Ingen-Schenau 1982).

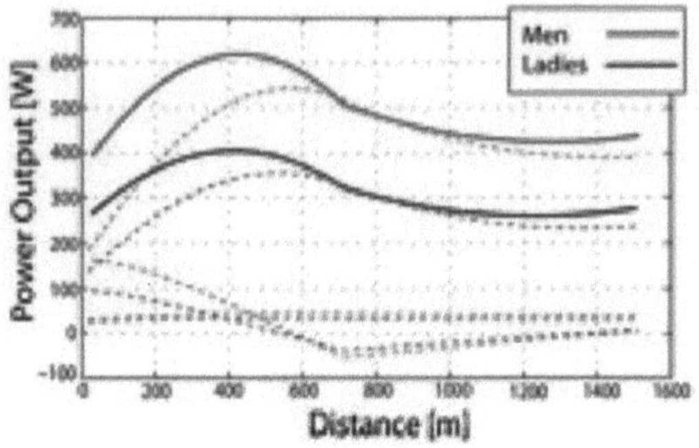

Figure 3. Reference Power outout calculated using the time intervals acquired during the Norwegian cup (Hamar 13-14 October 2007) for the 1500m competition (ladies and men).

3 Methods

3.1 Wind tunnel

For the experiments, the wind tunnel of NTNU (Norwegian University of Science and Technology) in Trondheim has been used. The contraction ratio is 1:4,23, and the test section of the wind tunnel is 12.5 meters long, 1.8m high, and 2.7m wide. The wind tunnel is equipped with a 220kW fan that can produce a variation of speed between 0.5 - 30m/s. The balance (Carl Schenck AG) used is a six components balance capable to measure the three forces and the three momentums around the three axes. Variations of forces and momentum are measured using strain gauges glued to the balance body. The voltage outputs are measured by a LABVIEW based PC program

3.2 Suits test

6 different suits have been tested in order to evaluate the effect of each suit in a 1500m speed skating competition. The suits tested have been numbered from 1 to 6 and some of them are currently used by top level athletes. Each suit has been mounted on a doll and the doll has been mounted in the wind tunnel. The posture of the doll has been chosen following the data previously acquired about the average posture of a speed skater in a 1500m competition ($\theta_2 = 25^o, \theta_3 = 115^o$). The drag of each suit has been measured in a range of speed (s) from 8m/s up to 18m/s and a CdA-Speed curve has been then plotted for each suit. The CdA-Speed

curves acquired have been introduced in a Power balance model and a final time in the performance has been estimated for each suit.

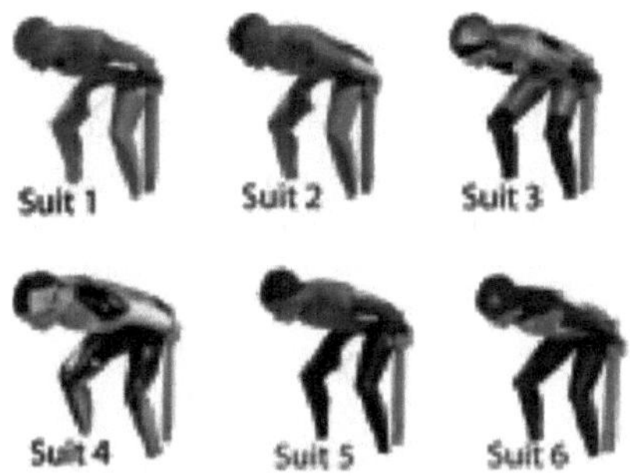

Figure 4. A picture of the 6 suits tested. All suits have different patterns in different parts of the body. Suits 1,3,5,6 have a moderate rough textile on their legs while suit 2 has an extreme rough pattern and suit 4 is totally smooth.

The suits tested are all different and produced by different manufactures. A level of smoothness (s) has been assigned for each textile. The smoothness level varies from 1 to 5 (1 = very smooth, 5 = very rough) and it has been assigned to each part of the suit (lower legs, upper legs, trunk, hat, arms). Smoothness and roughness levels are important in order to reduce the drag. Considering the human body as composed by a series of spheres and cylinders and knowing that it is possible to shift the transition to turbulent regime (in spheres and cylinders) at different speeds by using rough textiles, it is then possible to sensible reduce the drag choosing the right textile (Oggiano et al 2007).

	Suit 1	Suit 2	Suit 3	Suit 4	Suit 5	Suit 6
Lower leg	3	5	3	1	3	3
Upper leg	2	1	2	1	2	2
Trunk	1	1	1	1	1	1
Head	3-1	1	3-2	1	3-2	3-2
Arms	1	3	2	2	1	2

Figure 5. Suits descrirtion

In order to give a more quantitative idea of the smoothness factor used, it is possible to correlate the smoothness factor with a roughness coefficient calculated for cylinders. Based on structure width and depth, a surface parameter can be defined:

$$k_{surface} = \sqrt{wd} \tag{8}$$

The surface curvature (given by the cylinder diameter *Diam*) determines the pressure gradient that influences the flow separation conditions. It is possible to combine the two parameters in a dimensionless roughness coefficient.

$$r = \frac{k}{Diam} \times 10^3 \tag{9}$$

s	*r*
1	*r<10*
2	*10<r<30*
3	*30<r<60*
4	*60<r<90*
5	*r>100*

Figure 6. Relation between smoothness coefficients and roughness parameter

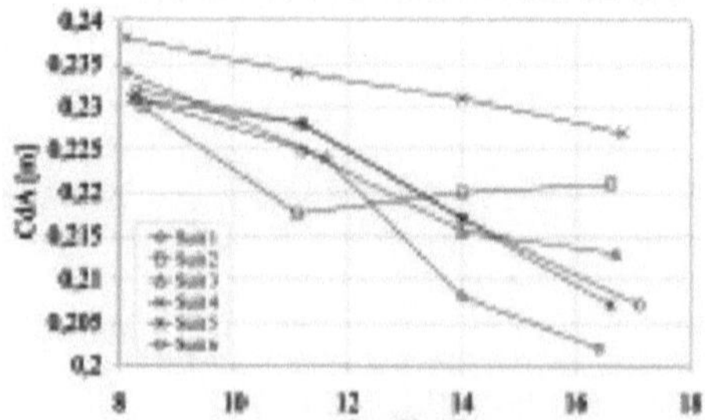

Figure 7. $CdA_{suit} - Speed$ curves for 6 different suits. Suit 2 has a rough textile and reach the transition at lower speed, while suit 4 having a smooth textile on the legs does not reach transition before 16m/s, which is the maximum speed for a speed skater

4 Results

Equation 7 has been used to calculate the instant speed of the skater during the race using different skating. The explicit formula can be written as:

$$V^{i+1} = \sqrt{\frac{2}{m}\left[\left(P^i_{uR} - P^i_D - P^i_f\right)\Delta t + \frac{1}{2}mV^{i2}\right]} \tag{10}$$

Where V^{i+1} is the speed calculated at the instant step $i+1$ and V^i is the speed calculated at the previous step i. All the quantities at the step i are known. The power output is the reference power output P^i_{uR} shown in Figure 3 and calculated as described in paragraph 2. P^i_D can be written as follow:

$$P^i_D = \frac{1}{2}\rho V^{i3} C_d A_{suit} \tag{11}$$

The power dissipated through ice friction can be written as:

$$P^i_f = \mu m g V^i \tag{12}$$

$C_d A_{suit}$ is now a function of the speed and it changes for each suit. A friction coefficient $\mu = 0.005$ is used. The final time for a 1500m race is then estimated:

	Suit 1	Suit 2	Suit 3	Suit 4	Suit 5	Suit 6	Hybrid
Ladies	125,84	126,24	126,34	128,34	126,84	126,84	124,54
Boys	107,54	109,34	108,74	110,64	108,44	108,54	107,64

Figure 8. Final times race calculated for the different suits

Suit 1 showed the best results while suit 4 showed the worst. This is due to the fact that the textiles adopted on suit 4 are too smooth and it has been impossible to reach the transition to turbulent regime on the legs and then reduce the drag. Suit 2 showed a different behaviour for girls and boys. This is due to the fact that the textiles used is on the legs are extremely rough and shift the transition at lower speed than any other of the suits tested, but at the same time increase the $C_d A_{min}$ at higher speed. Considering that the maximum speed (during a 1500m competition) for the women is around 14m/s and for men is around 16 m/s suit 2 gives better performances if used by girls and permits to gain almost half of second. It is interesting to notice that the difference between the performances simulated with suit 1 and suit 4 is 3.31s for the men and 2.31 second for the women.

Considering the Olympic Games in Torino 2006, a time difference of 3.1s for the men and 2.5s difference for the women was the difference between the 1st classified and the 23rd classified for the men and the 1st classified and the 5th classified for the women.

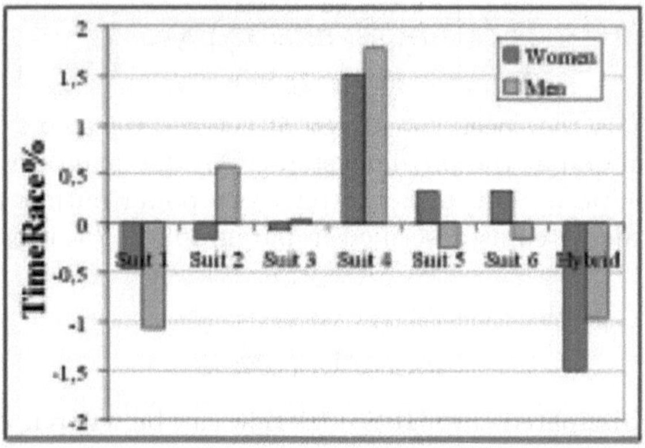

Figure 9. Comparison between each suit and the average time race.

In order to evaluate the effect of the textiles on the legs a hybrid suit has been created. The legs of suit 2 has been covered with a smoother textile (s = 4) and the Drag has been acquired. A comparison between the C_dA curves for suit 2 and the hybrid suit shows that the introduction of a smoother textile on the legs gives better performances in terms of drag reduction. The final time calculated the hybrid suit is 1.7s lower than the one calculated using the suit 2. A comparison has been made calculating the average time race.

$$TimeRace\% = \frac{Timerace_{suit} - Timerace_{average}}{Timerace_{average}} \times 100 \qquad (13)$$

5 Discussion

The results presented above gives an indication of the effect, of the seven suits tested, on speed skaters performances on the 1500m race. It has been noticed that the effect of the textile used on the legs is the most important in terms of drag reduction. A rough textile can reduce the drag if compared to a smooth one but, if the smoothness coefficient is too high, it can affect the performances negatively. In order to understand how different parts of the body moves during a skating action, a deeper look into the skaterś

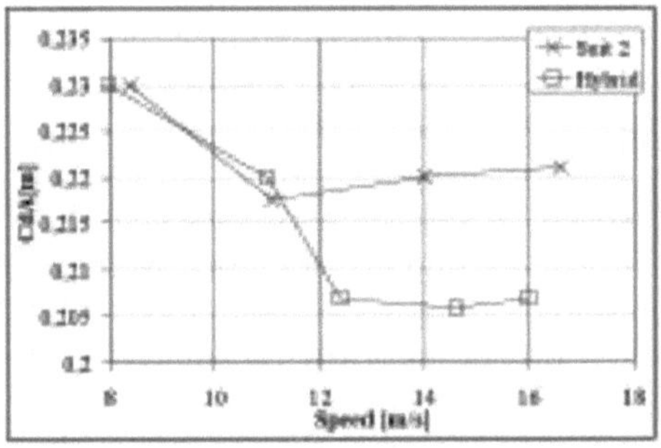

Figure 10. $C_dA - Speed$ curve for hybrid suit and suit 2.

position assumed during a race is needed. The posture tested in the wind tunnel is a static position but, during the race, a speed skater changes his posture and this can affect the aerodynamic parameters. The posture of a skater during a race can roughly be described with three parameters: the angles between the different parts of the body, the diameter of those parts and the change in distance between the legs.

By dividing a lap in two different parts (the curve and the straight part), it is possible to notice some difference in the posture assumed during the straight and the posture assumed during the curve. During the curve the skater has to compensate the centrifugal force by bending his body and shifting his centre of mass. In this phase the right arm is often used in order to find the right balance and increase the thrust. During the straight part the skater keep both the arms attached to the body in order to reduce the frontal area and then reduce the aerodynamic resistance. In short distance competitions and during the final sprint, the arms are not kept attached to the body but used in order to increase the thrust.

The lower part of the legs, from the feet up to the knees is kept almost always perpendicular to the flow ($80^o < \theta_1 < 90^o$) while the trunk is kept as parallel as possible to the flow ($90^o < \theta_3 < 120^o$). Arms, when used in order to increase the balance can assume positions from parallel to perpendicular to the flow. The head is kept parallel to the flow ($45^o < \theta_4 < 90^o$). The higher the θ_4 is, the higher the drag of the skater is. The distance between the lower parts of the legs varies during the race due to the pushing phase and this affect the drag.

Size of different body parts and average speed are also two important parameters that should be considered in order to decide which kind of roughness gives the best results in terms of drag reduction. All these parameters should be considered in order to improve speed skating suits.

6 Conclusions

Skin suit do affect speed skaters performances. A final time difference of about 3 seconds has been estimated (for both women and men) when suit 1 and suit 4 are compared.

The test performed on the hybrid suit showed that textiles present on the legs have a central role in order to reduce the drag.

Increasing the roughness of the textile means, at the same time, decreasing the drag at low speed but increasing the drag at high speed. The strict correlation between roughness, drag coefficient and speed impose the use of different suits for different disciplines in order to have the best results. As shown in Figures 3, 4, and 5, the average speed of a speed skater changes not only during the race but also from ladies to men competitions and it decreases when distance increases.

A rougher pattern on the legs should then be used for girls and for long distance competitions while a smoother textile is needed in men short distance competitions, where the average speed is higher. Possible suggestions are the use of a textile with $s = 3$ or $s = 2$ for short distance competition and in general when the average speed is larger than 12m/s and the use of a textile with s=4 for long distance competitions where the average speed is lower than 12m/s. These limits are not certain and future work is needed in order to have more precise values. However, a correlation between the smoothness coefficient and the average speed is certain and it has been exposed.

The use of totally smooth textiles on the legs (s=1) is not recommended in any case while they can be used for the trunk and the hat. The use of extremely rough textiles (s=5) is not recommended neither since they increase the drag at high speeds.

Bibliography

Di Prampero PE, Cortili G, Mognoni P, and Saibene F. *Energy cost of speed skating and efficency of work against air resitance; J Appl Physiol* 40, 4: 584591, 1979

Van-Ingen-Schenau-GJ. *The influence of air friction in speed skating; Journal of Biomechanics* 15, 6. (1982): 449-458

P.E. Di Prampero, G. Cortili, P. Mognoni and F. Saibene *Equation of motion of a cyclist. J Appl Physiol* 47: 201206, 1979

Pugh. 1970. *J of Physiology* 207, 823-835

Achenbach E. *The effects of surface roughness and tunnel blockage on the flow past spheres; Journal of fluid mechanics* Vol. 65 Pt.1 1964

P.W. Bearman and J.K. Harvey, *Control of Circular Cylinder flow by the use of dimples; AIAA-Journal* Vol.31 No.10 October 1993

J.J. De Koning, G. De Groot and G.J. Van Ingen Schenau, *Ice friction during speed skating. Journal of Biomechanics* 25, 6. (1992): 565-71

J.J. De Koning, G. De Groot and G.J. Van Ingen Schenau, *A power equation for the sprint in speed skating. Journal of Biomechanics* 25, 6. (1992): 573-580.

J.J. De Koning, G. De Groot and G.J. Van Ingen Schenau,*Experimental evaluation of the power balance model of speed skating; J Appl Physiology* 98, 6. (2005): 227-233.

Oggiano L, Sætran L R, Løset S and Winther R, *Reducing the Athletes Aerodynamical Resistance; J of Computational and Applied Mechanics*, vol 5, no 2, pp 1-8, (2007)

Cross-Country Skiing

Helge Nørstrud
Department of Energy and Process Engineering,
Norwegian University of Science and Technology,
Trondheim, Norway

Abstract The present chapter gives a theoretical analysis of the forces acting on a cross-country skier at various course gradients. A formulae for the human power output for given endurance and athletic characteristic is also presented. In addition, numerous examples are calculated to explain the force balance on a skier on a flat, an uphill and a downhill ski course.

1 Introduction

Walking on a pair of skies has been a mean of transportation since the ancient time, see e.g. the rock carving from Rødøy, Norway (Figure 1) which dates back to the Stone Age (about 3000 B.C.).

Figure 1. Rock carving of a skier (Photo: Tromsø Museum).

The word ski could be related to the Norwegian name for a sliced piece of wood, i.e. "skive". Hence, it is easy to understand that the skis in the early days were made out of wood. Nowadays the skis are mainly made out of plastic or composites. The technique of skiing and the binding style has also gone through an evolution and here we can mention the Norwegian Sondre Nordheim (1825–1897) and Mathias Zdarsky (1856–1940) of Austria which patented the Lilienfeld-binding in 1896. Also the ski wax and the skiers clothing (see Figure 2) made similar developments.

However, as time went on the cross-country skiing became a sport competition and was introduced in the first Winter Olympic Games in Chamonix, Switzerland in 1924. Today we can observe a world wide interest in cross-country skiing both as a leisure activity and a hard demanding competition between athletes.

Figure 2. Svein H. Brekke (1939–2008) from Morgedal, Telemark in a traditional cross-country ski outfit (Photo: Unknown).

We will in the following analyse a cross-country skier traversing various courses and give results for velocity and time for given atmospheric and snow condition, see Figure 3.

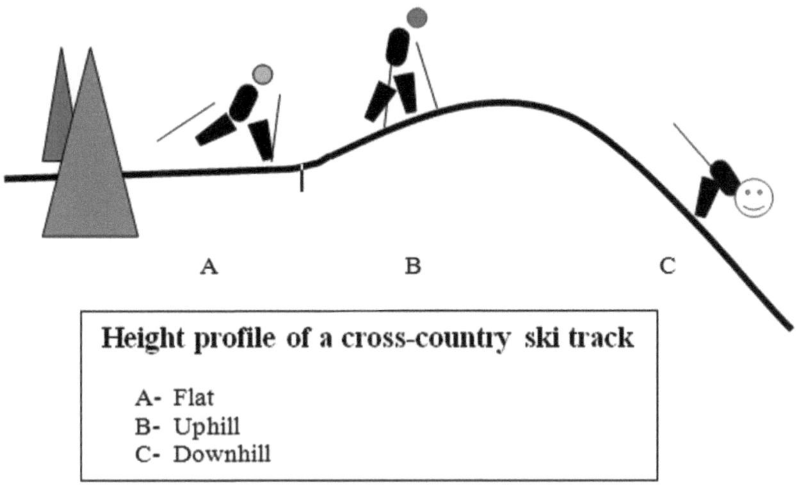

Figure 3. A cross-country ski course of various demands to the skier.

2 Flat ski courses

A force balance in the horizontal direction (see Figure 4) gives

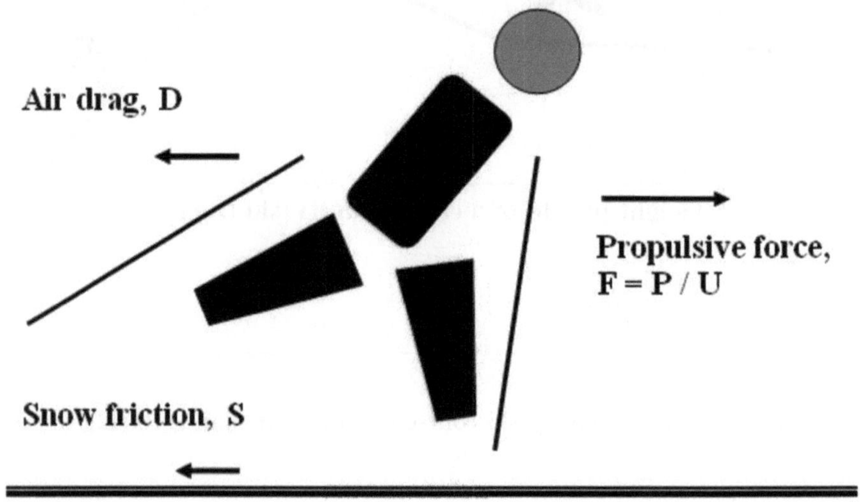

Figure 4. Force balance on a skier on a flat course.

$$F\,[N] - D\,[N] - S\,[N] = 0 \tag{1}$$

Here the propulsive force F can be expressed as

$$F = \frac{P}{U} \tag{2}$$

where P [W] is the human aerobic power and U [m/s] is the constant velocity of the skier over a distance d [m] performed during the time T [s] $= d/U$. From the "Bioastronautics Data Book", NASA SP-3006, 1964, p.182, we can formulate the power output over time as

$$P = \frac{\mathtt{A}^2}{4\ell nT} \tag{3}$$

The athletic factor $\mathtt{A}\,[-]$ is defined as

$$\begin{aligned} \mathtt{A} &= 100 \quad \text{(Athlete)} \\ &= 90 \quad \text{(Sporty)} \\ &= 80 \quad \text{(Normal)} \end{aligned}$$

and yields a measure of the performance level of the person under consideration. A super-athlete can be classified as $\mathtt{A} > 100$, see Figure 5.

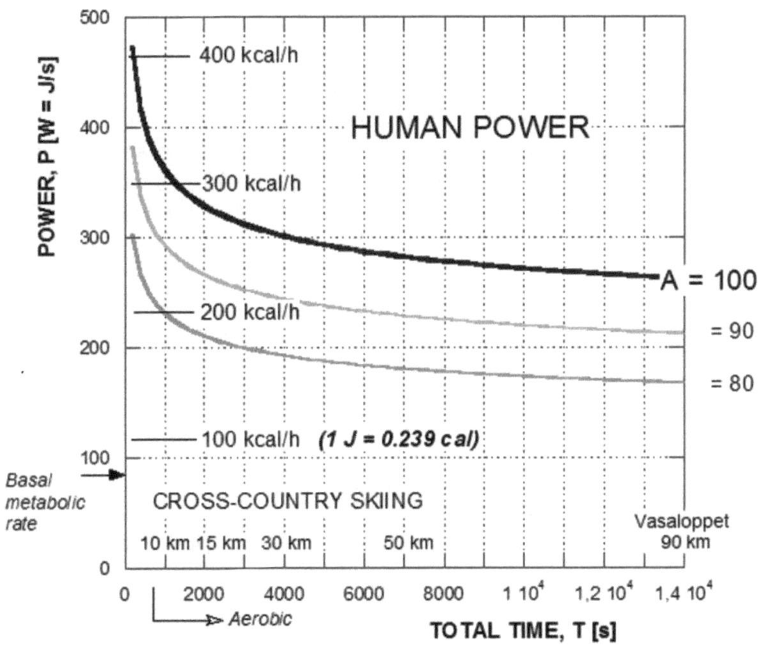

Figure 5. Human power output versus time.

The propulsive force F has to overcome the aerodynamic drag force D and the friction force S acting on the skis surface from the snow, i.e.

$$D = \frac{1}{2}\rho U^2 c_D A \tag{4}$$

$$S = \mu m g \tag{5}$$

where the air density is ρ [kg/m^3] and the drag area $c_D A$ [m^3] is the product of the dimensionless drag coefficient $c_D[-]$ and the frontal area A [m^2] of the skier. Furthermore, $\mu[-]$ designates the friction coefficient, m [kg] is the mass of the skier with ski gear and g $[\mathrm{m/s^2}] = 9.81$ is the gravity constant. Equations (2), (4) and (5) can be inserted in Eq. (1) to yield

$$\frac{P}{U} = \frac{1}{2}\rho U^2 c_D A + \mu m g \tag{6}$$

or with the substitution of $U = d/T$ and Eq. (3) in Eq. (6) we obtain

$$T = \ell n T \frac{4d}{\mathtt{A}^2}\left[\frac{1}{2}\rho\frac{d^2}{T^2}c_D A + \mu m g\right] \tag{7}$$

For given parameters (d, $\mathtt{A}$, ρ, $c_D A$, μ, m and g) Eq. (7) can now be iterated to find the time T required to ski the horizontal distance d. The iteration consists of the selection of a starting value for T and then solve for the right hand of Eq. (7). This process has to be repeated until both sides of the equation give the same value, i.e. the sought value for the time T which satisfies Eq. (7).

An alternative form for Eq. (7) can be formulated for $T > 1000$ s through a linear curve fit of the ratio $T/lnT \approx 0.1T + 71.22$ (see Figure 6) and is written as

$$T = \frac{40d}{\mathtt{A}^2}\left[\frac{1}{2}\rho\frac{d^2}{T^2}c_D A + \mu m g\right] - 712.2 \tag{8}$$

Again iteration is required to solve for the time T for otherwise given parameters. However, if we seek a first estimate for $T > 1000$ s we can eliminate the air drag by setting $c_D = 0$ and rewrite Eq. (8) to yield

$$\begin{aligned} T &= \frac{40d}{\mathtt{A}^2}\mu m g - 712.2 \\ &\approx \frac{40d}{\mathtt{A}^2}\mu m g \end{aligned} \tag{9}$$

The disregarding of the value 712.2 leading to Eq. (9) is done in order to compensate for the elimination of the air drag which otherwise would have increased the time T. Hence, Eq. (9) gives a reasonable estimate for T in order to start the iteration process.

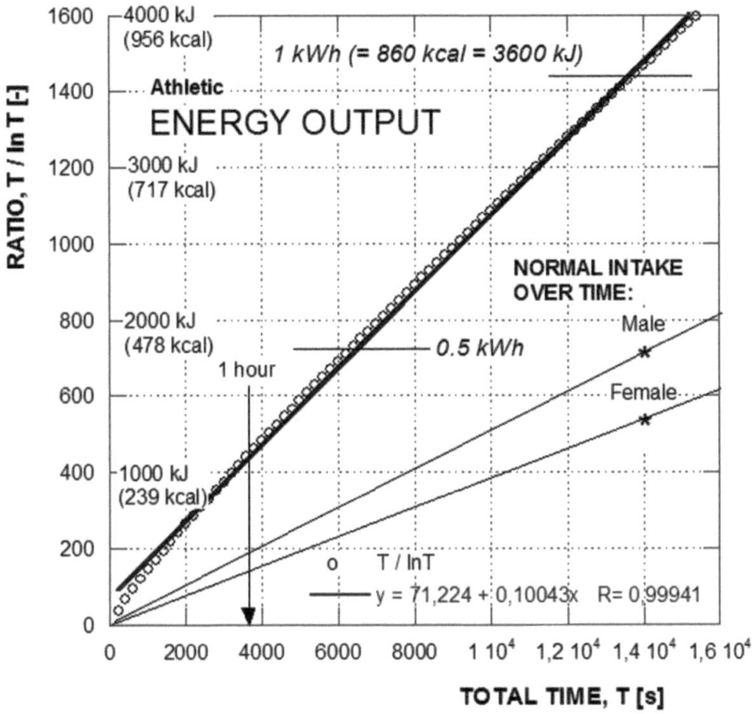

Figure 6. Energy output over time for an athlete.

NOTE: The ratio T/lnT, see Eq. (3), is also a measure of the energy output PT [Ws = J] over a given time T [s] and for an athletic parameter of 100 we will obtain 1 kJ = $2.5T/lnT$.

Average daily intake of calories for adolescents is 2700 kcal for male and 2000 kcal for female. This corresponds to 0.50 kWh and 0.37 kWh respectively over 3.89 hours (=14 000 s), see previous figure. This means that an performing athlete has to consume more than twice the normal calorie intake in order to keep the energy balance. A lumberjack has to double again this value.

2.1 Case study - Vasaloppet in Sweden

The world largest classic cross-country ski competition covers a distance of 90 km from Mora in Dalarne to Sälen and it dates back to 1922 in its present form. Record performance was achieved in 1998 by Peter Göransson (SWE) with the time 3:38:57 (T =13 137 s).

In 2004 over 14 000 people participated in Vasaloppet which got its name from the Nobelman Gustav Vasa (later the King of Sweden) who fled in the opposite direction from the Danish king Kristian II (Kristian the Tyrant) in 1521.

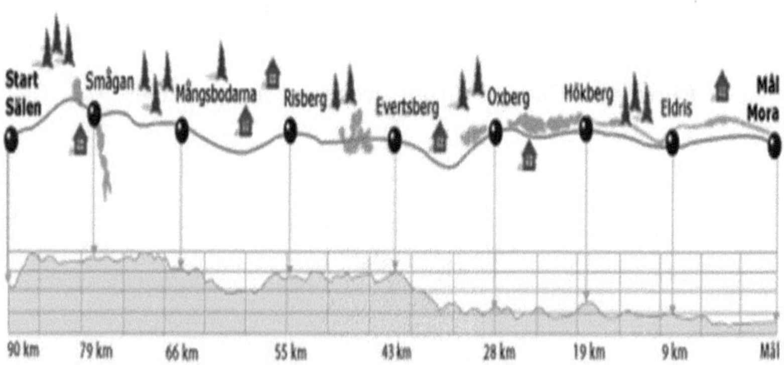

Figure 7. Course profile of Vasaloppet.

Since **Vasaloppet** covers a long distance over a relative flat terrain (see Figures 7 and 8) it can serve as an example for the flat course case. The skier

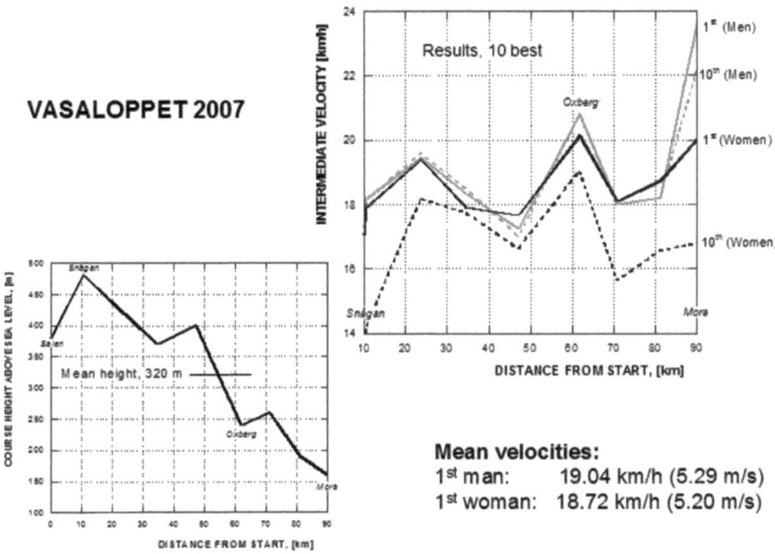

Figure 8. Results from Vasaloppet 2007.

is in an almost upright position and the velocity is low. Hence, the snow friction will be the dominant part of the resistance to the skier's movement. The selected reference values are

Athletic factor = 100	$\mu = 0.05$
$m = 70$ kg	$d = 90\ 000$ m
$c_D A = 0.8$ m^2	$\rho = 1.1878$ kg/m^3 (at 320 m altitude)

A first guess for the time T is 12 360 s which comes from Eq. (9) and an iteration of Eq. (7) gives finally the results for $T = 16\ 800$ s as shown in the first row in Table 1.

Table 1. Simulation results for Vasaloppet.

T **[s] =**	$\mathbf{c_D A}$ **[m²] =**	μ **[-] =**	**Drag,** ***D*** **[kg]**	**Friction,** ***S*** **[kg]**
16 800	0.8	0.05	1.39 (28 %)	3.50 (72 %)
17 020	0.8	0.0513	1.35 (27 %)	3.59 (73 %)
17 020	0.85	0.05	1.45 (29 %)	3.50 (71 %)
14 303	0.8	0.033	1.92 (45 %)	2.32 (55 %)

As stated the friction force represents 72 % of the total resistance. Furthermore, rows 2 and 3 represent simulation for the winner Oskar Svärd (SWE) of Vasaloppet in 2007. In row 2 the friction coefficient is increased by 2.6 % while keeping the drag area the same as for the reference case (row 1). Row 3 shows the result when the drag area is increased by 6.3 % and the friction coefficient is the same as for the reference value.

Oskar Svärd also won Vasaloppet in 2003 with the time 03:58:23 (14 303 s) and the simulation result is shown in row 4. We have kept the drag area as the reference area, but improved the gliding condition for the skis.

NOTE: The presented analysis of a cross-country skier on a flat course is also valid e.g. for a speed skater and as an example we will simulate the World Champion Svein Kramer (NED) when he won the 5000 m distance in Calgary (1128 m above sea level), Canada in March 2007 with the time 6:07,48 ($T = 367.48$ s).

Assuming the parameters:

athletic factor	=	100
m	=	70 kg
$c_D A$	=	0.23 m^2
μ	=	0.012
ρ	=	1.198 kg/m^3

Figure 9. Statue of Hjalmar Andersen in Trondheim, Norway.

we will by iteration of Eq. (7) for the distance of $d = 5000$ m obtain the final time of $T = 370.08$ s when assuming constant velocity during the run, i.e. we will obtain $U_{mean} = 13.51$ m/s. The power output will be $P = 422.7$ W which is needed to overcome the air drag (74 %) and the ice friction (26 %).

Sven Kramer also obtained a world record in Salt Lake City (1320 m), USA in the same month on the 10 000 m distance with the time 12:41,69 ($T =$

761.69 s). Replacing the air density with $\rho = 1.077$ kg/m^3, but otherwise keep the same parameters as listed above, our simulation for d =10 000 m will yield $T = 772$ s. It is up to the reader to change the input data in order to beat the World Champion on 10 000 m. One hint is to reduce the air drag by 36 gram (1.7 %) and then you are the hero with the final time of 12:41,00 ($T = 761$ s).

3 Uphill courses

Due to the inclination of an uphill course with respect to the horizontal reference (see Figure 10) the force balance on the skier will now read, see Eq. (1)

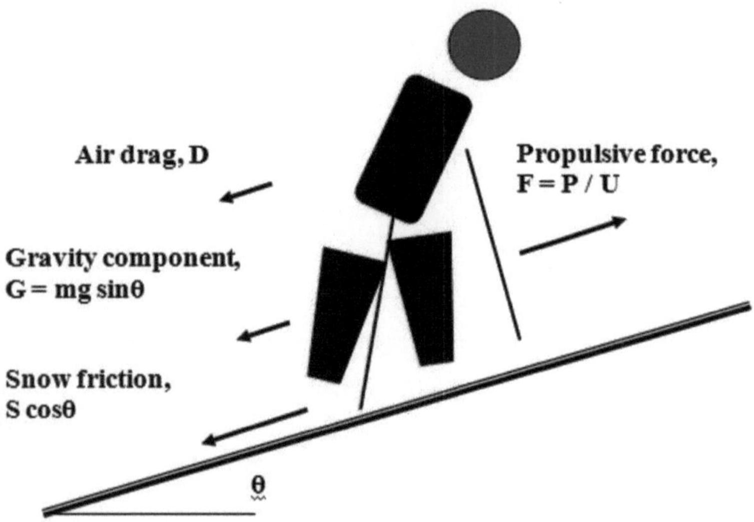

Figure 10. Force balance on a skier on an uphill course.

$$F - D - S\cos\theta - G[N] = 0 \tag{10}$$

Here the snow friction force S has been reduced by the factor cos θ where θ [deg] represents the slope angle and G is the gravity force component in the direction of the skier's motion, i.e.

$$G = mg\sin\theta \tag{11}$$

For a flat course we will have $\theta = 0$ (or sin $\theta = 0$) and $G = 0$. Inserting Eqs. (2)–(5) and (11) in Eq. (10) will give

$$T = \ln T \frac{4d}{\mathrm{A}^2}\left[\frac{1}{2}\rho\frac{d^2}{T^2}c_D A + mg\left(\mu\cos\theta + \sin\theta\right)\right] \tag{12}$$

Eq. (12) is again to be solved by iteration to yield the time T for otherwise given values for the parameters involved. Note that a slight increase of the athletic factor will reduce the final time T and this is often the case for the winner in the end spurt.

For steep hills (say $\theta > 5$ degrees) the aerodynamic drag (low velocity) tends to be negligible compared to the other resistive forces and Eq. (12) can be written as

$$\frac{T}{\ln T} = \frac{4d}{\mathtt{A}^2} mg\,(\mu\cos\theta + \sin\theta) \tag{13}$$

or with the linearization Eq. (7) of T/lnT for $T > 1000$ s we will obtain

$$T = \frac{40d}{\mathtt{A}^2} mg\,(\mu\cos\theta + \sin\theta) - 712.2 \tag{14}$$

In the limit of $\theta = 90$ degrees Eq, (14) reduces to

$$T = \frac{40h}{\mathtt{A}^2} mg - 712.2 \tag{15}$$

where the distance d has been replaced by the height h [m]. Hence, Eq. (15) gives the virtual time T which the skier has to spend in order to "lift" her/himself a height h vertically against gravity.

3.1 Case study – 1000 m Uphill course

We will use the same reference person from Vasaloppet , but change the air density to $\rho = 1.1$ kg/m^3, the distance to $d = 1000$ m and evaluate Eq. (12) for 4 course gradients θ. The results are provided in Table 2.

Table 2. 1000m uphill course.

	θ [deg]	T [s]	U [m/s]	U [km/h]	Drag [%]	Friction [%]	Gravity [%]
1	**0** (0 %)	122	8.2	29.5	46	54	0
2	**5** (8.7 %)	223	4.5	16.1	9	33	58
3	**10** (17.6 %)	370	2.7	9.7	2	22	76
4	**15** (26.8 %)	533	1.9	6.8	1	16	83

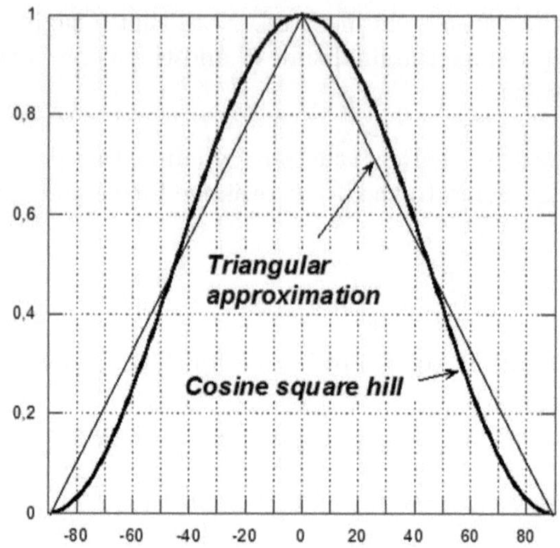

Figure 11. This figure shows a geometric hill produced by a cosine-square function.

3.2 Case study – Final climb in Val di Fiemme.

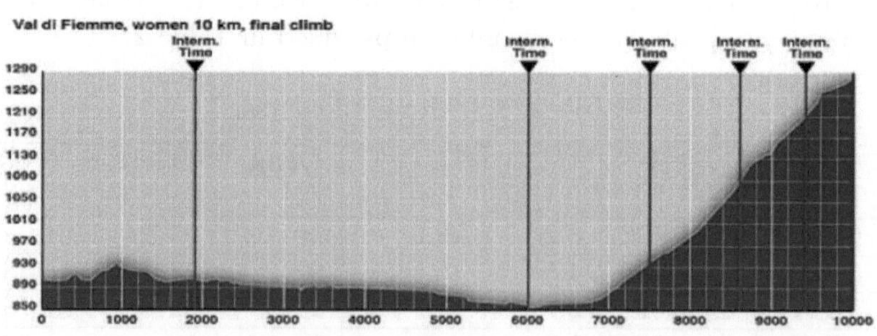

Figure 12. Val di Fiemme ski course.

Figure 12 shows that the final climb in Val di Fiemme, Italy (FIS Tour de Ski, 2007) has an average slope of $\theta = 7.7$ deg (13.5%) $= \arctan \Delta h/\Delta d =$

arctan 420/3100 were the vertical height is $\Delta h = 420$ m. Writing Newton´s second law as F [N]$= mg$ or rewritten with the velocity $\Delta h \Delta T$ gives P [W] $= mg\Delta h\Delta T$. Furthermore, the relation

$$P\Delta T[J] = \Delta hmg$$

will give the approximate energy output needed for climbing up the hill. Inserting the value $m = 70$ kg in the last equation will yield $P\Delta T = 288.4$ kJ (= 68.9 kcal). An athlete can produce this energy output over a time of $\Delta T = 766$ s (12:46,0), see Eq. (3). To perform this task under the mentioned time requires a superathlete (see Figure 13).

Figure 13. A superathlete (Photo: Richard Skagen, Adresseavisen).

4 Downhill ski courses

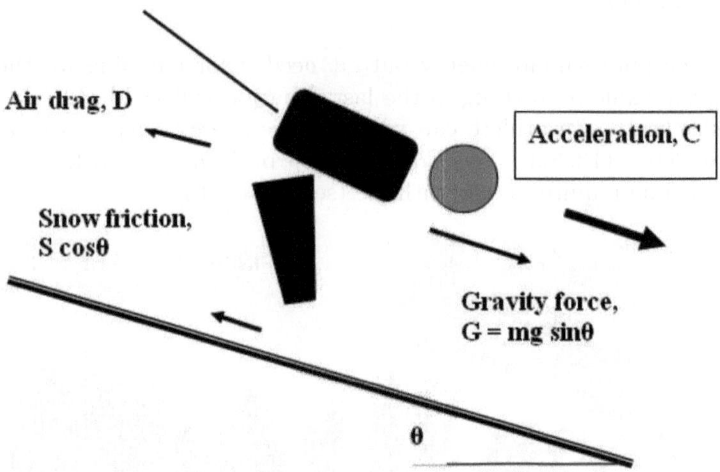

Figure 14. Force balance on a skier on a downhill course.

In the case of a downhill ski course (Figure 14) the skier can relax (after a possible push in the start) and let the gravity force take over as the force needed for propulsion. It also means that an acceleration force C [N] has to be included in the force balance on the skier, i.e.

$$C = G - D - S\cos\theta > 0 \tag{16}$$

where

$$C = m\frac{du}{dt} \tag{17}$$

Here u [m/s] designates the instant velocity $u = u(t)$ of the skier as function of the time t [s]. Inserting Eqs. (11), (4), (5) and (17) in Eq. (16) will result in the ordinary differential equation

$$m\frac{du}{dt} = mg\sin\theta - \frac{1}{2}\rho u^2 c_D A - mg\mu\cos\theta \tag{18}$$

When $C = 0$ we have reached the terminal velocity U_{term} since the acceleration du/dt is equal to zero, i.e. from Eq. (18) we will obtain

$$U_{term} = \sqrt{\frac{2mg(\sin\theta - \mu\cos\theta)}{\rho c_D A}} \tag{19}$$

The expression under the square root sign has to be positive which means that the $sin\theta > \mu cos\theta$ or $\mu < sin\theta / cos\theta$, see Figure 15. Or in other words, the gravity force G must be greater than the friction force $Scos\theta$ in order to achieve an acceleration force in the downhill direction. When the gravity and friction forces are of equal magnitude, no downhill pull on the skier is possible (without the skiers help) and no terminal velocity U_{term} will exist, see Eq. (19). Rearranging Eq. (18) will lead to

$$\frac{du}{dt} = -\frac{1}{2m}\rho c_D A u^2 + g\left(\sin\theta - \mu\cos\theta\right) \tag{20}$$

Firstly, we rewrite the acceleration term in Eq. (20) as

$$\frac{du}{dt} = u\frac{du}{dt}\frac{dt}{d\ell} = u\frac{du}{d\ell} = \frac{1}{2}\frac{d\left(u^2\right)}{d\ell} \tag{21}$$

and will obtain the differential equation

$$\frac{d\left(u^2\right)}{d\ell} = 2g\left(\sin\theta - \mu\cos\theta\right) - \frac{\rho c_D A u^2}{m} \tag{22}$$

Introducing the terminal velocity U_{term} from Eq. (19) into Eq. (22) we will obtain

$$\frac{d\left(u^2\right)}{d\ell} = 2g\left(\sin\theta - \mu\cos\theta\right)\left[1 - \left(\frac{u}{U_{term}}\right)^2\right] \tag{23}$$

Introducing the following new dimensionless variables

$$\begin{aligned} \hat{u} &= \tfrac{u}{U_{term}} \Rightarrow d\left(u^2\right) = U_{term}^2 d\left(\hat{u}^2\right) \\ \hat{\ell} &= \tfrac{2g(\sin\theta - \mu\cos\theta)}{U_{term}^2}\ell \Rightarrow d\left(\ell\right) = \tfrac{U_{term}^2}{2g(\sin\theta - \mu\cos\theta)} d\left(\hat{\ell}\right) \end{aligned} \tag{24}$$

Eq. (23) will be rewritten to the separable differential equation as

$$\frac{d\left(\hat{u}^2\right)}{1 - \hat{u}^2} = d\left(\hat{\ell}\right) \tag{25}$$

or with the substitution

$$\hat{v} = \hat{u}^2$$

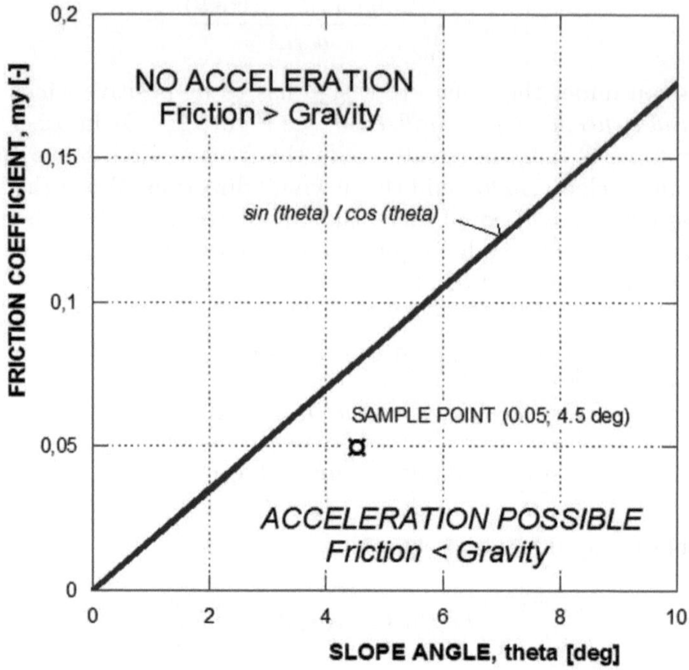

Figure 15. Acceleration requirement.

we obtain

$$\frac{d\hat{v}}{1-\hat{v}} = d\hat{\ell} \tag{26}$$

An integration of Eq. (26) between the limits

$$\begin{aligned} \hat{v} &= \hat{v}_0\left(u_0\right), \hat{v}\left(U\right) \\ \hat{\ell} &= \hat{\ell}_0\left(0\right), \hat{\ell}\left(d\right) \end{aligned}$$

will yield

$$\hat{\ell} = \ln\left(1-\hat{v}_0\right) - \ln\left(1-\hat{v}\right) \tag{27}$$

where we have used the integral relation

$$\int \frac{dx}{1-x} = -\ln(1-x) + C$$

Here C is an integration constant. Inserting the original variables back into Eq. (27) yields

$$\frac{2g(\sin\theta - \mu\cos\theta)}{U_{term}^2} d = \ln\left[\frac{1-\left(\frac{u_0}{U_{term}}\right)^2}{1-\left(\frac{U}{U_{term}}\right)^2}\right] \tag{28}$$

and Eq. (28) rewritten with exp $(lnx) = x$ gives the final result for the velocity $U = U(d)$, i.e.

$$U = U_{term}\left\{1-\left[1-\left(\frac{u_o}{U_{term}}\right)^2\right]\left[\exp\left(\frac{2g(\sin\theta-\mu\cos\theta)}{U_{term}^2}d\right)\right]^{-1}\right\}^{\frac{1}{2}} \tag{29}$$

Returning now to Eq.(20) we can write

$$\frac{du}{dt} = g(\sin\theta - \mu\cos\theta) - \frac{1}{2m}\rho c_D A u^2 \tag{30}$$

or by introducing again the terminal velocity U_{term} into Eq. (30) we will obtain

$$\frac{du}{dt} = g(\sin\theta - \mu\cos\theta)\left[1-\left(\frac{u}{U_{term}}\right)^2\right] \tag{31}$$

By introducing the new variables

$$\begin{aligned} \hat{u} &= \tfrac{u}{U_{term}} \\ \hat{t} &= \tfrac{g(\sin\theta-\mu\cos\theta)}{U_{term}} t \end{aligned} \tag{32}$$

we will obtain the separable differential equation

$$\frac{d\hat{u}}{1-\hat{u}^2} = d\hat{t} \tag{33}$$

Integration of Eq. (33) between the limits

$$\begin{aligned} \hat{u} &= \hat{u}_0(u_0), \hat{u}(U) \\ \hat{t} &= \hat{t}_0(0), \hat{t}(T) \end{aligned}$$

will lead to the solution

$$\hat{t} = \frac{1}{2}\left[\ln\frac{1+\hat{u}}{1-\hat{u}} - \ln\frac{1+\hat{u}_0}{1-\hat{u}_0}\right] \tag{34}$$

Inserting the original variables in Eq. (34) we finally obtain the result for $T = T(U)$ where

$$T = \frac{U_{term}}{2g(\sin\theta - \mu\cos\theta)}\left[\ln\frac{U_{term}+U}{U_{term}-U} - \ln\frac{U_{term}+u_0}{U_{term}-u_0}\right] \tag{35}$$

Introducing the constants

$$\begin{aligned} k_1 &= 2g\,(\sin\theta - \mu\cos\theta) \\ k_2 &= \tfrac{\rho c_D A}{m} \end{aligned} \tag{36}$$

will yield

$$U_{term} = \sqrt{\frac{k_1}{k_2}} \tag{37}$$

and these values inserted in Eq. (29) gives

$$U = \sqrt{\frac{k_1}{k_2}}\left\{1 - \left(1 - \frac{k_2}{k_1}u_0^2\right)\exp\left(-k_2 d\right)\right\}^{\frac{1}{2}} \tag{38}$$

or

$$U = \sqrt{\frac{k_1}{k_2}}\left[1 - \exp\left(-k_2 d\right)\right]^{\frac{1}{2}} \tag{39}$$

where Eq. (39) is valid for $u_0 = 0$. Likewise we can introduce the reference constants into Eq. (35) and obtain the result

$$T = \frac{1}{\sqrt{k_1 k_2}}\left[\ln\frac{1+U\sqrt{k_2/k_1}}{1-U\sqrt{k_2/k_1}} - \ln\frac{1+u_0\sqrt{k_2/k_1}}{1-u_0\sqrt{k_2/k_1}}\right] \tag{40}$$

or

$$T = \frac{1}{\sqrt{k_1 k_2}}\ln\frac{1+U\sqrt{k_2/k_1}}{1-U\sqrt{k_2/k_1}} \tag{41}$$

where Eq. (41) is valid for $u_0 = 0$.

Example – FIS regulates that a cross-country downhill slope of $\theta = 4.5$ deg shall not have a vertical drop larger than 9 m. Hence, as a simulation with the parameters

$$
\begin{aligned}
m &= 70 \text{ kg} \\
c_D A &= 0.4 \text{ m}^2 \\
\mu &= 0.05 \\
\theta &= 4.5 \text{ deg} \\
d &= 100 \text{ m} \\
\rho &= 1.1 \text{ kg/m}^3
\end{aligned}
$$

will with Eqs. (29) and (35) yield the results:

$U =$ **6.46 m/s (= 23.3 km/h)**
$T =$ **28.11 s**

4.1 Case study – Downhill courses of various gradients

We will assume the following input values:

$$
\begin{aligned}
\rho &= 1.1 \text{ kg/m}^3 \\
m &= 70 \text{ kg} \\
c_D A &= 0.4 \text{ m}^2 \\
\mu &= 0.0 \\
u_0 &= 0 \text{ m/s}
\end{aligned}
$$

and the results for various skiing lengths and gradients are shown below in figures 16 – 18. It must be noted that a cross-country course of $\theta = 15$ degrees is not practical, but we can transfer it to a downhill or a speed-skiing course.

Figure 16 shows that the gravity and friction forces are constants as function of the distance d for given slope θ, see also Figure 14. Since the gravity force is larger than the friction force an acceleraton force is obtained (Figure 15). This acceleration force will be reduced when the drag force increases with increased velocity U. This velocity approaches the terminal velocity (Figure 17) at which the acceleration force is zero. Finally, the lower the velocity U is at a given distance d, the higher will the required time T be in order to reach that point, see Figure 18.

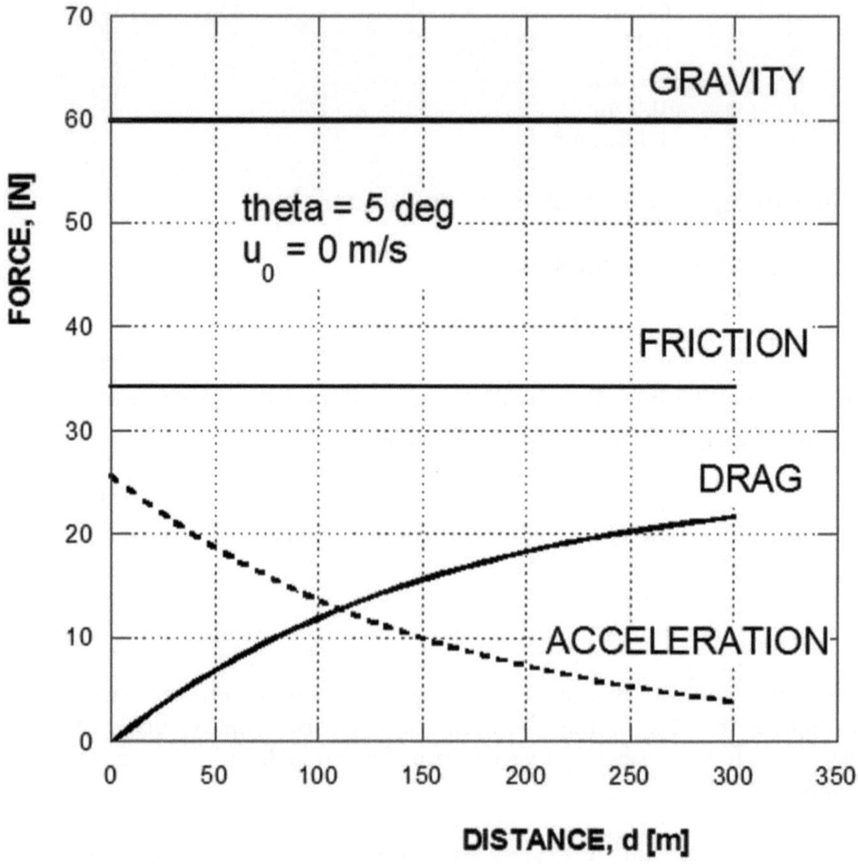

Figure 16. Forces as function of skiing distance.

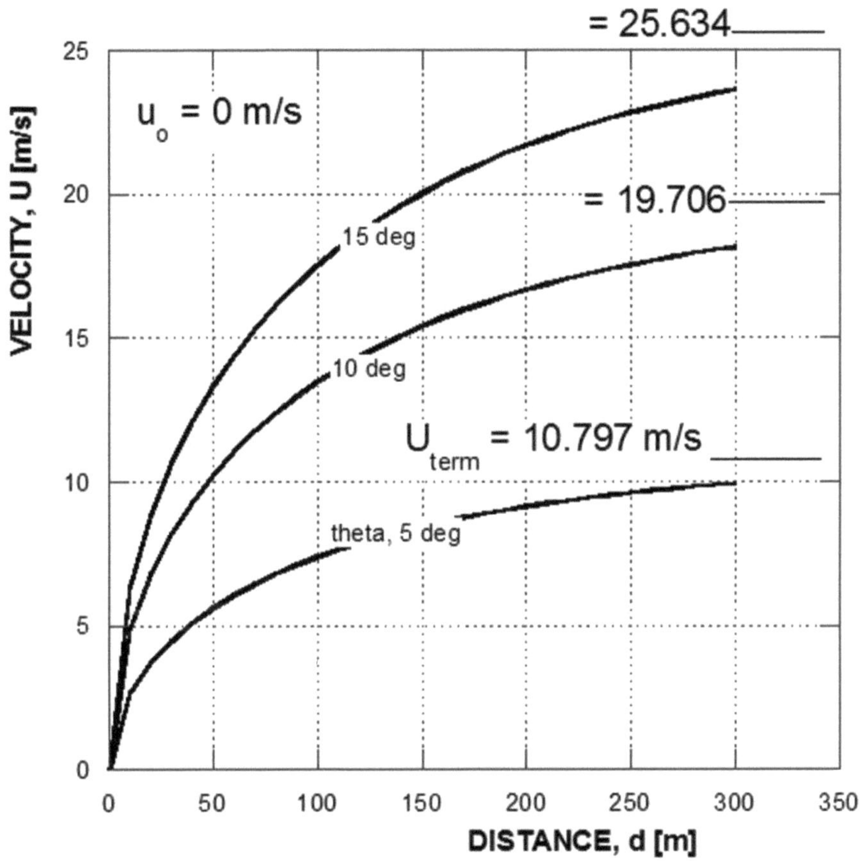

Figure 17. Velocity as function of skiing distance.

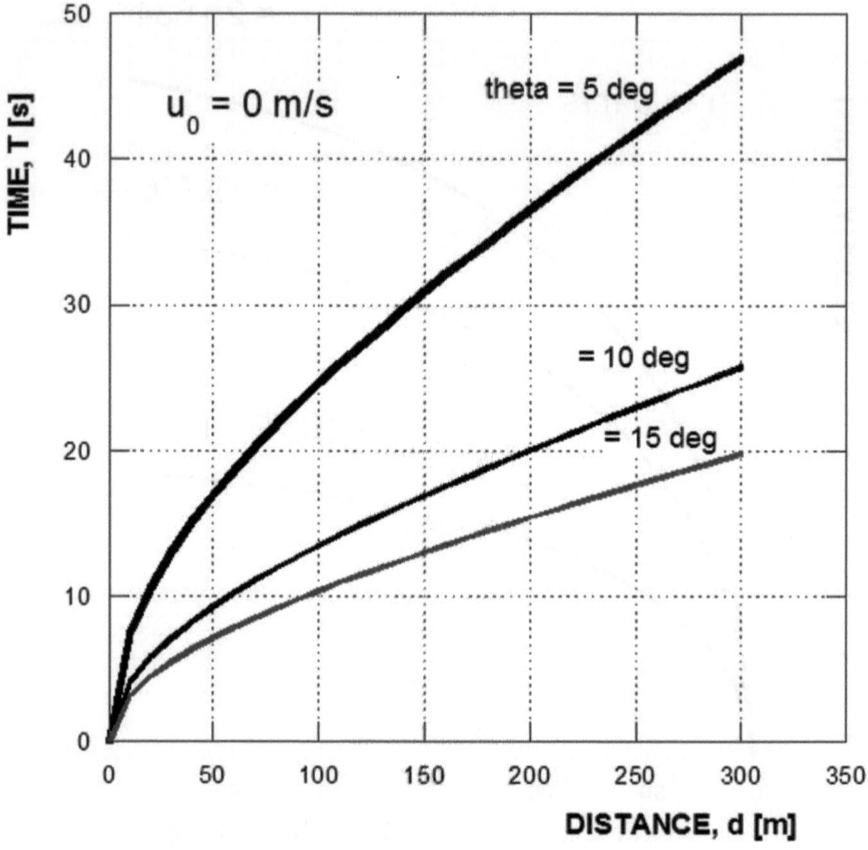

Figure 18. Time as function of skiing distance.

Alpine Downhill and Speed-Skiing

Helge Nørstrud
Department of Energy and Process Engineering,
Norwegian University of Science and Technology,
Trondheim, Norway

1 Introduction

The equations given in the previous chapter for a downhill cross-country skier are also applicable to an alpine downhill skier. However, we have ignored the lift force on the cross-country skier since the velocity is relatively low. For an alpine skier and especially a speed-skier the aerodynamic lift on the body will be important. Since the lift will indirectly reduce the ski friction, a reduction of the friction coefficient μ could be applied. For the speed-skier at velocities around 250 km/h, the lift force will seriously influence the stability of the athlete.

The next paragraph will introduce the Brachistochrone problem which describes the geometry of the downhill slope for the fastest descend when friction and aerodynamic drag is ignored. This also means that for given downhill course a large initial velocity of the skier (or e.g. of a bobsleigh) is advantageous.

2 The Brachistochrone problem

The great mathematicians of the 17th Century (Newton, Leibniz and Bernoulli among others) were engaged in finding the two-dimensional curve for a body which should move from point A to a lower point B under the influence of gravity and with the shortest time (Brachistochrone in Greek). Ignoring friction in the problem, the solution is related to the cycloid path (see Figure 1) which is given in a parametric form as:

$$X/R = (2\pi/360)\varphi - sin\varphi$$

$$Y/R = 1 - cos\varphi$$

Figure 1. The generation of a cycloid (Illustration: Trondheim Science Centre).

Here the parameter φ [deg] defines the cycloid in a Cartesian space X,Y and R is the radius of the rolling circle. If we invert the cycloid about the horizontal X-axis we will obtain the inverted cycloid (or the Brachistochrone) as shown in Figure 2.

The time and length given below are in dimensionless form with the inverted half-cycloid representing the reference path, see Table 1. The arc of the cycloid will always be the Brachistochrone when compared to the straigth lines from A to B_1, B_2 or B_3. It is interesting to note that even when the straight line is shorter than the inverted half-cycloid, the time it takes to descend on a straight line is 18.5 % longer.

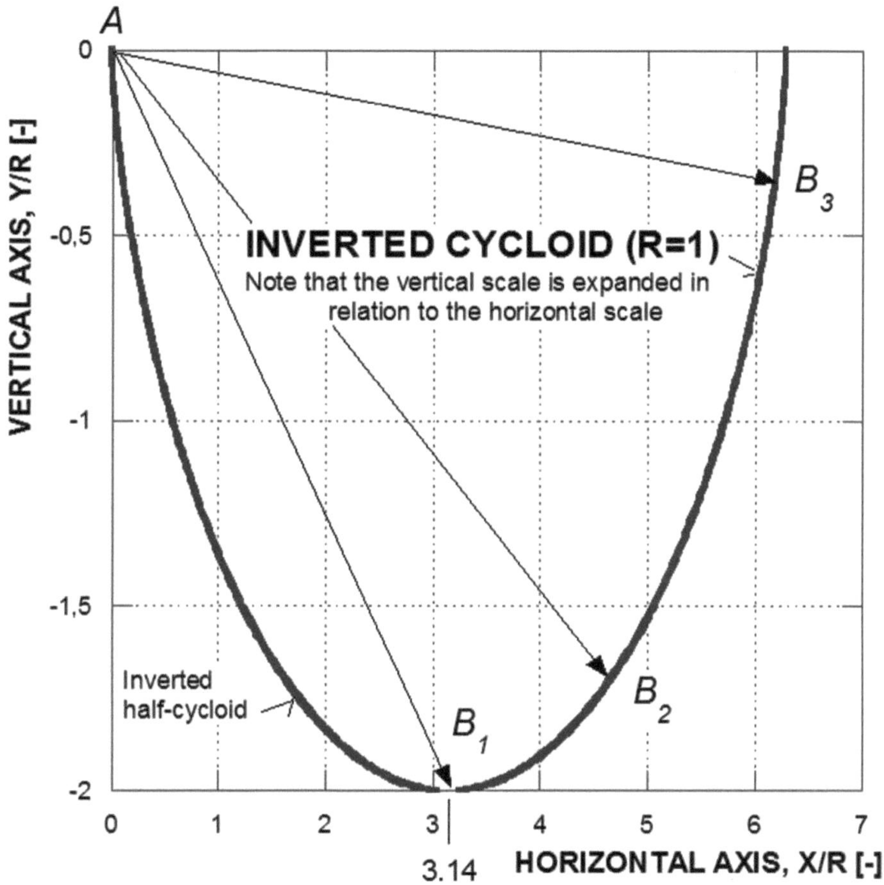

Figure 2. The inverted cycloid (or the Brachistochrone).

Table 1. Length and time comparison for various paths.

PATH	Inverted half-cycloid	Ellipse	Parable	Sine	Straight line
LENGTH	1.0	1.023	0.964	0.955	0.931
TIME	1.0	1.004	1.043	1.071	1.185

3 Alternative analysis of downhill skiing

Rewriting Eq. (18) from the previous chapter "Cross-Country Skiing", we will obtain the alternative form

$$a = g \sin\theta - D/m - g\mu \cos\theta \tag{1}$$

where

$$a = \frac{du}{dt}$$

$$u = \frac{d\ell}{dt}$$

Combining the last two relations leads to

$$udu \equiv \frac{1}{2} d(u^2) = a d\ell \tag{2}$$

Integration of Eq. (2) for a = constant (Note that $g = 9.81$ m/s^2 = constant) between the limits

$$\begin{aligned} \ell = 0 : u = u_0 ; t = 0 \\ \ell = L : u = U ; t = T \end{aligned} \tag{3}$$

gives

$$U^2 = u_0^2 + 2aL \tag{4}$$

Eq. (4) combined with Eq. (1) will give the result

$$U = \left\{ \left[u_0^2 + 2Lg(\sin\theta - \mu\cos\theta)\right] \left[1 + \frac{L}{m}\rho c_D A\right]^{-1} \right\}^{\frac{1}{2}} \tag{5}$$

From $du = adt$ we obtain $U - u_0 = aT$ or $a = (U - u_0)/T$ and this inserted in Eq. (1) will yield

$$T = (U - u_0) \left[g(\sin\theta - \mu\cos\theta) - \frac{1}{2m} \rho c_D A U^2 \right]^{-1} \tag{6}$$

Eqs. (5) and (6) yields explicit solutions for U and T when the other parameters are given. Since the course angle θ can vary with the total length L, we will divide L up in n intervals of length $\Delta\ell$ and this with Eqs. (5) and (6) yields

$$u_0 = \left\{ \left[u_{n-1}^2 + 2\Delta \ell g \left(\sin\theta - \mu \cos\theta \right) \right] \left[1 + \frac{\Delta \ell}{m} \rho C_D A \right]^{-1} \right\}^{\frac{1}{2}} \tag{7}$$

and

$$T_n = (u_n - u_{n-1}) \left[g(\sin\theta - \mu\cos\theta) - \frac{1}{2m} \rho c_D A u_n^2 \right]^{-1} \tag{8}$$

The total time used on the course is obtained from $T = \Sigma T n$ after Eqs. (7) and (8) are evaluated.

3.1 Case study – Downhill Åre 2007

Figure 3. Aksel Lund Svindal in action (Photo: Tor Richardsen, SCANPIX).

The downhill men competition during FIS Alpine World Ski Championships in Åre, Sweden (February 2007) was won by Aksel Lund Svindal (NOR) with the time 1:44,68 ($T = 104.68$ s). The course starts at 1240 m altitude and ends at 396 m, i.e. with a vertical drop of 844 m. The total length of the course was $L = 2922$ m with a gradient ranging between 7 % and 69 % (mean gradient 30 %). The ranking number 48 came in on 1:51,57 which is 6.89 s after Svindal. This shows the split seconds between the competitors.

Without any further information about the course profile, we will simulate the downhill run of Svindal by estimating the following parameters:

$$\begin{aligned} m &= 110 \text{ kg} \\ u_0 &= 2 \text{ m/s} \\ C_D A &= 0.2 \text{ m}^2 \\ \rho &= 1.1071 \text{ kg/m}^3 \text{ (at the mean altitude of 1042 m)} \\ \mu &= 0.04 \end{aligned}$$

Inserting these values together with the length $L = 2922$ m and the mean value of $\theta = 16.79$ degree in Eqs. (5) and (6) given above, we will obtain the results:

$U = 45.69$ m/s (= 164.5 km/h)
$T = 122.53$ s

3.2 Case study – Speed skier in Kulm, Austria

The FIS Speed Ski World Cup took place in Bad Mittendorf, February 2007. The course was situated on the ski jump Kulm and a simulation is based on the following values:

$$\begin{aligned} m &= 95 \text{ kg} \\ C_D A &= 0.13 \text{ m}^2 \\ \mu &= 0.08 \\ \rho &= 1.1 \text{ kg/m}^3 \\ \theta &= 23 \text{ deg} \\ \ell &= 400 \text{ m} \\ u_0 &= 2 \text{ m/s} \end{aligned}$$

With the use of Eq. (5) we will obtain the result as
$U = 39.44$ m/s (142 km/h)

The World Cup was won by Ludwig Gewessler (AUT) with the speed of 135.85 km/h. Since the vertical drop of the course is $h = 156$ m, an idealistic estimate for the final time will yield $U = \sqrt{(2gh)} = 55.32$ m/s (=199.2 km/h).

3.3 Case study – Speed skier in Velocity Peak

Figure 4. A speed-skier in the NTNU wind tunnel in Trondheim.

The speed-ski course Velocity Peak lies close to Silverton, Colorado and starts at an altitude of 3810 m and has a vertical drop of 365 m. The total length L of the course is 643 m.

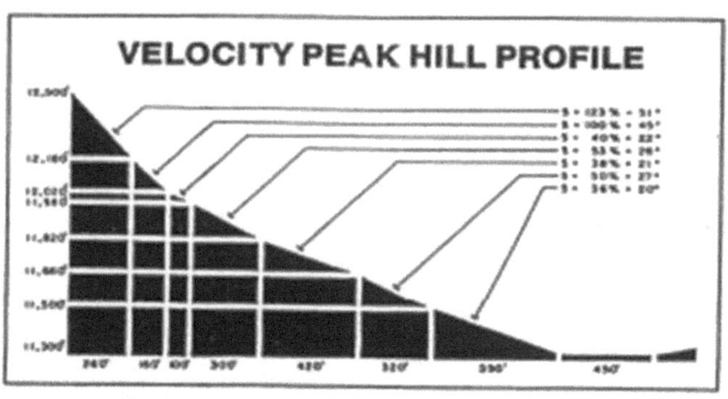

Figure 5. Height profile of Velocity Peak.

The selected parameters for this case are:

$$
\begin{aligned}
m &= 90 \text{ kg} \\
C_D A &= 0.13 \text{ m}^2 \\
\mu &= 0.06 \\
u_0 &= 6 \text{ m/s}
\end{aligned}
$$

We will simulate the downhill course by dividing the total length in 7 intervals and using Eqs. (7) and (8). The results are shown in the Table 2. The total time for reaching the end velocity $U = 206.8$ km/h is $T = \Sigma\Delta t = 16.26$ s.

Table 2. Simulation result for a speed-skier.

Length interval	**Slope angle,** θ [deg]	**Length,** $\Delta\ell$ **[m]**	**Air density,** ρ **[kg/m³]**	**Velocity, u [km/h]**	**Time,** Δt [s]
1	51	79	0.840	118.7	4.72
2	45	49	0.848	145.9	1.33
3	22	30	0.850	151.4	0.73
4	26	91	0.852	169.9	2.04
5	21	128	0.856	183.6	2.60
6	27	98	0.860	198.0	1.85
7	20	168	0.865	206.8	2.99

Franz Weber (AUT) world speed record from 1982 was achieved in Velocity Peak with the final velocity of 203.16 km/h, i.e. our simulation is *too good*!

Performance factors in ski jumping

Wolfram Müller*

* Human Performance Research^{Graz}, Karl-Franzens University and Medical University of Graz, Graz, Austria

Abstract Predominant performance factors in ski jumping are: High in-run velocity, high linear momentum perpendicular to the ramp at take-off due to the jumping movement and the utilisation of aerodynamic lift, accurate timing of the take-off jump with respect to the edge of the ramp, appropriate angular momentum at take-off in order to obtain an aerodynamically advantageous and stable flight position as soon as possible, choice of advantageous body and equipment configurations and angles of attack during the entire flight in order to obtain optimum lift and drag values and the ability to control the flight stability. After the transition from take-off to the flight phase the pitching moment has to be balanced close to zero (1) in order to avoid tumbling accidents; this can become extraordinary difficult when gust or other disturbing effects occur. During the flight, the gravitational force F_{g}, the lift force F_{l}, and the drag force F_{d} act upon the athlete and his equipment and determine the flight path of the centre of gravity of a ski jumper with a given set of initial conditions and parameters. During the flight phase the athlete's position changes. The athlete can strongly influence the aerodynamic forces by changing his posture. He can affect the drag force, the lift force and the torque; the latter enables him to change his flight position and angle of attack with respect to the air stream.

1 Introduction

1.1 Historical development

The ski jumping technique has changed several times since the first Olympic Winter Games in Chamonix, France, 1924, when Jacob Tullin Thams (Norway) won the ski jumping competition. Many authors have analyzed the in-run and take-off techniques as well as the flight styles in the various phases of the sports historical development.

World Cup ski jumping events of today are held on three types of hills: "normal hills" are designed for jump lengths up to 110 m, "large hills"

Figure 1: Jacob Tullin Thams (NOR) at the first olympic ski jump competition in Chamonix, France, 1924. Photo: *IOC* homepage, with permission.

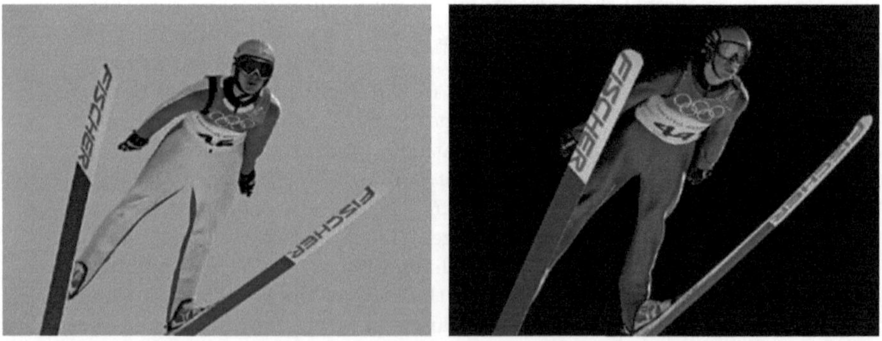

(a) T. Morgenstern (AUT) (b) A. Kofler (AUT)

Figure 2: (a) Olympic gold medalist of Turin (2006): Thomas Morgenstern (AUT). (b) Olympic silver medalist of Turin (2006): Andreas Kofler. The score difference between 1st and 2nd place was only 0.1 points, which is not a significant difference: One meter is 1.8 points, in other words, the 0.1 points difference would correspond to a jump length difference of 5 cm which is an order of magnitude below the jump length measurement resolution of 0.5 m. Photos: Alois Furtner, with permission.

for jumps above 110 m, and "ski flying hills" for flights above 185 m (2). By means of computer simulations possible now (3) landing slopes can be optimized in terms of landing impact characteristics, height above ground, and jump length as a function of in-run velocity. The new landing profile for Innsbruck, for instance, has been designed by means of computer simulations in order to obtain a moderate increase of landing impact as a function of jump length when compared to conventional hill designs (4). The ski jumping world record has continuously increased over the years. The first to jump further than 100 m was Sepp Bradl (Austria) in 1936, and in 1994 the 200 meter line was exceeded for the first time by Andreas Goldberger (Austria), and B. E. Romören (Norway) reached 239 m in Planica (Slovenia) in the year 2005. The mean increase of the world record since 1936 is 1.9 m per year (5).

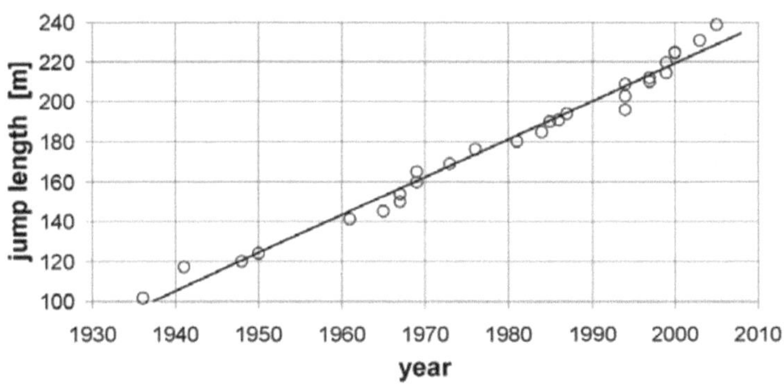

Figure 3: World record development, 1936 (B. Bradl, Austria) to 2007, current record: 239 m (B. E. Romören, Norway, 2005).

1.2 The V-Style or Boklöv-Style and the limitation of front ski percentage

All world class athletes of today use the V-technique which was pioneered by Jan Boklöv (Sweden) in 1985. During the flight phase the skis are not held parallel to each other. Associated with this flight style is an increase in jumping length at a given in-run velocity, thus the aerodynamic features of this flight style are advantageous when compared to the old parallel style. This is the case because lift and drag forces are advantageous and also because this flight style enables the athlete to lean forward in a more pronounced way (1) which leads to an additional aerodynamic improvement. The take-off movement, the stabilization phase after take-off, and the fine torque balance during the flight phase are very difficult sensory-motor tasks which have to be solved in "real time". This explains why it is so difficult for an athlete to remain on a top performance level over longer periods. Reasons for "loosing the feeling" are manifold. After the introduction of the *Boklöv*-style the athletes soon found out, that they could lean forward in a more pronounced way when mounting the binding further back on their skis. This led to extreme flight styles. Some athletes had used a front ski to total ski length relation of up to 60%. These extreme postures increased the tumbling risk enormously because the pitching moment can suddenly become unbalanced, e.g. due to a gust, and many world class athletes had severe tumbling accidents.

Based on the measurement of the aerodynamic forces and pitching moments associated with various flight positions in the large scale wind tunnel of Railtec Arsenal, Vienna (1), the authors have suggested to limit the maximum percentage of front ski to total ski length (6) and the *FIS* has followed this suggestion in 1994. From this season on, the front ski to total ski length percentage was limited to 57%. As a consequence the pitching moment balance has been eased and only one tumbling accident occurred during the 1994/95 World Cup, compared with 10 in 1993/94.

1.3 Decreasing weight of ski jumpers and the solution of the underweight problem

Wind tunnel measurements have shown that both lift and drag have strongly increased due to flight style and equipment changes (1). The development towards larger aerodynamic forces has increased the importance of low weight as a performance factor (7) and this resulted in a substantial decrease of athletes' body weight, starting out from a mean Body Mass Index (*BMI*) of 23.6 in the years 1970-75 (5; 8; 9). The *BMI* is the body mass m in kilograms divided by h^2, with h being the body height in meters. A dangerous disease associated with extremely low weight is anorexia nervosa (10; 11; 12). Several severe cases among ski jumpers have come to light. At the Olympic Games 2002 in Salt Lake City 22.8% of the ski jumpers (5) had a *BMI* below the *WHO* underweight border line of 18.5 kgm^{-2} (13). The problematic development toward extremely low weight was predicted in 1995 (6) and in this publication the solution strategy has been sketched already: Shorter skis, i.e. "smaller wings", compensate for the advantage of extremely low weight and thus the attraction for athletes to be underweight is removed. A research project funded by the *IOC* and

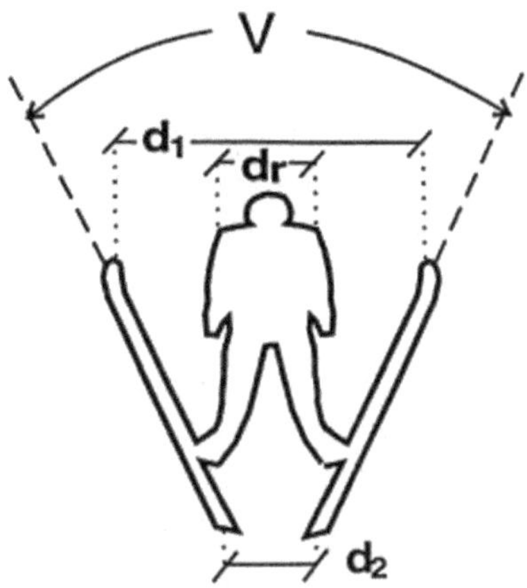

(a) V-style parameters

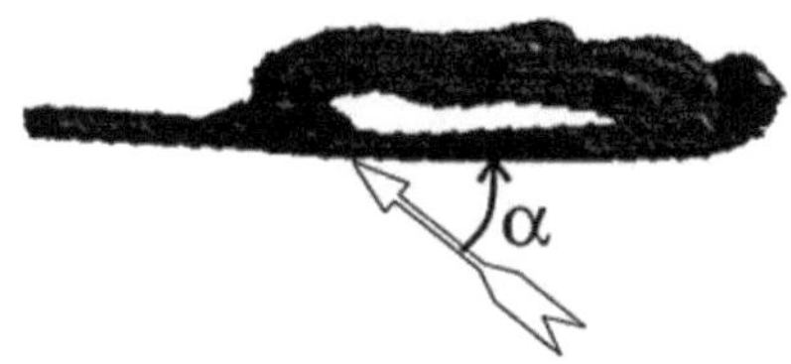

(b) V-style by C. Duffner

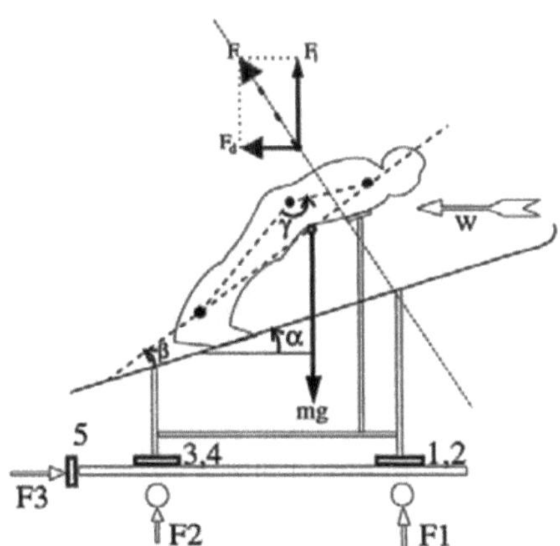

(c) Nomenclature of angles

Figure 4: V-style or Boklöv-style (a) and an example of an extreme flight style (b) as performed in 1993/94 by C. Duffner (GER). (c) Schematics of the wind tunnel positioning apparatus and angle nomenclature. Müller et al. 1996 (1), with permission.

the *FIS* clearly showed how dramatic the situation had become until 2002 (5). The evaluation of the complete World Cup data set from Hinterzarten in the year 2000 (the participation rate was 100%) showed that 16.3% of the athletes had a *BMI* < 18.5 kgm^{-2}, and 22.8% of the athletes investigated during the Olympic Games 2002 were below 18.5 kgm^{-2} (Fig. 5a). The lowest *BMI* found among competing world class ski jumpers was 16.4 kgm^{-2}.

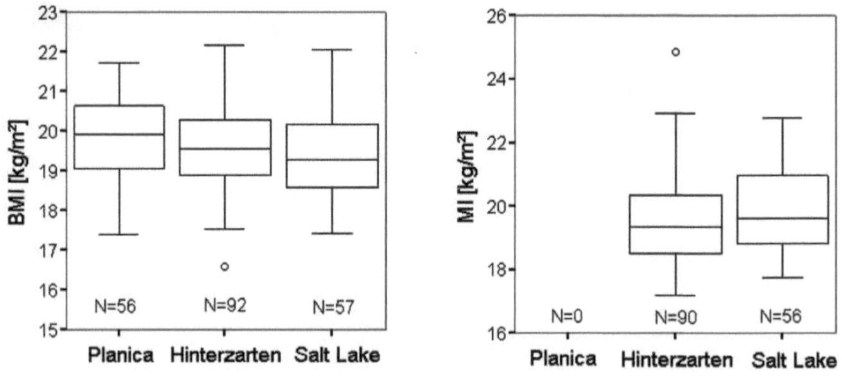

(a) Body mass index (*BMI*) (b) Mass index (*MI*)

Figure 5: Low weight of ski jumpers. Comparison of *BMI* and *MI* of the athletes at different competitions. W. Müller et al. 2006 (5), with permission.

In 2004, the FIS officials followed the concept to solve the problem by relating the weight - in terms of *BMI* - to the ski length (14). The new regulations allow a ski length of 146% of body height for athletes with a weight to squared body height ratio of 20 kgm^{-2} or above, with the weight being measured with jumping suits and boots immediately after the competition. This value corresponds to a *BMI* slightly above 18.5 kgm^{-2}. Every 0.5 units below 20 kgm^{-2}, the maximum ski length percentage is reduced by 2%.

1.4 Measurement of relative body weight: *BMI* and *MI*

However, the *BMI* ignores different body properties. For individual assessments, according to the *WHO* (15), "care should be taken in groups with unusual leg length to avoid classifying them inappropriately as thin or overweight". The *BMI* definition, the *Cormic index* $C = s/h$ (s being the sitting height in meters), and anthropometric data presented by N. Norgan (16) were the starting points for the deduction of a new measure for relative body weight: the Mass Index *MI*, which takes the relative leg length l of the individual into consideration (5). The *MI* is a modification of the *BMI* according to the definition equation

$$MI = BMI \cdot \left(\overline{C}/C\right)^k \tag{1}$$

with C being the individual *Cormic index* s/h and $\overline{C}$ is a value to be chosen in the middle of the *Cormic index* continuum: $\overline{C} = 0.53$ was used. The intention was, to define the new measure for relative body weight *MI* such that it is independent of the relative leg length. Using the regression coefficient of 0.9 kgm^{-2} per 0.01 C, as found by Norgan, k could be determined: $k = 2.015 \approx 2$, and thus the simple formula for the *MI* results as (5):

$$MI = 0.53^2 \cdot \frac{m}{s^2} \approx 0.28 \cdot \frac{m}{s^2} \tag{2}$$

The unit is the same as for the *BMI*: kgm^{-2}. With $k = 2.0$, the body height h does not appear in the final equation for the Mass Index *MI*. Fig. 5b shows the *MI* values of ski jumpers investigated in Hinterzarten (2000) and in Slat Lake City (2002).

In case, further studies would imply to use anthropometric data sets resulting in a value for the exponent $k \neq 2.0$, both, h and s would remain in the formula and the general term *MI** would result:

$$MI^* = \frac{m}{h^2} \cdot \left(\frac{\overline{C}}{s/h}\right)^k = \frac{m}{h^{2-k} s^k} \cdot \overline{C}^k \tag{3}$$

For practical purposes this formula could be approximated by a simple term of the type $MI^* = BMI + f \cdot (MI - BMI)$, with f being an approximately chosen constant factor. For $k = 2.0$ (which was used for the definition of the Mass Index *MI*) the factor f would equal 1.0 and $MI^* \equiv MI$ would result.

2 Physics of Ski Jumping

2.1 The in-run

During the in-run the athlete maximises acceleration by minimising both the friction between skis and snow and the aerodynamic drag. The obtained in-run speed v_0 has a high impact on the jump length (1; 17). The reduction of aerodynamic drag in the in-run phase is primarily a question of the athlete's posture and his dress and can be optimized by means of wind tunnel measurements and feedback training forms.

Figure 6: Wind tunnel measurements of the time course of the jumping force (Kistler force plate, wind speeds from 0 to 30 m/s). In order to simulate the low friction between the ice of the in-run track and the skis, rollers were mounted underneath the ski dummies or underneath real skis. Wind tunnel: Graz University of Technology. Compare to Fig. 7. Photo: W. Müller 2001.

The friction between skis and snow is not well understood and the theoretical as well as the empirical basis for these complex problems are not sufficiently developed for a scientifically guided friction minimisation approach. Ski preparation is still a field of practical experience. Obviously, the position of the athlete on the skis – which is determined by the given biomechanical geometry of the athlete's feet, legs, hips, the jumping boots, and the binding – is of relevance, and also, how the athlete guides the skis in the track is important. In the curved path of the in-run the force perpendicular to the ground was found to be 1.65 times the weight (18). This mechanical demand posed on the athlete influences the initial conditions for the take-off movement.

2.2 The take-off

Due to the curved form of the in-run just before the ramp the athlete has to counteract the centrifugal force acting on him (as seen from his point of view) and immediately after this phase the athlete's acceleration perpendicular to the ramp due to the muscular forces exerted follows (19; 20; 21). During this jumping phase of about 0.3 s duration the athlete has to produce a high momentum $\int \vec{F}(t) \cdot \mathrm{d}t = \vec{p}_{\mathrm{p0}} = m \cdot \vec{v}_{\mathrm{p0}}$ perpendicular to the

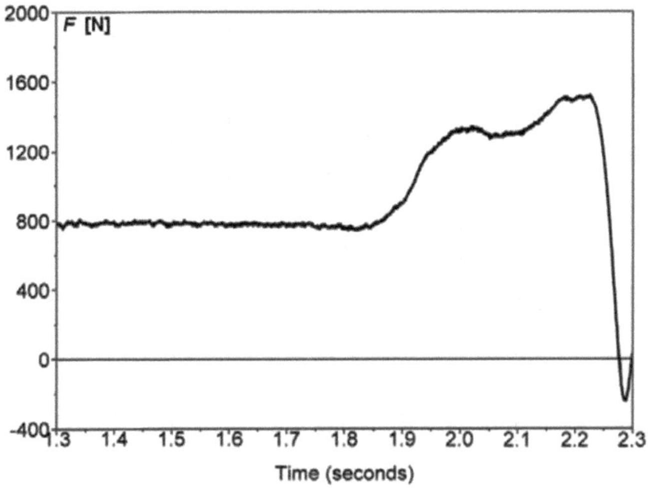

(a) Athlete *SH*: take-off force

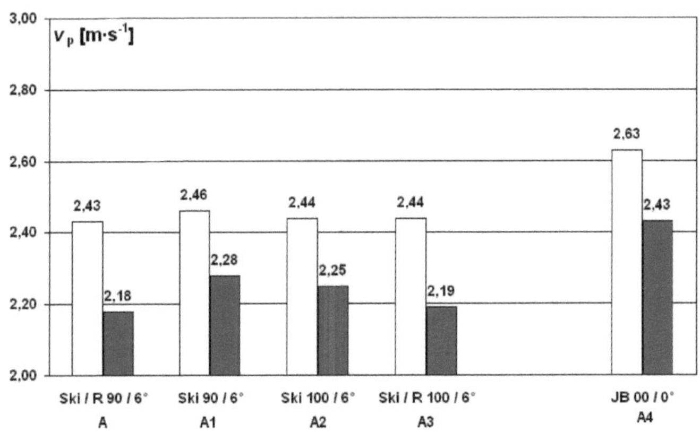

(b) Athlete *SH*: velocity perpendicular to ramp mockup

Figure 7: Athlete *SH*. Measurements of take-off forces in the wind tunnel (wind velocities of 80, 90, or 100 km·h^{-1}) and the according velocities perpendicular to the ramp mockup (calculated from the linear momentum obtained from the force curve evaluation). The measurements were made with jumping skis (2.60 m), with the exception of A4 and B4 which were made with jumping boots but without skis. For the latter case a theoretically determined velocity with skis was derived by considering the additional mass of 6 kg for the skis (filled bars A4 and B4) in order to make jumps with and without skis comparable.

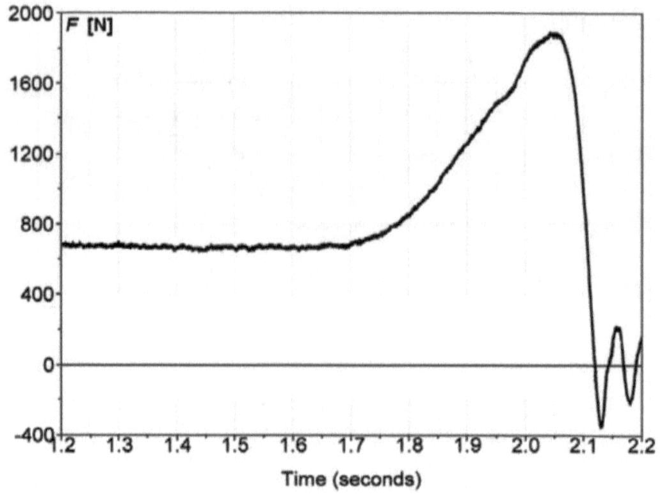

(c) Athlete *WL*: take-off force

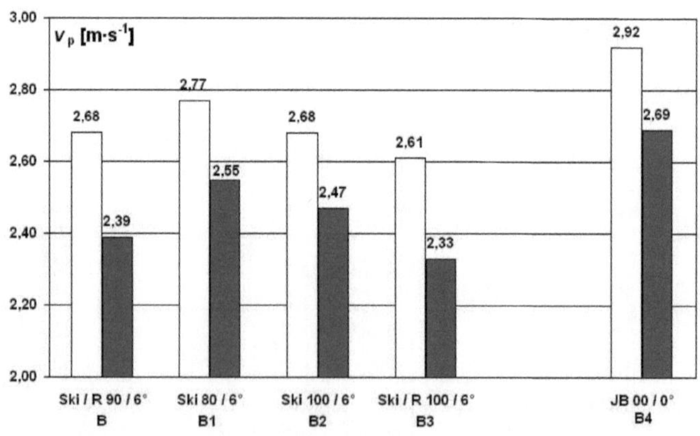

(d) Athlete *WL*: velocity perpendicular to ramp mockup

Figure 7: (continued) Athlete *WL*. For all measurements except *A4* and *B4* it was done the other way round. The real jumps on the inclined (6°) forceplate were made with skis (labeled *Ski*) or with skis and rollers underneath (*Ski/R*) in order to obtain a friction comparable to the situation on a real ramp. The velocities obtained with the use of rollers were all lower than those without, which indicates the increased difficulty to produce a maximum jump when tangential forces cannot be applied by the athlete. Linear momentums were highest with jumping boots (*JB*). Masses of the athletes with equipment: *SH* 73.3 kg, *WL* 70.0 kg. (Measurements: W. Müller et al., 2001).

ramp (m: mass of the athlete plus equipment; $\vec{F}$: accelerating force due to the athlete's muscles and additionally due to the aerodynamic lift force acting on the athlete) through which an advantageous take-off angle of the centre of gravity can be obtained. Values of $\vec{v}_{\mathrm{p}0}$ measured in a wind tunnel (Fig. 6) are shown in Fig. 7. The take-off velocity vector $\vec{v}_{00}$ is given by: $\vec{v}_{00} = \vec{v}_0 + \vec{v}_{\mathrm{p}0}$, with $\vec{v}_{\mathrm{p}0}$ being the velocity perpendicular to the ramp due to the jump.

Simultaneously, the athlete must produce an angular momentum $\vec{L}$ for the forward rotation in order to obtain an advantageous angle of attack as soon as possible. The athlete must anticipate the magnitude of the backward torque that will occur due to the air-stream in the initial flight phase so that his forward rotation will be stopped at the right moment.

The accurate timing of the muscle groups involved in the jumping movement has to occur such that the take-off jump is completed as close at the edge of the ramp as possible. Due to the glide path, with respect to the profile of the landing hill, a too early take-off would result in a substantially decreased jump length and the same is true when the timing is too late. In the latter case, the leg extension is not completed in time and the linear momentum perpendicular to the ramp $\vec{p}_{\mathrm{p}0}$ is too far from the athlete's maximum, and the angular momentum will not be optimal too. The flight trajectory is sensitive to both initial conditions, the angular momentum which largely determines the lift and drag forces after take-off, and to the linear momentum. It is not at all only the magnitude of the linear momentum which determines the jump length (1; 17; 22), even a maximum linear momentum cannot lead to the success without accurate timing and production of the associated angular momentum, which depends on the body configuration and movement of the athlete during the jump. The jumping forces determined for two world class athletes (Fig. 7) can serve as an example for this: Although the jumping force measurements with ski jumper *WL* showed higher linear momentums produced by this athlete it was the other jumper shown in this figure, *SH*, who won two Summer Gran Prix competitions a few day after these wind tunnel measurements and trainings (after more than two years in which *SH* did not win any competition).However, a very high jumping potential eases the athlete to correct for smaller mistakes before take-off and can therefore help stabilizing a high performance level.

Take-off forces have been measured in the field (20; 21). The force curves of different subjects showed different patterns (20) and this goes hand in hand with the curves shown in Figs. 7a and 7c and with results of kinematic

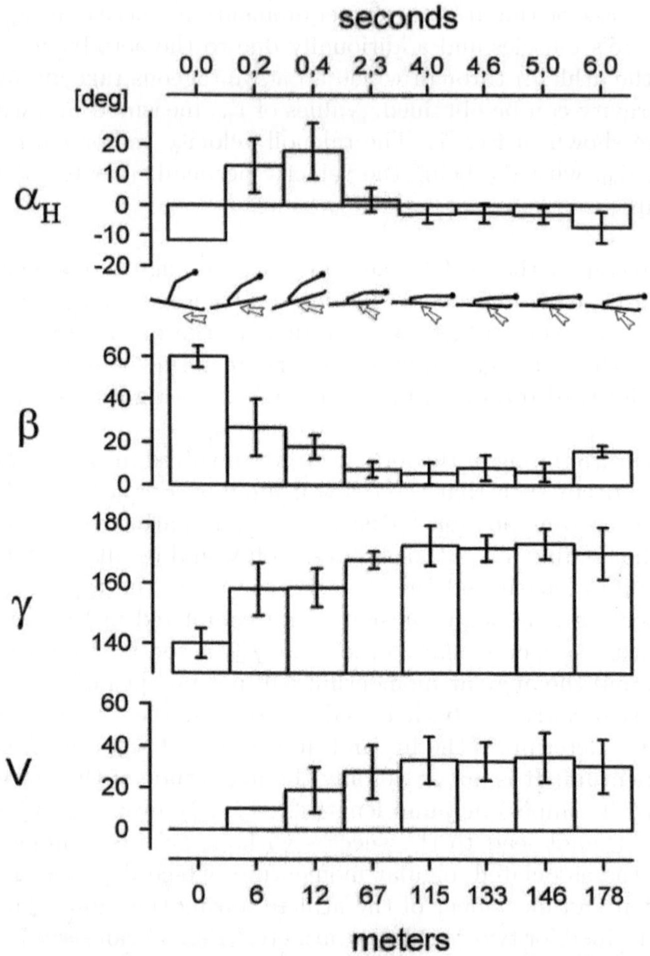

Figure 8: Field research results obtained during the 13th World Championships in Ski Flying (Planica 1994). The histograms show the average values and standard deviations of position angles during 15 excellent jumps. Angles α_H, β and γ were measured from slides taken from abeam at $l = 0$, 6, 12, 67, 115, 133, 146, and 178 m. Mean values for the flight times corresponding to the observer positions are given on top of the histogram. The angle V of the skis to each other was determined from digitized video images taken from approximately 200 to 300 m distance (Professional *SVHS*, *JVC KY-17 FIT* camera with *JVC TV* lens UM, 1:1.4, 7.5 - 97.5). W. Müller et al. 1996 (1), with permission.

analyses of the take-off movement where significant differences between individuals were found (23; 24). The athletes also utilize the aerodynamic lift force during take-off: force plate measurements in the wind tunnel have shown that the vertical momentum was larger than was found from the calculation of the area under the force-time curve, due to the effect of the lift force. This additionally acting force also decreased the take-off time in the wind when compared to calm conditions (19). It has been described that the correlation between maximum force and maximum jumping height differs considerably between different ski jumpers (25) and it is also still being debated as to which load maximises power output during various resistance exercises and how training at maximum power influences the jumping performance (26).

2.3 Aerodynamic characteristics of flight styles

For the analysis of flight styles the flight position angles of ski jumpers have to be measured from take-off on, during the entire flight phase until landing. Such measurements of V-style ski jumping orientation angles in the field have been made for the first time during the World Championships in ski flying in Planica, 1994 (Fig. 8). The position angles of the athletes and the angle of attack of the skis (projected angles) have been measured from slides taken from abeam at l = 0, 6, 12, 67, 115, 133, 146 and 178 m. The angle nomenclature used (see Fig. 4c): α is the angle of attack of the skis, β is the body to ski angle, γ is the hip angle; the angle V of the skis to each other (see Fig. 4a) was determined from digitized video images taken from in front. $V = 2 \cdot arcsin\left(\frac{d_1 - d_2}{2 \cdot L}\right)$, L being the known length of the ski and d_r being the distance reference (0.45 m). Fig. 4b shows an extreme position performed with 1993/94 equipment and flight style. The tumbling risk is high using such postures, because the pitching moment can suddenly become unbalanced, e.g. due to a gust.

Series of wind tunnel measurements of athletes in 54 different positions (see Fig. 9) in a 5 $m \times 5$ m cross section wind tunnel (Railtec Arsenal, Vienna) have led to a model of ski jumping which enables the investigator to take the very important positional changes of the athletes into account (1; 6). Fig. 4c shows the schematics of the positioning apparatus for the wind tunnel measurements , which enabled almost all postures of athletes and skis imaginable. In order to minimize aerodynamic interaction, the positioning apparatus was constructed of slim steel profiles (2 $cm^2 \times 3$ cm^2 and 1 cm diameter). Two pairs of dynamometers (1, 2 and 3, 4 respectively) measured the vertical forces F_1 and F_2, dynamometer 5 measured the horizontal force F_3. This setup allowed the determination of the lift force F_l

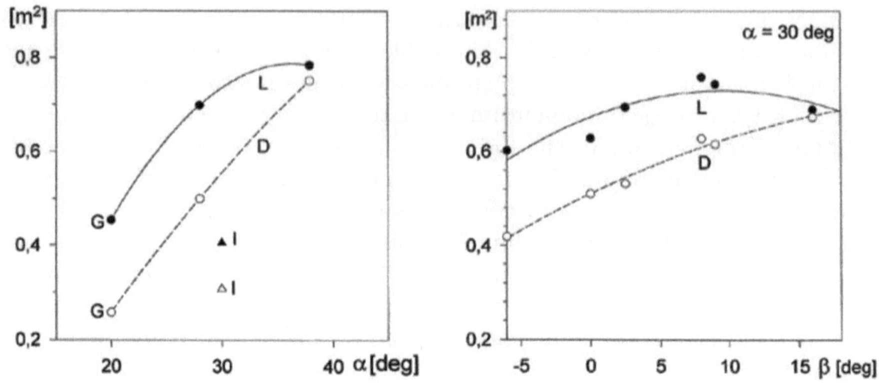

(a) L and D with constant body to ski angles

(b) L and D depending on the body to ski angle β

Figure 9: Circles label the measurements with athlete G (Andreas Goldberger, World Cup winner 1994/95, 1.70 m, 59 kg). Examples of lift area L and drag area D values are shown. (a): The body to ski angles were kept constant at $\alpha = 7°$, $\gamma = 164°$, $V = 35°$. The L and D values at a given $\alpha = 30°$ with $\beta = 9°$, $\gamma = 159°$, $V = 0°$ (old flight style, skis held in parallel), for athlete I (Anton Innauer, world record 1976, 1.73 m, 59 kg) with his equipment of 1976, are labelled with triangles. (b): Examples of L and D values depending on the body to ski angle β (athlete: 59 kg, 1.70 m). The shown values have been taken at $\alpha = 30°$ and $V = 35°$ with 1993/94 equipment. While the drag value still rises above $\beta = 10°$, the lift value saturates and is slightly decreased at $\beta = 15°$. W. Müller et al. (1), with permission.

and the drag force F_{d}. According to $F_{\mathrm{l}} = \rho \cdot L \cdot \frac{w^2}{2}$ and $F_{\mathrm{d}} = \rho \cdot D \cdot \frac{w^2}{2}$ the values for the lift area L and the drag area D were determined, w being the wind velocity. The air density ρ was calculated according to $\rho = \frac{p}{R \cdot T}$, p being the actual air pressure, T the absolute temperature and R the gas constant (288.3 $\mathrm{JK^{-1}kg^{-1}}$). Athletes, in various flight positions, have been investigated at wind velocities between 25 and 30 $\mathrm{ms^{-1}}$. In order to follow effects resulting from equipment advances over the last two decades, both today's world class athletes and the world champion of 1976 with their respective equipment and flight styles have been investigated. The lift area (L) and drag area (D) values for many positions of the athletes have been evaluated. All data obtained have been corrected for blocking effects although these were minimal (25 m^2 tunnel cross-section). Many of the measurements were performed at 2 wind velocities (25 and 27.8 ms^{-1}). The obtained D- and L-values varied typically by ±2%, and singularly up to 5%, owing to unavoidable vibrations and small movements of the athlete. Using a model of the athlete would (partly) avoid the latter, but this would ignore the very interesting individual posture differences between athletes. A selection of wind tunnel data obtained with 1993/94 equipment is shown in Figs. 9. 9a shows the lift L and drag D values of the athlete G at different angles of attack . Circles mark results with athlete G (1.70 m, 59 kg) with 1993/94 equipment. The body position of G was held constant ($\beta = 7°$ and $\gamma = 164°$) and the angles of attack were 20°, 28°, and 38° respectively with the opening of the skis held constant at $V = 35°$. These body position angles equal the mean values during the flight phase obtained in the field (Fig. 8). Triangles in Fig. 9a mark the results obtained with athlete I (1.73 m, 59 kg; 176 m world record in 1976, Olympic gold medalist 1980) using his equipment and flying position from 1976 ($V = 0°$, i.e. skis positioned parallel) at $\alpha = 30°$. L was 54% and D was 56% of the values obtained with 1994/95 equipment and flight style (compare with G). For the flight configurations compared here, the L/D-ratio remained almost unchanged. Fig. 9b gives additional information on the L- and D-dependencies on the body-to-ski angle β. While D at β-values over 10° still significantly increases, the lift curve levels out. This is qualitatively easy to understand because at $\beta = 10°$ and $\alpha = 30°$ the body's angle of attack is already 40°. In all, the athlete was measured in 54 postures (Fig. 9 shows only some examples) in order to obtain a grid of L and D values according to a variety of postures useable as input data for computer simulation studies of ski jumping.

This first approach to a realistic mapping of V-style ski jumping has meanwhile been developed further (compare to Fig. 10) using field data sets from several World Cup events (27) and from the Olympic Games 2002 (28). Five runs during Olympic competitions were investigated (150 jumps) and from the data sets obtained (1800 images) the best 10 athletes of each competition were selected for detailed analysis and the mean of these 10 flight position angles at each observation point was used for the definition of a reference jump P which was used for comparisons with results obtained previously by Schmölzer and Müller with reference jump A (27) and as a starting point for the computer simulations of the flight paths at the Olympic venue in Park City, Utah (see Fig. 11).

The angle measurements have been made by means of a computer graphics program (*Corel Draw 10.0*). In the case of skis not appearing exactly

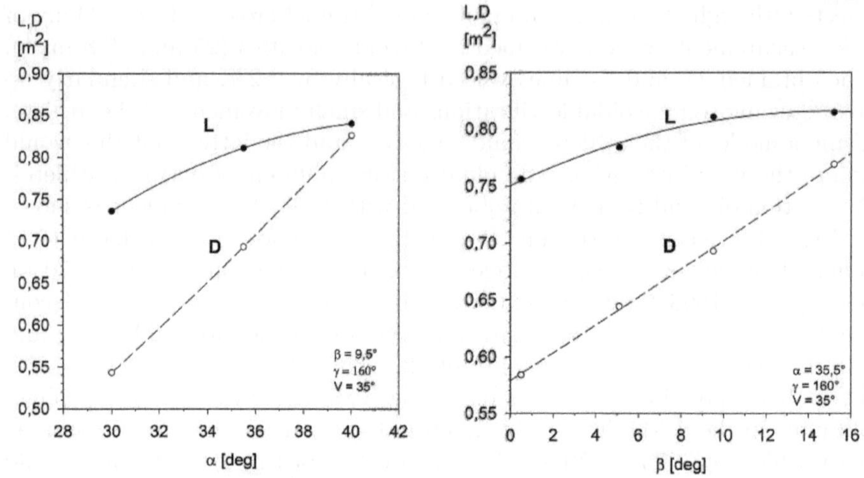

(a) L and D depending on angle of attack α

(b) L and D depending on body to ski angle β

Figure 10: Examples of lift and drag areas according to (27). The figures show the L and D values for reference jump A measured with a $1:1$ model of a ski jumper (height $h = 1.78$ m). (a) shows the dependency to different angles of attack α. The body position was held constant ($\beta = 9.5°$ and $\gamma = 160°$) and the angles of attack were 30°, 35.5°, and 40°. The opening of the skis was held constant at $V = 35°$. (b) shows L and D values depending on the body to ski angle β. The values shown have been taken at $\alpha = 35.5°$, $\gamma = 160°$ and $V = 35°$. B. Schmölzer and W. Müller 2002 (27), with permission.

one behind the other in the image and projected angles between the skis less than 6° a line in the middle between the two skis has been used for the determination of the α_H- and β-angles. 85 images where skis appeared under angles larger than 6°to each other (i.e. 15.5% of a total of 550 images) have been excluded from further evaluation. Wind was almost calm during all field studies, maximal gust values measured officially were less than 1.8 ms^{-1}. In order to maximise positioning accuracy, series of additional measurements with $1:1$ models of ski jumpers in the large scale wind tunnel (27) as well as measurements with a model on a smaller scale (in the 1.5 $m \times 1$ m wind tunnel of the Technical University of Graz; Fig. 12) have been conducted in order to receive a very dens grid of data for all

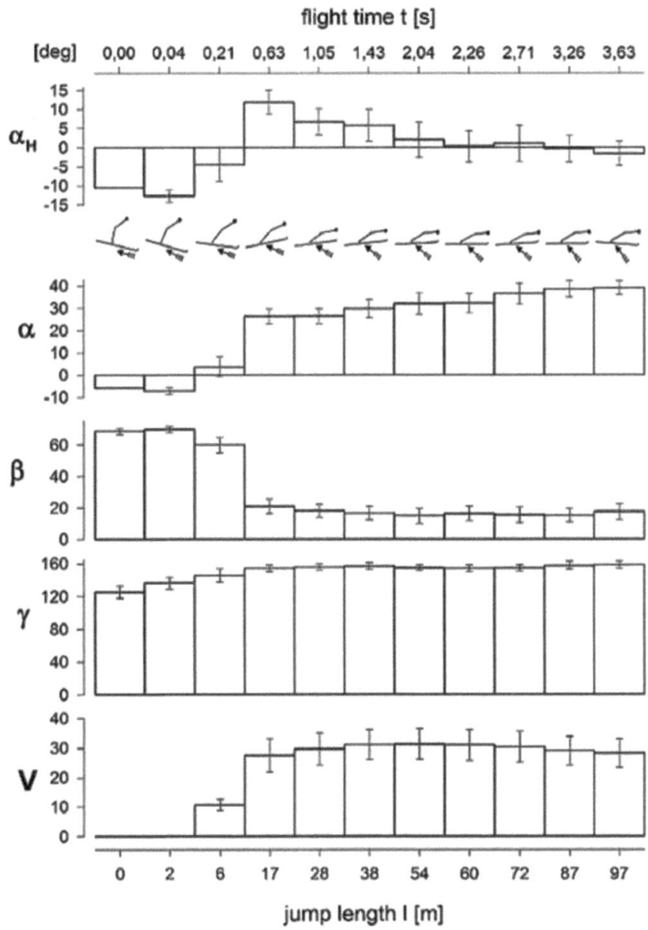

Figure 11: Mean flight position angles according to (28). The mean angles found during the Olympic Games competitions 2002 (venue altitude: 2000 m) show remarkable deviations when compared to results from field studies at competitions performed at lower altitudes (27; 28). The histograms show the average values and standard deviations of position angles from the best 10 athletes in each of the 5 runs at the $K = 120$ m jumping hill. The number of angle measurements ranged from 18 to 50 at each position. The angle V of the skis to each other was determined from digitised images taken from at the end of the run out. B. Schmölzer and W. Müller 2005 (28), with permission.

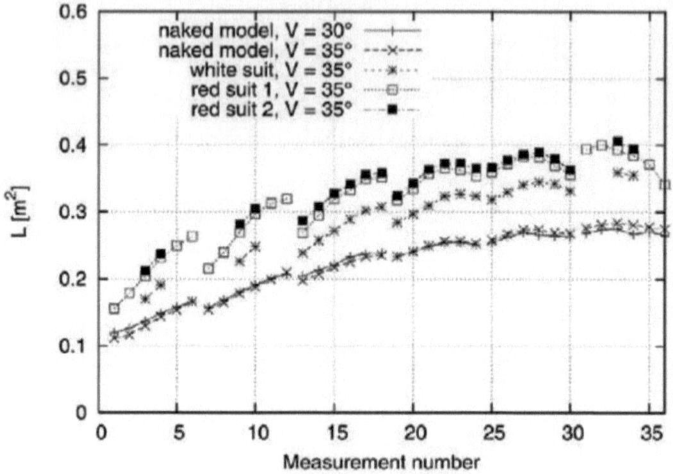

(a) L values of model

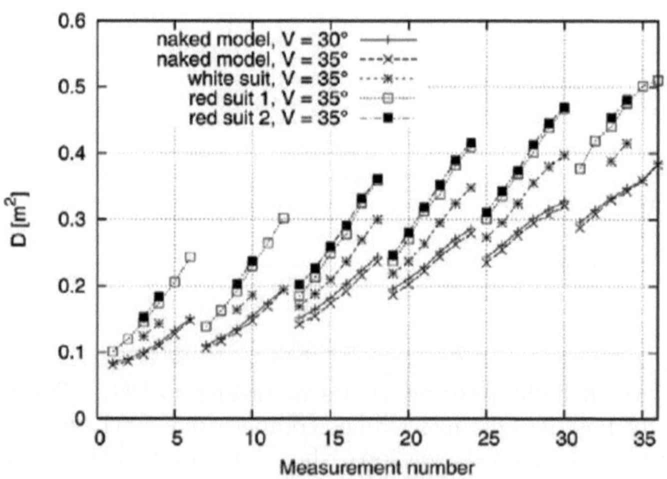

(b) D values of model

Figure 12: Examples of lift and drag areas obtained with a model scaled down by $1 : \sqrt{2}$ (29). (a) shows the lift area L dependency on the angle of attack (with respect to the skis) α and on the body to ski angle β. The β - angle was varied from 0 to 25 degrees in steps of 5 degrees and α - angles were 20 degrees (measurement numbers 0 to 6), 25 degrees (7 to 12), and so on until $\alpha = 45$ degrees (31 to 36). (b) Drag area D data determined as described for L. W. Meile et al. 2006 (29), with permission.

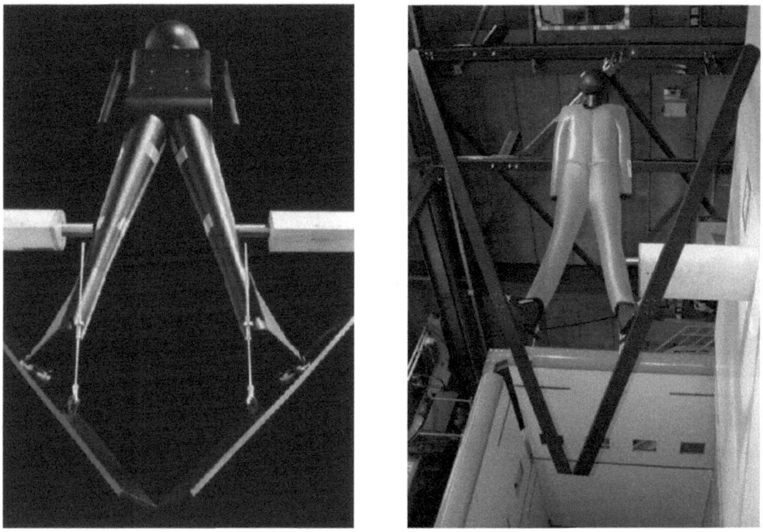

(c) Naked model (d) Model with jumping suit

Figure 12: (continued) (c) The naked scaled down model ($1 : \sqrt{2}$) used for L and D measurements (see Fig. 12a and 12b) is shown in the photo. (d) The photo shows the model in a red jumping suit with dimensions according to the *FIS* ski jumping regulations.

imaginable postures (29). The wind tunnel measurements with the scaled down model have also been designed for comparative studies with *CFD* approaches (29) which are discussed in a separate chapter in this book (*Ski Jumping Aerodynamics: Model-Experiments and CFD-Simulations* by W. Meile, W. Müller, and E. Reisenberger). Changing postures of the athlete during the flight have pronounced effects on the flight trajectory and thus on the performance. The solution of the nonlinear coupled equations of motion describing the flight of a ski jumper cannot be worked out in one's head and necessitates a computer simulation approach. The computer model and practically relevant results of simulation studies are discussed in the chapter on *Computer Simulation of Ski Jumping Based on Wind Tunnel Data* by W. Müller.

Acknowledgements Supported by the Medical Commission of the International Olympic Committee (*IOC*), The International Ski Federation

(*FIS*), and the Austrian Research Funds (*FWF* 14388 Tec, 15130 Med). I would like to thank C. Cagran for proof reading and for the layout of the manuscript.

Bibliography

[1] W. Müller, D. Platzer, and B. Schmölzer. Dynamics of human flight on skis: improvements on safety and fairness in ski jumping. *Journal of Biomechanics*, 29(8):1061–1068, 1996.

[2] H.-H. Gasser. *Grundlagen für die Projektierung einer Schisprungschanze.* FIS, available from http://www.fis-ski.com, 2005.

[3] W. Müller. Biomechanics of ski-jumping - Scientific jumping hill design. In E. Müller, H. Schwameder, E. Kornexl, and C. Rascher, editors, *Science and Skiing*, pages 36–48. E & FN Spon, 1997.

[4] W. Müller and B. Schmölzer. The new jumping hill in Innsbruck: Designed by means of flight path simulations. In *World Congress Biomechanics (WCB), University of Calgary, Canada, Proceedings CD*, 2002.

[5] W. Müller, W. Gröschl, R. Müller, and K. Sudi. Underweight in ski jumping: The solution of the problem. *International Journal of Sports Medicine*, 27:926–934, 2006.

[6] W. Müller, D. Platzer, and B. Schmölzer. Scientific approach to ski safety. *Nature*, 375:455, 1995.

[7] W. Müller and T. T. J. DeVaney. The influence of body weight on ski jumping performance. In S. Haake, editor, *The Engineering of Sport*, pages 63–69. Balkema, 1996.

[8] F. Vaverka. Somatic problems associated with the flight phase in ski-jumping. *Studia i monografia AWF we Wroclawiv*, 40:123–128, 1994.

[9] F. Vaverka. Research reports, Olomouc. personal communication, 1987 and 1995.

[10] A. E. Becker, S. K. Grinspoon, A. Klibanski, and D. B. Herzog. Current concepts: eating disorders. *New England Journal of Medicine*, 340:1092–1098, 1999.

[11] K. Sudi, Karl Öttl, Doris Payerl, Peter Baumgartl, Klemens Tauschmann, and Wolfram Müller. Anorexia Athletica. *Nutrition*, 20:657–661, 2004.

[12] P. F. Sullivan. Mortality in anorexia nervosa. *American Journal of Psychiatry*, 152:1073–1074, 1995.

[13] WHO Expert Committee. Physical status: the use and interpretation of anthropometry. In *WHO Technical Report Series*, number 854, page 364. WHO, 1995.

[14] W. Müller. A scientific approach to address the problem of underweight athletes: a case study of ski jumping. *Medicine & Science in Sports and Exercise*, 35(5 Supplement 1: 6th IOC World Congress on Sport Sciences 2002):124, 2002.

[15] WHO Expert Committee. Physical status: the use and interpretation of anthropometry. In *WHO Technical Report Series*, number 854, page 355. WHO, 1995.

[16] N. G. Norgan. Population diferrences in body composition in relation to the BMI. *European Journal of Clinical Nutrition*, 48:10–27, 1994.

[17] W. Müller and B. Schmölzer. Computer simulated ski jumping: The tightrope walk to high performance. In *World Congress Biomechanics (WCB), University of Calgary, Canada, Proceedings CD*, 2002.

[18] G. J. C. Ettema, S. Braten, and M. F. Bobbert. Dynamics of the in-run in ski-jumping: a simulation study. *Journal of Applied Biomechanics*, 21(3):247–259, 2005.

[19] M. Virmavirta, J. Kiveskäs, and P. V. Komi. Take-off aerodynamics in ski jumping. *Journal of Biomechanics*, 34(4):465–470, 2001.

[20] M. Virmavirta and P. V. Komi. Measurement of take-off forces in ski-jumping. Part I and II. *Scandinavian Journal of Medicine & Science in Sports*, 3(4):229–243, 1993.

[21] P. Kaps, H. Schwameder, and C. Engstler. Inverse dynamic analyses of take-off in ski-jumping. In E. Müller, H. Schwameder, E. Kornexl, and C. Rascher, editors, *Science and Skiing*, pages 72–87. E & FN Spon, 1997.

[22] J. Denoth, S. M. Luethi, and H. Gasser. Methodological problems in optimisation of the flight phase in ski jumping. *International Journal of Sport Biomechanics*, 3:404–418, 1987.

[23] F. Vaverka, M. Janura, M. Elfmark, J. Salinger, and M. McPherson. Inter- and intra-individual variability of the ski-jumper's take off. In E. Müller, H. Schwameder, E. Kornexl, and C. Rascher, editors, *Science and Skiing*, pages 61–71. E & FN Spon, 1997.

[24] A. Arndt, G.-P. Brüggemann, M. Virmavirta, and P. V. Komi. Techniques used by olympic ski jumpers in the transition from take-off to early flight. *Journal of Applied Biomechanics*, 11(2):224–237, 1995.

[25] S. Bruhn, A. Schwirtz, and A. Gollhofer. Diagnose von Kraft- und Sprungkraftparametern zur Trainingssteuerung im Skisprung. *Leistungssport*, 5(2):34–37, 2002.

[26] J. Cronin and G. Sleivert. Challenges in understanding the influence of maximal power training on improving athletic performance. *Sports Medicine*, 35(3):213–234, 2005.

[27] B. Schmölzer and W. Müller. The importance of being light: aerodynamic forces and weight in ski jumping. *Journal of Biomechanics*, 35(8):1059–1069, 2002.

[28] B. Schmölzer and W. Müller. Individual flight styles in ski jumping: results obtained during olympic games competitions. *Journal of Biomechanics*, 38(5):1055–1065, 2005.

[29] W. Meile, E. Reisenberger, M. Mayer, B. Schmölzer, W. Müller, and G. Brenn. Aerodynamics of ski jumping: experiments and CFD simulations. *Experiments of Fluids*, 41(6):949–964, 2006.

Computer simulation of ski jumping based on wind tunnel data

Wolfram Müller*

* Human Performance Research^{Graz}, Karl-Franzens University and Medical University of Graz, Graz, Austria

Abstract The computer model of ski jumping described here considers the changing flight postures of the athlete. It is based on sets of wind tunnel data associated with the changing postures of the athlete during the flight. The computer modelling approach allows studying the impact of all variables, parameters, and initial conditions which determine the flight path – and thus the jump length – of a ski jumper. In addition, wind u (wind vector) in the vertical plane can be selected with any direction and speed. Insertion of the construction parameters of a given hill allows calculating jump lengths and also the landing impact (component of the moment perpendicular to the landing slope) as a function of jump length for the particular hill. The model enables the calculation of the flight path; the inverse approach starts out from flight path data and results in the time-funtions of the lift of drag forces and areas which can be used for a quantitative flight analysis.

1 The modelling approach

During the flight, the gravitational force F_{g}, the lift force F_{l}, and the drag force F_{d} act upon the athlete and his equipment (see Fig. 1) and determine the flight path of the centre of gravity of a ski jumper with a given set of initial conditions and parameters:

$$\begin{aligned} F_{\mathrm{g}} &= m \cdot g \\ F_{\mathrm{l}} &= \frac{\rho}{2} \cdot w^2 \cdot c_{\mathrm{l}} \cdot A \\ F_{\mathrm{d}} &= \frac{\rho}{2} \cdot w^2 \cdot c_{\mathrm{d}} \cdot A \end{aligned} \tag{1}$$

w is the value of the relative wind verctor ($\vec{w} = \vec{u} - \vec{v}$, $\vec{u}$ being the velocity of external wind, $\vec{v}$ the velocity of motion), c_{l} and c_{l} are the lift and drag coefficients, respectively, and A is the reference area. $L = c_{\mathrm{l}} \cdot A$ and $D = c_{\mathrm{d}} \cdot A$

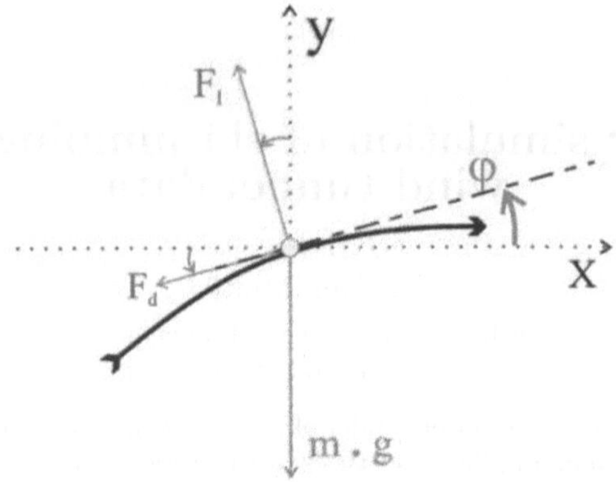

Figure 1: Forces acting on a flying object (2D-model): The lift force F_{l}, the drag force F_{d}, and the weight weight $m \cdot g$. At calm wind conditions $L = \frac{F_{\mathrm{l}}}{(\frac{\varrho}{2} \cdot v^2)}$ and $D = \frac{F_{\mathrm{d}}}{(\frac{\varrho}{2} \cdot v^2)}$ with v being the the value of hte velocity of motion and ρ the air density. In case of wind $\vec{u}$ the relative wind $\vec{w} = \vec{u} - \vec{v}$ has to be considered.

are called the lift and drag areas which can be measured in a wind tunnel. The air density ρ decreases with increasing altitude and temperature: $\rho = p/RT$ (p being the air pressure which decreases exponentially with increasing altitude, T is the absolute temperature, and $R = 288.3\ \mathrm{JK^{-1}kg^{-1}}$ is the gas constant).
The equations of motion consider all forces acting during the flight. These equations have already been used in the pioneering work by Straumann in 1927 (1) and later in several analytical studies (2; 3; 4). They can be solved numerically for a given set of initial conditions with any desired accuracy:

$$\begin{aligned} \dot{v}_{\mathrm{x}} &= (-F_{\mathrm{d}} \cdot cos\varphi - F_{\mathrm{l}} \cdot sin\varphi)\ \frac{1}{m} \\ \dot{v}_{\mathrm{y}} &= (-F_{\mathrm{d}} \cdot sin\varphi + F_{\mathrm{l}} \cdot cos\varphi)\ \frac{1}{m} - g \\ \dot{x} &= v_{\mathrm{x}} \\ \dot{y} &= v_{\mathrm{y}} \end{aligned} \tag{2}$$

During the flight phase the athlete's position changes. The athlete can strongly influence the aerodynamic forces by changing his posture. He can affect the drag force, the lift force and the torque; the latter enables him to change his flight posture and angle of attack with respect to the air stream. The real problem with simulation studies of the flight path of a ski jumper is the difficulty to obtain accurate lift and drag area functions $L(t)$ and $D(t)$, respectively, which correspond to the changing postures of the athlete during the flight and to the time functions of the angles of attack of the body parts and the skis. Based on a simplified model which constrained the motion to that of a rigid body, it has already been shown by Remizov (4) that maximisation of the jump length necessitates an increase of the angle of attack during the flight (α, angle of attack of the skis). Hubbard et al. (5) developed a four-segment dynamic model of ski jumping based on a Lagrangian formulation of the equations of motion of the body segments, in which the jumper is modelled as a collection of planar, rigid bodies. However, a satisfying prediction accuracy of the set of muscle joint torques of the athlete as a function of time in order to position himself in the air stream in a desired way also necessitates sophisticated wind tunnel measurements of the pressure distribution on all body segments and on the skis in all positions the athlete goes through during the flight, which still now days are not available.

Both time functions $L(t)$ and $D(t)$ of the whole system (athlete in his gear with skis) depend in a most complicated way on the body configuration of the athlete with respect to the plane of the skis, on the angle V between the skis to each other, and on the skis' angle of attack α with respect to the relative wind vector $\vec{w}$. Even small changes of the posture can have noticeable effects on L and D and thus on the flight trajectory. The often made comparison of a ski jumper with a wing does not work at all: angles of attack of a wing range from about 0° to about 12° whereas, a ski jumper's body angle of attack is typically around 50° (6; 7; 8) a wing would stall at such high angles of attack and it makes more sense to compare the ski jumper to a flat plate(9; 10). Since the calculation of the aerodynamic forces acting on a ski jumper by means of computational fluid dynamics (*CFD*) is far from the necessary accuracy for relevant predictions, the only way to obtain accurate L and D values are measurements of the aerodynamic forces in a wind tunnel with a large cross section area (in order to keep the blocking effect low).

The computer model based on the sets of wind tunnel data allows studying the impact of all variables, parameters, and initial conditions which determine the flight path – and thus the jump length – of a ski jumper. In addition, wind $\vec{u}$ (wind vector) in the vertical plane can be selected with any direction and speed ($\vec{w} = \vec{u} - \vec{v}$), with $\vec{v}$ being the velocity of motion along the path and $\vec{w}$ the relative wind. Insertion of the construction parameters of a given hill allows calculating jump lengths and also the landing impact (component of the moment perpendicular to the landing slope) as a function of jump length for the particular hill.

2 Obtainable simulation accuracy

Figure 2: Measurement of position angles in the field (Olympic Games 2002) (8). The figure shows the angle nomenclature. B. Schmölzer and W. Müller 2005, with permission.

The obtained simulation accuracy of this approach for calculating the trajectory of the centre of gravity of the ski jumper is only determined by the limits of the experimentally obtainable accuracy for the simulation in-put values. Recently, a detailed failure analysis has been presented for both absolute obtainable simulation accuracy and for comparative studies of effects of parameter and initial value variations as well (7):
The simulation output accuracy depends primarily on the accuracy of $L(t)$ and $D(t)$, which is determined by:

1. The inexactness of the determination of the position angles. We assume, for example, a measurement error for α of $+2°$. A simulation using the associated values for L and D for the main flight phase (from $t = 1.0$ s on) resulted in a 3.5 m reduction in jump length (109 m compared to 112.5 m using the Sapporo $K = 120$ m jumping hill profile). Analogously, an increased value of β of $+2\%$ resulted in a 2.3 m reduction in jump length in the simulation.
2. The inaccuracy of the lift and drag measurements in the wind tunnel due to the inexactness of the model position (angular error $<1°$). An increase in both L and D by $+2\%$, whereas an increase of D alone by 2% would reduce it from 112.5 m to 109.6 m.
3. Blocking effects: An inaccuracy due to insufficient blocking corrections always remains when series of objects with different geometry (different athlete positions) are to be investigated. The failure can be kept insignificant by using large scale wind tunnels with cross section areas of 20 m^2 or above.

2.1 Reliability of the comparative simulation studies

The very high reliability of comparative simulation studies can be illustrated by assuming a large error in the input L and D values, which are the limiting parameters for the accuracy of the simulation. If, for instance, the effects of different masses are to be investigated, we obtain the following results (Sapporo $K = 120$ m profile, reference jump A): Reducing the mass from 65 kg to 63 kg results in a jump length increase from 112.5 m to 115.0 m (+2.5 m).When using the same protocol, but with L and D values increased by 3% (assumption of a large L and D measurement error), we find a jump length of 115.0 m with mass being 65 kg and 117.7 m with mass set to 63 kg. The predicted jump length increase of 2.7 m due to the 2 kg reduction in mass is, in this case, almost the same as above. A simulation protocol with only L set higher by 3% throughout the flight resulted in +2.7 m jump length increase due to 2 kg less mass, and finally only D set 3% higher resulted in a +2 m jump length increase due to the 2 kg mass reduction. The prediction error for the worst of these cases was less then 0.3 m (for comparison: the length measurement resolution in a real competition is 0.5 m).

3 Computer simulation studies

Computer simulations using jumping protocols based on accurately measured aerodynamic in-put data (L- and D-functions) can be applied for the quantitative interpretation of field study, laboratory, wind tunnel, or theoretically obtained data, e.g. their effects on the jump length or on the flight velocity; the predictions can also be used to answer many questions arising in the discussion of training methods, safety and health considerations, fairness in the sport, optimized jumping hill design, development of the sport, and changes to the regulations.

3.1 Computer simulation in-puts: choice of initial conditions and parameters

The simulation results obtained depend on the complete set of initial conditions and parameters chosen: the in-run velocity, the velocity perpendicular to the ramp (due to the athlete's jumping force), the ramp angle, the time functions $L(t)$ and $D(t)$ through the entire flight which mirror the flight style (the athlete's posture during the flight) and the equipment tuning used, the wind speed and direction, and the weight of the athlete (with his equipment). Additionally, the air density (which depends on the altitude and temperature) and the ramp angle as well as the parameters of the landing slope have to be chosen with respect to the real-world situation to be mapped.

3.2 Computer simulation out-puts

The 2-D computer-modeling approach described in the preceding paragraph enables the investigator to predict the trajectory during the flight phase and to investigate the effects of parameter of initial value variations. Due to the high precision of simulation outputs obtainable when performing comparative studies, many practically relevant questions can be answered reliably. Each simulation run results in the following outputs: jump length, landing velocity, landing velocity component perpendicular to the landing slope which results in the "equivalent landing height", height above ground (for every chosen jumping hill profile), velocity of motion, horizontal and vertical components of the velocity of motion, the flight trajectory as a function of the horizontal axis or of flight time, and the lift and drag forces acting on the athlete and his equipment during the entire flight. Of course, further variables, as for instance, the linear momentum $\vec{p}(t)$ or the kinetic energy $E_{\text{kin}}(t)$ can also be calculated and displayed graphically.

3.3 Questions concerning performance optimization

Of very high practical importance is the utilization of comparative computer simulations for the determination of all kinds of factors which influence the flight path. So, for instance, it is not possible to interpret any wind tun-

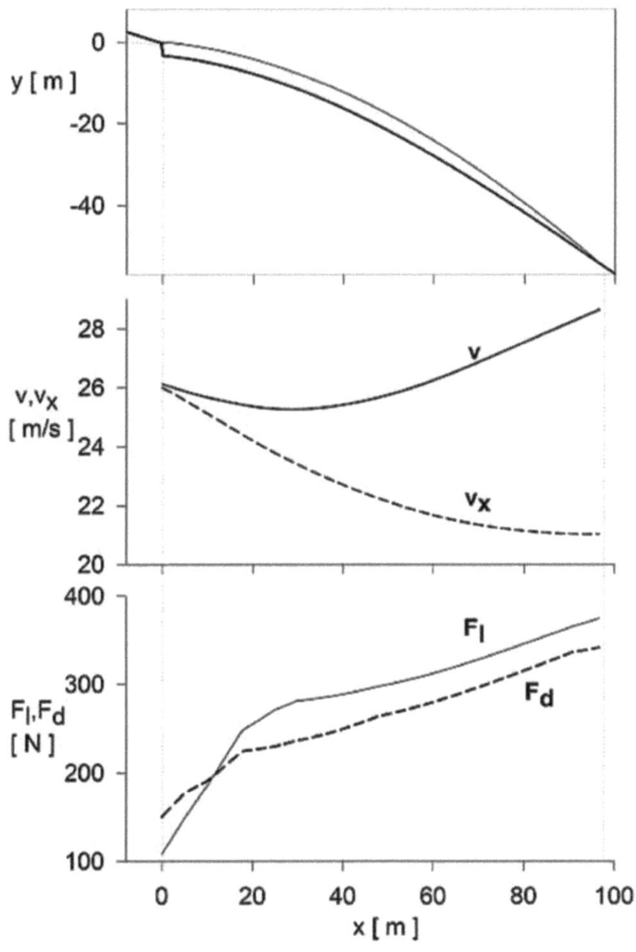

Figure 3: Flight trajectory, velocity, and lift and drag forces during a simulated jump (reference jump A) (7). The figure on top of the panel shows the profile of the jumping hill in Sapporo and the trajectory $y = y(x)$. The velocity of motion v (solid line) and the horizontal component of this velocity v_{x} (broken line) are shown in the figure in the middle. The lower figure shows the lift force F_{l} and drag force F_{d} acting on the athlete and his equipment. The air density was set to 1.15 kgm^{-3}; the mass of the athlete with equipment was 65 kg. B. Schmölzer and W. Müller 2002, with permission.

nel measurements associated with flight style or equipment changes correctly without applying the measured results to an appropriate computer simulation protocol: The equations of motion cannot be worked out in one's head, and, additionally, the profile of the jumping hill has to be considered too. This holds also true for all other factors which determine the jump length. Each phase of a ski jump (in-run, take-off jump, early flight phase, stabilized flight phase, and landing) has an impact on the subsequent phase. As has been shown already by Remizov 1984 (4) and by Denoth et al. 1987 (2) optimization approaches which focus on separate phases of a ski jump may not be relevant: the discussion of the "optimum style" has to include all parts of a jump. This is also supported by the field study results of Schmölzer and Müller (8), obtained during the Olympic Games 2002 , which demonstrate that different athletes, e. g., S. Amann (SUI) and A. Malysz (POL), used distinctly different styles which both resulted in a top performance. The optimum flight style for one athlete might be disadvantageous for others, due to different motor abilities and different anthropometrical and aerodynamic characteristics of individual athletes, which have an important impact on the difficulty for the athlete to stabilize the flight, i.e. to regulate the net pitching moment close to zero (6) as soon as possible after take-off.

However, based on the simulation of the entire flight trajectory, it may be very useful to change an input value (or a set of values) only in a distinct phase of the flight in order to get a general idea of the impact of this performance factor in dependence on the phase of flight. For instance, it has been shown that an increase of the drag D in the first third of the flight diminishes jump length dramatically whereas the same increase in D has only minor effect in the last third of the flight (11; 12), and high lift forces are important through the entire flight. This emphasizes the predominant importance of a rapid transition from take-off to a flight position associated with low drag (13). The athlete can balance the lift to drag ratio individually by means of his flight posture and should take care to keep the absolute value of drag small after take-off, although the associated lift value is below maximum in this case.

3.4 Simulation of the effects of flight posture variations – wind tunnel data as in-puts

It is very important to notice that the drag area D of a ski jumper increases continuously with increasing angle of attack α and also with increasing body to ski angle β, whereas, the lift area L shows a plateau (6; 7). Sets of aerodynamic data of V-style ski jumping obtained in a large scale wind tunnel with athletes (6; 14), or with 1 : 1 – models of athletes (7) can be applied to investigate all combinations of the time-functions of position angles imaginable. The wind tunnel data sets described in these publications above allow finding L and D measurement values (or interpolated values) for many simulation protocols of practical relevance (6; 14; 7; 8). The detailed sets of wind tunnel measurements with a scaled down model ($1:\sqrt{2}$) in a smaller wind tunnel (measured at reciprocally increased wind speed as demanded by aerodynamic similarity laws) (15) were primarily designed for comparing *CFD*-results to measured data, but, appropriate calibration of these values to results obtained from 1 : 1 measurements in a large scale wind tunnel would further enlarge the data sets available for detailed optimization studies.

3.5 Parameters and initial value variations: Simulation starting point

The simulations summarized here start out from the reference jump A (compare to Fig. 4) and the following protocol (7): approach velocity $v_0 = 26\ \text{ms}^{-1}$, $v_{p0} = 2.5\ \text{ms}^{-1}$, $m = 65$ kg, $\rho = 1.15\ \text{kgm}^{-3}$. This setting results in a jump length of 120 m, i.e. to the K-point of the hill profile in Innsbruck, which was used for this set of simulations.

3.6 In-run velocity v_0 and velocity v_{p0} perpendicular to the ramp

An increase of the in-run velocity v_0 of $0.1\ \text{ms}^{-1}$ (0.36 km/h) increases the jump length by 1.6 m (16). Notice the high sensitivity on the in-run velocity! The impact of the athlete's jump (in terms of the velocity perpendicular to the ramp v_{p0} at take-off) on the performance: An increase of v_{p0} of $0.1\ \text{ms}^{-1}$ results in 1.2 m jump length increase. Variations of v_{p0} in the simulation protocol show that a good jumping force of the athlete resulting in $v_{p0} \approx 2.5\ \text{ms}^{-1}$ is a necessity for a successful jump, however, a further increase of v_{p0} to a very high value of $2.8\ \text{ms}^{-1}$, for example, would increase the jumping distance by 3.6 m only.

3.7 Flight style and aerodynamic features of the equipment ($L(t)$ and $D(t)$ input functions)

When only the lift area $L(t)$ is increased or decreased by a constant percentage during the entire flight (with respect to the reference jump A): the slope is +1.8 m per 1 percent increase in L. When only the drag area D is increased or decreased by a constant percentage during the entire flight in the range of ±10% (with respect to the reference jump A): the slope is

-1.2 m per 1 percent increase in D.

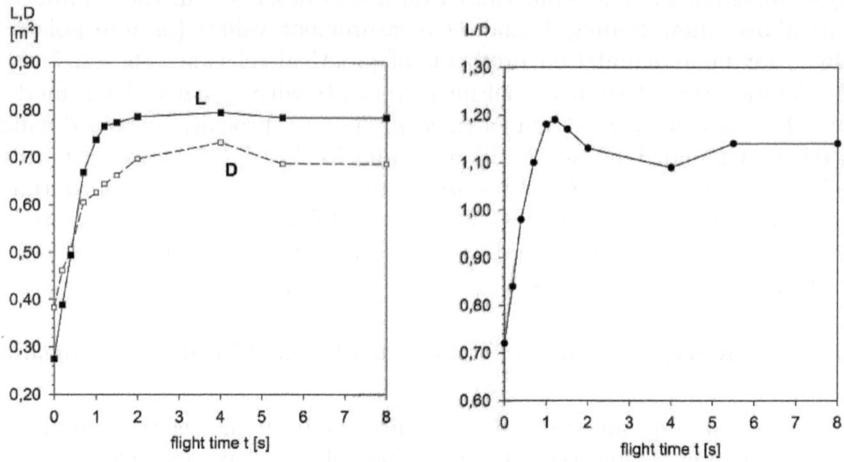

(a) L and D of reference jump A (b) L to D ratio of reference jump A

Figure 4: (a) shows the L and D values of reference jump A (7) which was designed by using the mean position angles of the 10 best athletes measured in each of a series of World Cup competitions. In (b) the respective ratio L/D is plotted. The values are functions of time, reflecting the athlete's position changes during the flight. B. Schmölzer and W. Müller 2002 (7), with permission.

Each change in the flight style or the equipment used can be expressed in a modification of the $L(t)$ and $D(t)$ input functions used, provided that the respective wind tunnel data is available.Additional to the angle of attack of the skis α and the body to ski angle β, the lift and drag areas also depend on the hip angle γ used by the athlete (7): the maximum lift L and a drag area D close to minimum have been measured at $\gamma = 160°$ in the wind tunnel (Fig. 5a). The mean γ-values found in the field when studying the 10 best athletes of several World Cup events were: 159°, 159°, 159°, 158°, and 161°at the flight times 1.0 s, 1.2 s, 1.5 s, 2.0 s, and 4.0 s, respectively (7). This excellent match of optimization prediction and field data, found from the best ski jumpers in the world, indicates the fascinating ability of top athletes to find out the optimum empirically, just guided by their proprioception and the coaches' advice.

The V-angle (angle between the skis to each other) most athletes use today is about 35°(7). Smaller V-angles would reduce the backward rotating torque of the skis (6). However, at the high altitude of Park City (2000 m) a mean V-angle of 30.5° was found (from $t = 1.0$ s to $t = 3.3$ s), and the mean γ-angle was only 155.5°. This result has to be seen in connection with the larger β-angles the athletes used at this altitude (8).

3.8 Altitude effects on the flight style

During the Olympic Games 2002, the effect of the low air density of 1.0 kgm^{-3} (due to the altitude of 2000 m of the venue in Park City) on the flight position during the entire flight has been investigated (8). Reduced aerodynamic forces are associated with lower values of the backward-rotating torque due to the air flow: the athletes cannot lean forward in such an extreme way as at lower altitudes. They have to use larger β-angles in order to avoid flight instability and tumbling accidents. In fact, the mean β-angles from $t = 1.0$ s to 3.6 s found in Park City were noticeably larger than the values found in field studies during World Cup competitions at lower elevation (7; 8). The mean β-angle during the flight phase from one second on was 16.1°, whereas the mean value found at lower elevations was 11.7°. The athletes did not only increase their β-angle in order to obtain a stable flight in thin air, the altered positions also lead to a better performance, and additionally, the landing velocity was considerably lower which eased the landing (8).

3.9 Individual flight styles

The flight position angles differed markedly from one athlete to the other; however, the analysis of flight posture data of the Olympic medallists of the Games in Salt Lake City 2002 showed that they were able to reproduce their individual flight style in an impressive way (8). The gold medallist Simon Amann used a low angle of attack of the skis α and additionally a low body to ski angle β in order to keep the drag force low through the entire flight until the preparation for the landing, whereas the silver medalist Adam Maysz used a noticeably higher angel of attack α and also a low body to ski angle β. Different athletes utilize the advantages associated with high lift and low drag in distinctly different way. These results support the opinion that "the optimum flight style" – applicable to all athletes – does not exist. Analogously to this finding, it has been pointed out by Vaverka (17) that also the take-off movements of various world class athletes deviate remarkably from each other.

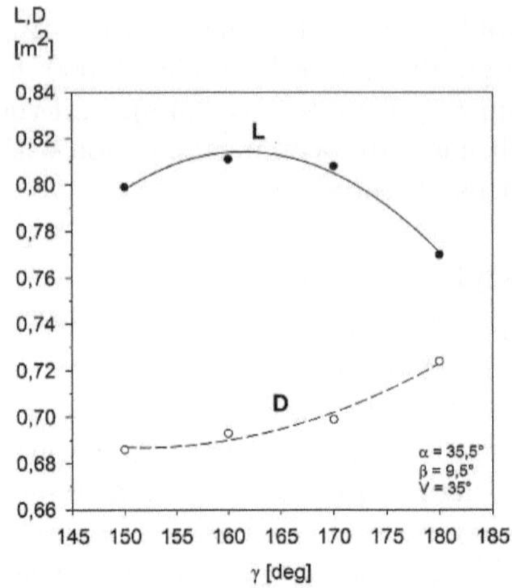

(a) Dependency of L and D on γ

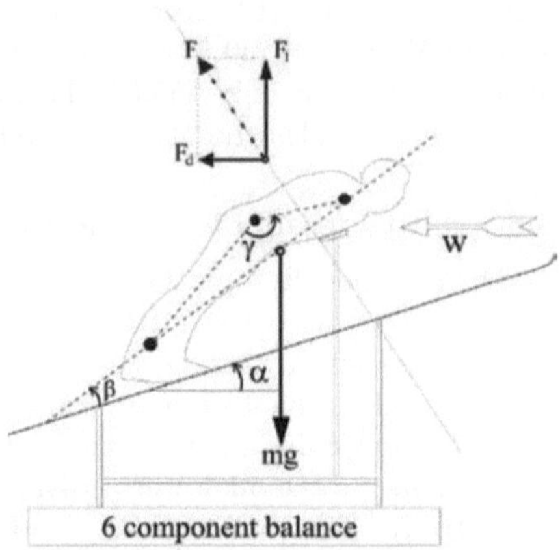

(b) Nomenclature of the position angles

(c) 1 : 1 model in wind tunnel

(d) A. Goldberger

Figure 5: Lift and drag areas as functions of the athlete's hip angle γ. The L value reaches a maximum at 160 degrees and the drag area D is small at this hip angle. In this case it is obvious that larger or smaller hip angles would reduce the jump length. Quantitative predictions can be obtained by means of computer simulations. The schematic drawing (b) shows the nomenclature of the position angles and the mounting of the 1 : 1 models of athletes for measurement purposes in the large scale wind tunnel (Railtec Arsenal, Vienna). All labeled angles could be varied for the measurement series. (c) Large scale wind tunnel measurements using a 1 : 1 model of a ski jumper. (d) Wind tunnel measurements with A. Goldberger (AUT): Series of measurements of 54 flight postures. Schmölzer and W. Müller 2002 (7), with permission; photos: W. Müller.

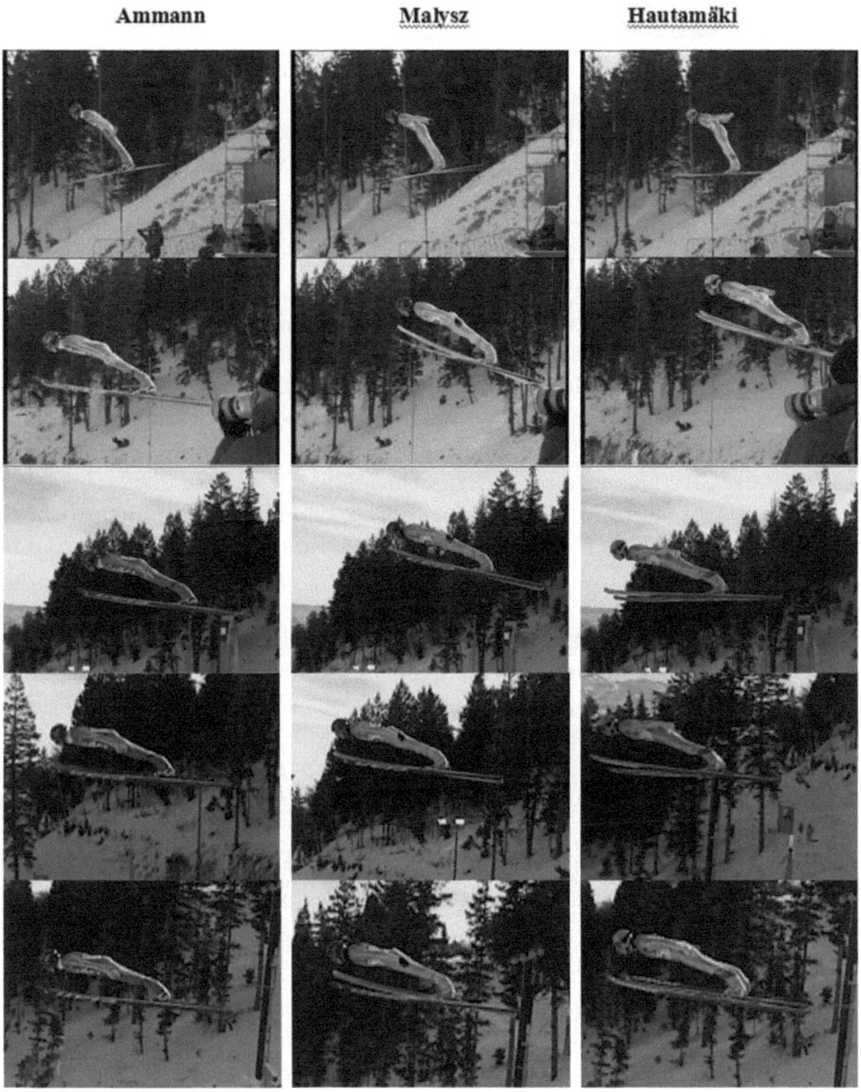

Figure 6: Individual flight styles of Olympic medaillists 2002. Series of digitised video images of the three Olympic medallists ($K = 120$ m jumping hill) at 6, 17, 28, 38, and 54 meters distance from the ramp. B. Schmölzer and W. Müller 2005 (8), with permission.

3.10 The influence of wind

The effect of wind blowing depends on the wind velocity u and on the direction the wind comes from (wind angle ζ, within the vertical plane). Wind blowing up the hill increases the jump length dramatically and decreases the landing velocity which eases the landing; wind from behind the jumper reduces the jump length and increases the landing velocity. A simulation series, with athletes' gear of 1994, and with wind directions from $\zeta = 0$ (wind assumed to blow horizontally in the positive x-direction, i.e. the direction of the run-out) to $\zeta = 350°$ in anti-clockwise steps of 10°(using the parameters of the ski flying hill in Planica) allowed to determine the most advantageous and the most disadvantageous direction of a constantly blowing wind (6): Wind blowing up the hill with $u = 3$ ms^{-1} resulted in an increase of jump length of 16 m, whereas wind blowing down the hill with $u = 3$ ms^{-1} reduced the jump length by 23.7 m.

For Park City ($K = 120$ m), with athletes in the gear of 2002, the effect of wind with $u = 3$ ms^{-1} blowing during the whole flight from an advantageous angle ($\zeta = 135°$) resulted in an increased jump length of $l = 120.0$ m (compared to 106.2 m at calm conditions; $m = 75$ kg), $l = 128.7$ m (compared to 115.9 m; $m = 65$ kg), and $l = 136$ m (compared to 125.7 m; $m = 55$ kg) (8). Using the same wind vector u in the simulation, a pronounced increase in jump length l occurred at all sizes of jumping hills. These simulation studies point out the enormous effect of wind on the jump length and also on the landing velocity.
Also in reality, jumps to or beyond the K-point of a hill are usually performed only the help of wind blowing up the hill. Changing wind velocities during a competition raise the question of fairness and it has to be emphasized that just 1 ms^{-1} (only 3.6 kilometers per hour!) difference in wind speed can easily decide between winning or loosing (16).

3.11 Weight as a performance factor

The importance of being light for ski jumping performance has been investigated in detail recently (6; 16; 7; 8; 18): the body mass profoundly influences the jump length and the velocity of motion. The lighter athlete has the advantage of flying further, and additionally the touch down is eased due to a lower landing velocity. The slope is -0.9 m per kilogram mass increase.

This results for the case that just the mass is increased in the simulation protocol. It has to be added, that lower weight allows the athlete to lean forward in a more pronounced way which results in an additional increase of jump length due to aerodynamic advantages associated with a lower body to

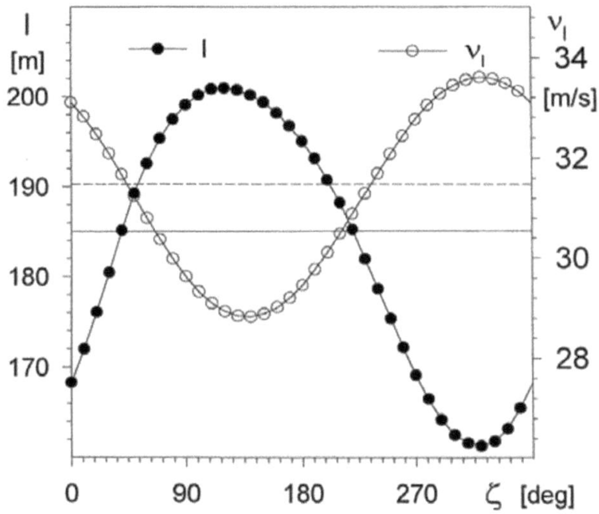

Figure 7: Influence of wind u on the jump length l and on the landing velocity v_l (6). $m = 70$ kg, $u_0 = 28.6$ ms^{-1}, $u_{p0} = 2.11$ ms^{-1}, wind blowing constantly with 3 ms^{-1} during the whole flight. Directions of wind vector u (in the x-y–plane) have been varied. The gust angle is counted positive in counter-clockwise direction, i.e. 0 in x-direction, 90° when blowing upwards, etc. The horizontal line and the broken horizontal line indicate the jump length l and the landing velocity v_l, respectively, at calm conditions: $u =$ 0. Both, l and v_l are dramatically influenced even at this moderate wind velocity of $u = 3$ ms^{-1}. W. Müller et al. 1996 (6), with permission.

ski angle β (6; 7) which might double the effect of low mass. This additional increase of jump length is difficult to quantify because of the complex flight stabilization questions associated; such an attempt would need a modeling approach including questions of flight stabilization and pitching moment balance, based on the knowledge of the air pressure distribution on all surfaces of the jumper-ski system.

Series of anthropometrical measurements of world class ski jumpers were made in Planica (World Cup 2000), Hinterzarten (Summer Grand Prix 2000), and Salt Lake City (Olympic Games 2002) (19). The mean *BMI*-values found in these studies were 19.8 kgm^{-2}, 19.6 kgm^{-2}, and 19.4 kgm^{-2}, respectively. A comparison of these results to results obtained by Vaverka in the time interval between 1970 and 1995 (20; 21) shows that the mean *BMI* has decreased by approximately 4 kgm^{-2} during the last 30 years.

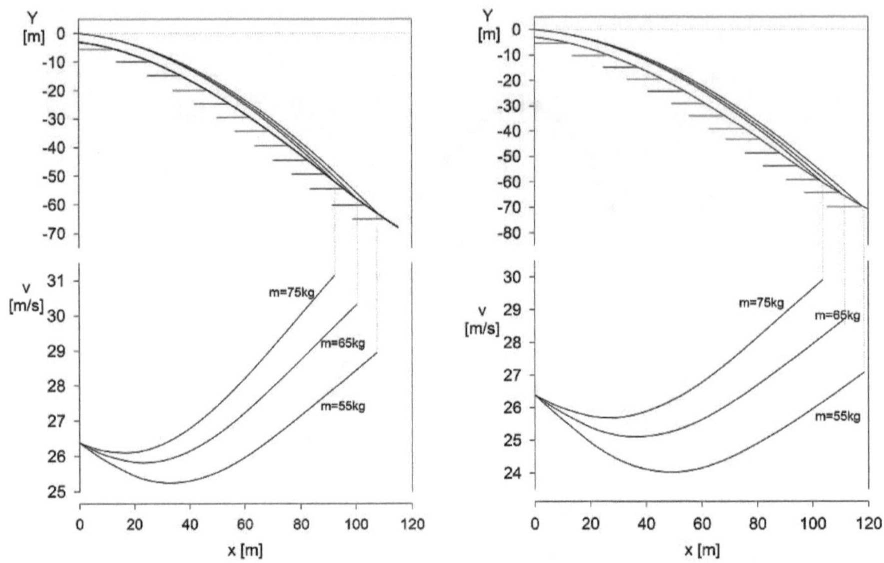

(a) Jump length at $u = 0$

(b) Jump length at $u = 3\ \mathrm{ms}^{-1}$ from $\zeta = 135°$

Figure 8: Simulated jumps using the L- and D-tables from the reference jump for model A and the profile of the jumping hill in Park City (7). The jumping hill parameters for Park City ($K = 120$ m) are: $\alpha = -10.5°$, $\beta = 35°$, $\beta_{\mathrm{P}} = 38°$, $\beta_{\mathrm{L}} = 37.77°$, $\gamma = 35°$, $h = 59.52$ m, $n = 103.51$ m, $r_1 = 93$ m, $r_2 = 105$ m, $r_L = 356.5$ m, $l_1 = 18.67$ m, $l_2 = 13.90$ m, $t = 6.7$ m, $s = 3$ m. For all jumping hills approved by the *FIS* the parameters can be found in the *FIS Certificates of Jumping Hills.* The trajectories and velocities for three different masses (55, 65 and 75 kg) are shown. The approach velocity v_0 was 26.27 ms^{-1}, the air density $\rho = 1.0\ \mathrm{kgm}^{-3}$ (corresponding to the altitude of 2000 m at the Olympic Games venue 2002) and $v_{\mathrm{p}0}$ was set to 2.5 ms^{-1}. The wind velocity u was set to 0 for the simulations shown in (a) and (c). Results in (b) are analogous to (a), however, in this case a wind blowing constantly with $u = 3\ \mathrm{ms}^{-1}$ from an advantageous direction ($\zeta = 135°$) during the whole flight was used. B. Schmölzer and W. Müller 2002 (7), with permission.

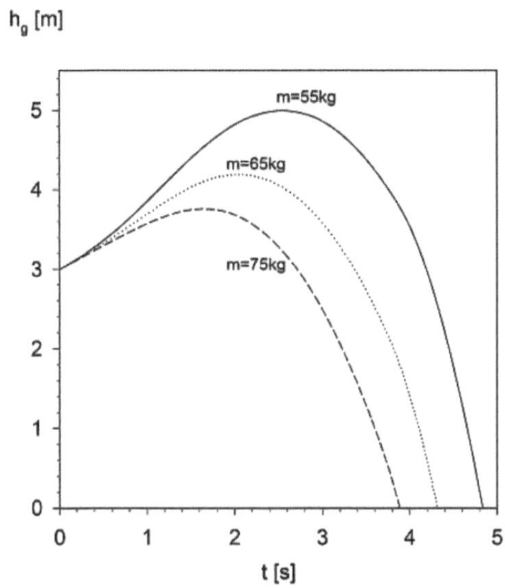

(c) Height above ground at $u = 0$

Figure 8: (continued) (c) shows the height above ground h_g for three different masses as a function of the flight time t with the wind velocity u set to 0 (compares to Fig. 8a). B. Schmölzer and W. Müller 2005 (8), with permission.

3.12 The solution of the underweight problem in ski jumping

A reduction of ski length for athletes who are too light can solve the underweight problem because then it is not attractive for the athletes to be underweight any more: shorter skis are smaller "wings" for the athlete and this aerodynamic disadvantage compensates for the advantage of low weight. Studies on the athletes' anthropometric status on one hand (19; 22) and on the importance of weight in the context of aerodynamic forces acting on a ski jumper on the other hand (6; 16; 7) formed the basis for the discussion of the final form of the changes to the regulations (18). The *FIS* has meanwhile followed the suggestion to solve the problem by relating maximum permitted ski length to relative body weight. In addition to the existing ski length limitation (maximum ski length is 146% of body height) this percentage is now reduced for those athletes whose relation of weight (including the jumping shoes and the suit) to body height squared is less than 20 kgm^{-2}. This corresponds to a correctly measured *BMI* of slightly

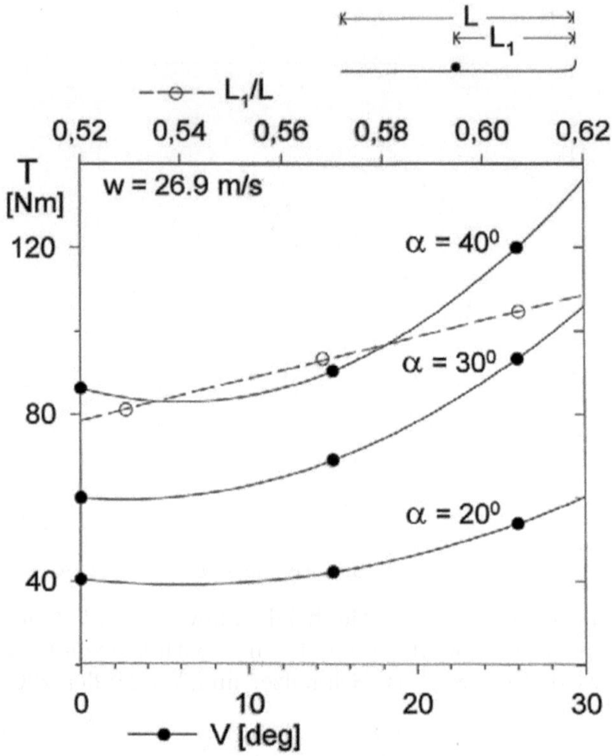

Figure 9: Torque (T) of skis alone: The figure shows two types of torque dependencies. *Solid line family*: Skis ($L = 2.56$ m) have been mounted with their centre of gravity (1.10 m from ski end) underneath an axle, i.e. a front ski to total ski length relation $L_1/L = 0.57$. The angle V of the skis to each other has been varied at different angles of attack α. A pronounced increase of T with increasing V has been found. *Broken line*: Here α (30°) and V (26°) have been held constant and the L_1/L relation was varied (from 0.52 to 0.62) by moving the skis relative to the axle. T increases with L_1/L. W. Müller et al. 1996 (6), with permission.

more than 18.5 kgm^{-2} (18.5 kgm^{-2} is the *WHO* underweight cut-off point). Every 0.5 units below 20 kgm^{-2} the maximum ski length percentage (of the athlete's height) is reduced by 2%. For example, a value of 19 kgm^{-2} would result in a ski length of only 142% of the body height. However, the *BMI* cut-off point chosen by the *FIS* is much lower than the value of 21 kgm^{-2} suggested by the author (18) and in this heuristic example the reduction of the ski length percentage was only 2% per unit *BMI*, whereas the current regulations use 4% per 1 kgm^{-2}. Two consequences from these deviations can be observed: Firstly, some athletes with *BMI* values below the *WHO* underweight cut-off point still perform in an outstanding way. Secondly, athletes who do not extend their body to maximum length, when their body weight is measured, can obtain a longer ski that way. Therefore the current regulations should be adapted.

The Body Mass Index $BMI = m/h^2$ (m body mass in kg; h body height in meters) is widely used to define appropriate body weight. The Mass Index *MI* is an improvement of the *BMI*, which takes the relative leg length l of the individual into consideration (19). It is determined by $MI = 0.53^2 \cdot m/s^2 \approx 0.28 \cdot m/s^2$, with m being the mass in kg and s the sitting height in meters.

3.13 The world record development

Despite the fact that the height difference between the edge of the take-off platform and the lowest point of the outrun is limited to $\Delta H = 130$ m, the competitive performance in terms of jump length has continuously increased, the present world record being 239 m. This was possible due to the development of aerodynamically improved equipment, particularly due to larger ski areas, thicker and stiffer dresses, an aerodynamically advantageous new flight style (the Boklöv- or V-style) and a substantial decrease of body weight. All these factors enabled flatter glide paths and thus longer jumps at a given maximum height between run-out and ramp. The recent changes to the regulations have reduced the possibilities for further increases of the flight path angle and it is to be expected that new records in ski flying will not occur so frequently any more. Without the height limitations, at larger and steeper hills than the ones available today, much longer jumps are imaginable because from approximately 200 m jump length on the glide path angle of a ski jumper with current equipment and flight style does not change noticeably any more (23). However, from the safety point of view such developments should not be supported for competitive ski jumping and the current hill height limitation should be maintained in the future for *FIS* events; this will also keep the costs of ski jumping hills within reasonable limits.

3.14 Inverse dynamics applied to ski jumping

The author has developed an inverse approach for analyzing a ski jumper's flight phase and the according software has been developed by M. Böck (24). When starting out from flight path data, the first and second time derivatives of $x(t)$ and $z(t)$ result in the velocity and acceleration components, respectively. This allows the calculation of lift and drag forces and areas as functions of flight time (compare to Fig. 1). This approach can be applied advantageously to training practice. It enables a quantitative analysis of the flight phase. In addition to $F_{\mathrm{l}}(t)$, $F_{\mathrm{d}}(t)$, $L(t)$, and $D(t)$ the height above ground $h(t)$, the velocity of motion $v(t)$, its components $v_{\mathrm{x}}(t)$ and $v_{\mathrm{z}}(t)$, as well as the landing impact in terms of the equivalent landing height h_{l} result. For demonstrating this inverse approach and the according software (24) the flight trajectory of Fig. 3 (top panel; $x(t)$ and $z(t)$ values were taken as inputs) has been used: As is to be expected, the lift and drag forces result (compare to Fig. 3, lower panel) as well as the lift and drag area functions shown in Fig. 4. When using flight path data from field measurements appropriate data processing should be applied in order to minimize noise artefacts.

Acknowledgements Supported by the Medical Commission of the International Olympic Committee (*IOC*), The International Ski Federation (*FIS*), and the Austrian Research Funds (*FWF* 14388 Tec, 15130 Med). I would like to thank C. Cagran for proof reading and for the layout of the manuscript.

Bibliography

[1] R. Straumann. Vom Skiweitsprung und seiner Mechanik. In *Jahrbuch des Schweizerischen Ski Verbandes*, pages 34–64. Selbstverlag des SSV, 1927.

[2] J. Denoth, S. M. Luethi, and H. Gasser. Methodological problems in optimisation of the flight phase in ski jumping. *International Journal of Sport Biomechanics*, 3:404–418, 1987.

[3] H. Koenig. Theorie des Skispringens angewandt auf die Flugschanze in Oberstdorf. In *Uhrentechnische Forschung*, pages 235–253. Steinkopf Verlag, 1952.

[4] L. P. Remizov. Biomechanics of optimal flight in ski-jumping. *Journal of Biomechanics*, 17(3):167–171, 1984.

[5] M. Hubbard, R.L. Hibbard, M. R. Yeadon, and A. Komor. A multi-segment dynamic model of ski-jumping. *International Journal of Sport Biomechanics*, 5(2):258–274, 1989.
[6] W. Müller, D. Platzer, and B. Schmölzer. Dynamics of human flight on skis: improvements on safety and fairness in ski jumping. *Journal of Biomechanics*, 29(8):1061–1068, 1996.
[7] B. Schmölzer and W. Müller. The importance of being light: aerodynamic forces and weight in ski jumping. *Journal of Biomechanics*, 35(8):1059–1069, 2002.
[8] B. Schmölzer and W. Müller. Individual flight styles in ski jumping: results obtained during olympic games competitions. *Journal of Biomechanics*, 38(5):1055–1065, 2005.
[9] E. Reisenberger, W. Meile, G. Brenn, and W. Müller. Aerodynamic behaviour of prismatic bodies with sharp and rounded edges. *Experiments of Fluids*, 37(4):547–558, 2004.
[10] O. Flachsbart. Messungen an ebenen und gewölbten Platten. In L. Prandtl and A. Betz, editors, *Ergebnisse der Aerodynamischen Versuchsanstalt zu Göttingen, IV. Lieferung*, pages 96–100. Oldenbourg Verlag, 1932.
[11] B. Schmölzer and W. Müller. The influence of lift and drag on the jump length in ski jumping. In *Book of Abstracts of the 1st International Congress on Science and Skiing*, page 274. Austrian Association of Sports Sciences (ÖSG), 1996.
[12] W. Müller, editor. *Proceedings of the 9th International Conference and Symposia,* Vancouver, Canada, August 21–24, 1996, 9th Biennial Conference: Canadian Society of Biomechanics. Canadian Society for Biomechanics, 1996.
[13] M. Virmavirta, J. Isolehto, P. Komi, G.-P: Brüggemann, Müller E, and H. Schwameder. Characteristics of the early flight phase in the olympic ski jumping competition. *Journal of Biomechanics*, 38(11):2157–2163, 2005.
[14] W. Müller, D. Platzer, and B. Schmölzer. Scientific approach to ski safety. *Nature*, 375:455, 1995.
[15] W. Meile, E. Reisenberger, M. Mayer, B. Schmölzer, W. Müller, and G. Brenn. Aerodynamics of ski jumping: experiments and CFD simulations. *Experiments of Fluids*, 41(6):949–964, 2006.
[16] W. Müller and B. Schmölzer. Computer simulated ski jumping: The tightrope walk to high performance. In *World Congress Biomechanics (WCB), University of Calgary, Canada, Proceedings CD*, 2002.
[17] F. Vaverka, M. Janura, M. Elfmark, J. Salinger, and M. McPherson. Inter- and intra-individual variability of the ski-jumper's take off. In E. Müller, H. Schwameder, E. Kornexl, and C. Rascher, editors, *Science and Skiing*, pages 61–71. E & FN Spon, 1997.

[18] W. Müller. Body weight and performance in ski jumping: the low weight problem and a possible way to solve it. In *Proceedings of the 7th IOC World Congress on Sport Sciences, Athens, Greece*, page 43D, 2003.

[19] W. Müller, W. Gröschl, R. Müller, and K. Sudi. Underweight in ski jumping: The solution of the problem. *International Journal of Sports Medicine*, 27:926–934, 2006.

[20] F. Vaverka. Somatic problems associated with the flight phase in ski-jumping. *Studia i monografia AWF we Wroclawiv*, 40:123–128, 1994.

[21] F. Vaverka. Research reports, Olomouc. personal communication, 1987 and 1995.

[22] W. Müller. A scientific approach to address the problem of underweight athletes: a case study of ski jumping. *Medicine & Science in Sports and Exercise*, 35(5 Supplement 1: 6th IOC World Congress on Sport Sciences 2002):124, 2002.

[23] W. Müller. Biomechanics of ski-jumping - Scientific jumping hill design. In E. Müller, H. Schwameder, E. Kornexl, and C. Rascher, editors, *Science and Skiing*, pages 36–48. E & FN Spon, 1997.

[24] M. Böck and W. Müller. Flight trajectory analysis software (unpublished).

Ski-Jumping Aerodynamics: Model-Experiments and CFD-Simulations

Walter Meile * and Wolfram Müller † and Ewald Reisenberger ‡

* Institute of Fluid Mechanics and Heat Transfer, Graz University of Technology, Graz, Austria

† Human Performance Research^{Graz}, Karl-Franzens University and Medical University of Graz, Graz, Austria

‡ Siemens VAI, Metals Technologies GmbH & Co, Linz, Austria

Abstract The importance of the predominant performance factors from the aerodynamic point of view, i.e. - besides the jumpers' weight - the position and posture during all phases of the inrun/flight, has been highlighted in the previous chapters. Here, the aerodynamic behaviour of a scaled down by $1 : \sqrt{2}$ model ski jumper is investigated experimentally at full scale Reynolds numbers as well as computationally, applying a standard RANS code. Particular focus is put on the influence of different postures on aerodynamic forces in a wide range of angles of attack with a high density of measurements, i.e., small angle increments from one posture to the next. The wind tunnel results form a fine grid of data which can be used for detailed studies on flight style and aerodynamic equipment optimization. The comparison of CFD results with the experiments reveals poor agreement. The studies clearly show simulation potentials and limits in contrast to the required accuracy of the predictions. Winning or loosing in ski jumping can be a question of a few percent difference in lift or drag; common turbulence models, however, do not provide results with an accuracy as required for analysis in top level ski jumping.

1 Introduction

The study presented in this chapter is one of the major parts of a research project on aerodynamics of sports, particularly on ski jumping. As outlined earlier, the flight path of a ski jumper is determined by the weight and the aerodynamic forces (lift and drag) acting on the jumper and his equipment, which influence the flight path and thus the jump length in a pronounced way. The athletes can strongly influence the forces by proper movement of parts of their bodies resulting in a changed posture. Ski jumpers have to solve extremely difficult optimization problems in real time in order to optimize their performance. Immediately after the take-off from the ramp, posture changes enable the extremely sensitive pitching moment to be balanced. In order to maximize the achievable jump length, the athlete has to vary the angle of attack during the flight (Remizov, 1984) and also the configuration of body parts and the skis. During the initial phase the drag acts particularly disadvantageous against the nearly horizontal motion in reducing the flight velocity. During later phases the flight path becomes steeper and the vertical component of the drag (against gravity) supports raising the flight path. Simultaneously, the lift component in horizontal direction increases, which is beneficial for larger jump lengths. Of course, the optimum flight style differs significantly from one athlete to the other due to anthropometrical differences and due to individually different motor abilities (Schmölzer and Müller, 2005). In general, high lift forces are advantageous throughout the flight, whereas the disadvantage associated with high drag is more pronounced during the first flight phase(Schmölzer and Müller, 1996). Thus, ski jumpers and their coaches spend much effort on improving the flight style and on aerodynamic improvements of the equipment, like binding settings, to optimize the individual angle between skis and the body.

Appropriate interpretation of the effects of lift and drag area variations on jump length due to changing flight positions necessitates a simulation model approach, as proposed by Müller et al. (1995, 1996); Müller and Schmölzer (2002). This model considers the posture changes during the flight phase by tabulated functions $L = L(t)$ and $D = D(t)$, L and D being the lift and drag areas, respectively. The functions have to be determined with athletes or models of athletes positioned in a wind tunnel in accordance with the postures found during competitions in the field, as described e. g. by Müller et al. (1996); Schmölzer and Müller (2002, 2005). Since body mass is an important performance factor, many ski jumpers were alarmingly underweight, and several cases of a very dangerous disease (anorexia nervosa) have come to light. The major aim of the investigations described here is focused towards better understanding of aerodynamical and biomechanical details of ski jumping. This in turn should enable measures to be taken counter obviously misleading developments and to improve fairness, safety, health, and performance in ski jumping (Müller et al., 1995, 2003; Schmölzer and Müller, 2002).

Measurements with ski jumpers in large-scale wind tunnels showed that even small changes in position as well as different jumping suits (characterized by shape and/or surface structure) may lead to distinct changes of the lift and drag areas. The consequences of changes of regulations, especially with respect to jumping suits, are in general not precisely predictable in advance. Comprehensive experiments in large-scale wind tunnels on very detailed modifications would lead to excessive measurement time and costs, and would require the disposability of several world-class athletes over unreasonably long periods.

This has led to the idea of performing basic studies on the aerodynamics of ski jumping. It was intended to find out if human bodies may be represented by combinations of geometrically simple bodies in Computational Fluid Dynamics (CFD). Furthermore, a way to estimate the interaction of individual parts of the body was to be developed. Both tasks require wind tunnel data of the aerodynamic characteristics of the objects within a wide range of orientations relative to the flow, which were provided in a preceding part of the project based on prismatic bodies (Reisenberger et al., 2004).

Wind tunnel measurement results obtained with a jumper model in reduced scale ($1 : \sqrt{2}$) are discussed. The selected dimensions represent an average measure out of a variety of today's world class athletes. Firstly, we concentrate on the "naked" model, both without and with the skis in position. During the first measurements, the arms and the head were held in a "fixed" position throughout the complete series. In addition, we investigated two different V-angles between the skis. Subsequently, several variations of arm and head positions were investigated, but in a reduced range of flight angles. In order to complete the investigations, three different jumping suits were tested: a narrow suit as used in downhill skiing (denoted white), one suit in normal dimensions according to present regulations (denoted red 1), and one suit with extreme width (denoted red 2).

In many cases, accurate predictions of consequences of intended changes to the regulations and possible innovations are urgently needed prior to a practical implementation. On the other hand, the aerodynamics of bluff bodies are yet a challenging task for modern simulation tools, and this is particularly true for the applied turbulence models. In order to evaluate the possibilities and/or limits of Computational Fluid Dynamics (CFD), a variety of selected postures of the naked model were simulated numerically and compared with the according measurements. Suitable simulations with jumping suits would require the consideration of fluid-structure interactions, which currently would highly exceed the available possibilities.

In the following section, the experimental facilities and procedures used for the present work are briefly presented, and section 3 provides a detailed description of the simulation details. The results are presented in section 4, and section 5 summarizes the studies and draws the conclusions.

2 Experimental setup and procedure

2.1 Wind tunnel, balances

The present experiments were performed in the Göttingen type wind tunnel of the Institute of Fluid Mechanics and Heat Transfer at Graz University of Technology. The most important characteristics can be given as follows: nozzle outlet cross-section 2 m × 1.46 m ($W \times H$), wind speed $v_{max} \approx 41$ m/s, non-uniformity $< 1\%$ of the mean velocity, longitudinal degree of turbulence $\approx 0.13\%$ along the jet axis. Comprehensive descriptions of the tunnel and the balances were given by Gretler and Meile (1991, 1993), while data acquisition, accuracy, and preliminary tests were specified by Reisenberger et al. (2004). A detailed description of the calibration procedure of the wind-tunnel balance is given by Reisenberger (2005).

2.2 Model design, dimensions

A model ski jumper in a typical flight position is drawn in Fig. 1. This figure exhibits the relevant angles (α, β, γ) which characterize posture and position relative to the airstream. Both posture and flight position are usually varied during the flight phase and differ between the athletes. In order to cover the entire range of relevant angles, we designed a suitable model allowing for the appropriate alignment. A welded steel frame forms the basis incorporating the necessary adjustment facilities. This frame is covered by different compound constructions to build the body's hull - a shell of fibre reinforced plastic forms the upper part, while head and legs are made from styrofoam with a laminated surface. The arms are made from hardwood, and separate pairs of arms were manufactured for each of the tested arm angles.

The model design started from the body dimensions of typically tall world class athletes, the height taken as an average value of $h = 1.78$ m with a corresponding ski length of 2.6 m. These values correspond to Model A in the work of Schmölzer and Müller (2002). Because of the restricted test section area, $1 : 1$ models including the skis were not suitable. With regard to similarity requirements, the model scale was set to the smallest value ($1 : \sqrt{2}$) according to the highest velocity attainable in the tunnel. This is justified by the following data: usual flight velocity ≈ 100 km/h, max. velocity in the wind tunnel test section ≈ 140 km/h.

Postures and positions during the flight are characterized by the angles depicted in Fig. 1. The ranges of angles covered during the present measurements are specified in Table 1 and, where appropriate, in the text. Present FIS regulations allow the characteristic circumferences of the suits to be oversized by 6 cm compared to circumferences of corresponding body parts. The size of the different suits was closely adapted to the reduced scale, while the fabrics and the structure of the material could not be modelled in reduced scale.

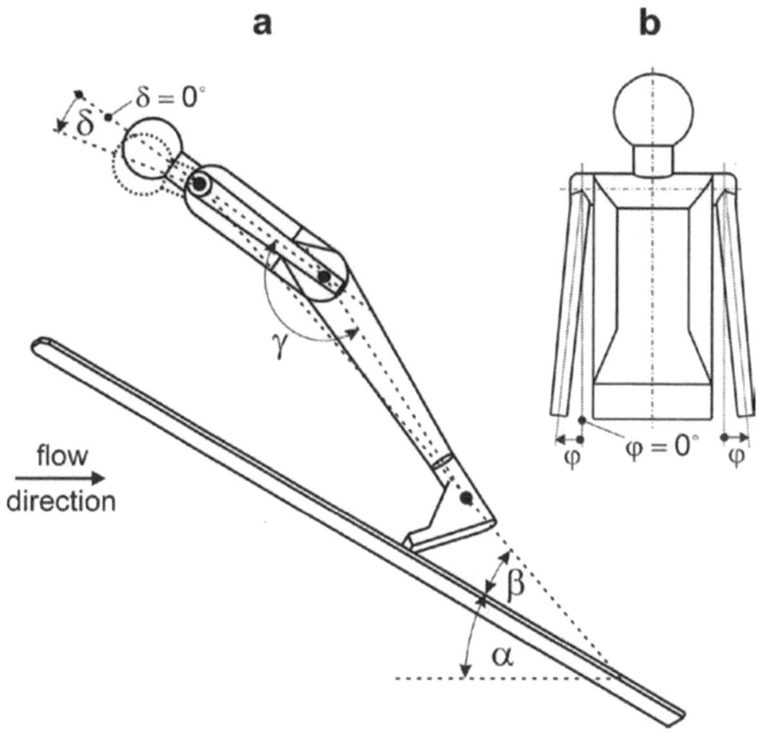

Figure 1: (a) Ski jumper in typical posture and flight position: α (angle of attack), β (body angle relative to the skis), γ (hip angle), δ (head angle), (b) Upper part of the model, definition of φ (arm angle). From Meile et al. (2006), with permission.

2.3 Model support

The measurement of aerodynamic forces requires an appropriate connection of the model to the balance. Even in the reduced scale, a configuration with model and skis in vertical arrangement would lead to a position of the model's head outside the airstream. Fortunately, the test section width of 2 m allows the model to be mounted horizontally, as depicted in Fig. 2.

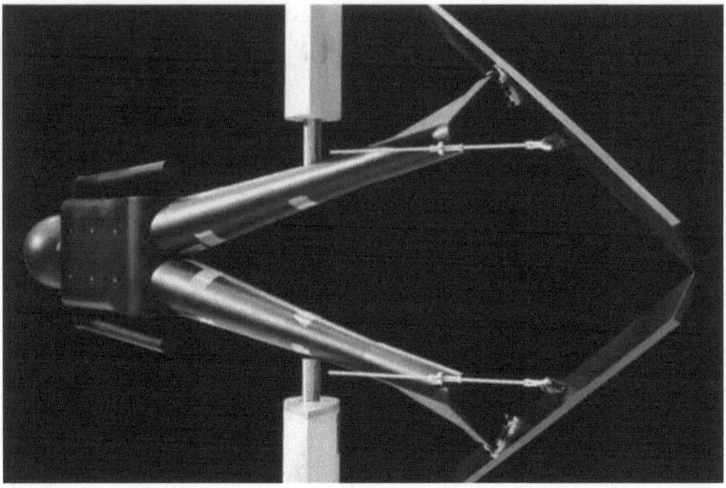

Figure 2: The model arrangement in the test section, view against the direction of the air stream. Lower strut: connection to the balance, upper strut: dummy support. From Meile et al. (2006), with permission.

The horizontal symmetry plane of the model coincides with the jet axis, i. e. it is placed 0.74 m above the test section floor. The connection to the external balance below the test section bottom plate is provided by a single vertical strut, which is shielded by a symmetric NACA profile to minimize the tare loads. In order to achieve symmetrical flow conditions, the support arrangement is mirrored in the upper region. This facility also serves for determination of the small, but unavoidable tare loads on the support strut. The interference between the support strut/wind shield and the model were estimated to be small and are, therefore, neglected.

The blockage ratio of the test section by the jumper model was between 6% and 14%. These extreme values correspond to measurement numbers 1 and 36, respectively. The related model areas were not measured, but estimated from the exact data calculated for cases corresponding to numbers 13 - 17, 19 - 23, and 25 - 29 (see Table 1).

2.4 Definitions

Drag and lift areas, as well as the coefficients of drag, lift and pitching moment, are deduced from the measured forces in the conventional manner as follows.

Number	$\alpha(°)$	$\beta(°)$	Naked $V = 30°$	Naked $V = 35°$	Red 1 $V = 35°$	White $V = 35°$	Red 2 $V = 35°$
1	20	0	x	x	x		
2		5	x	x	x		
3		10	x	x	x	x	x
4		15	x	x	x	x	x
5		20	x	x	x		
6		25	x	x	x		
7	25	0	x	x	x		
8		5	x	x	x		
9		10	x	x	x	x	x
10		15	x	x	x	x	x
11		20	x	x	x		
12		25	x	x	x		
13	30	0	x	x	x	x	x
14		5	x	x	x	x	x
15		10	x	x	x	x	x
16		15	x	x	x	x	x
17		20	x	x	x	x	x
18		25	x	x	x	x	x
19	35	0	x	x	x	x	x
20		5	x	x	x	x	x
21		10	x	x	x	x	x
22		15	x	x	x	x	x
23		20	x	x	x	x	x
24		25	x	x	x	x	x
25	40	0	x	x	x	x	x
26		5	x	x	x	x	x
27		10	x	x	x	x	x
28		15	x	x	x	x	x
29		20	x	x	x	x	x
30		25	x	x	x	x	x
31	45	0	x	x	x		
32		5	x	x	x		
33		10	x	x	x	x	x
34		15	x	x	x	x	x
35		20	x	x	x		
36		25	x	x	x		

Table 1: Range of measurements with assignment of measurement number; $\gamma = 160°$, $\delta = 0°$, and $\varphi = 5.3°$ during all measurements.

$$\text{Drag coefficient: } c_D = \frac{F_D}{\frac{\varrho}{2}v^2 A} \qquad \text{Drag area: } D = c_D A = \frac{F_D}{\frac{\varrho}{2}v^2}$$

$$\text{Lift coefficient: } c_L = \frac{F_L}{\frac{\varrho}{2}v^2 A} \qquad \text{Lift area: } L = c_L A = \frac{F_L}{\frac{\varrho}{2}v^2}$$

$$\text{Coefficient of moment: } c_M = \frac{M}{\frac{\varrho}{2}v^2 A l}$$

The moment is calculated from the readings of different load cells of the balance with the underlying coordinate system located at the intersection point of the jet axis and the axis of the model support. Because of the complicated model shape, we apply a reference area $A = 1\ \text{m}^2$ for simplicity. This procedure implies that drag/lift area and the corresponding coefficients exhibit equal numerical values. In deriving the moment coefficient, the thickness of the upper part of the body is taken as the reference length ($l = 0.126$ m), as in our previous investigations (Reisenberger et al., 2004). On the other hand, lift and drag area values are of interest for further processing in flight path simulations, and dimensionless coefficients are practically not available from full scale measurements. For this reason we shall present lift and drag areas (L and D) rather than coefficients in all diagrams. In addition, we also provide the dimensionless coefficients c_D and c_L in Table 2 for selected cases where the reference area A was evaluated from CAD drawings.

In order to achieve results significant for the full-scale flow, similarity requirements must be taken into account. They imply that the Reynolds number in the experiments must be equal to the value of the athletes during the flight. With the reference length as given above ($\bar{l} = l$), and the undisturbed flow velocity v as the reference velocity, the Reynolds number reads (ν being the kinematic viscosity of the air)

$$Re = \frac{v\bar{l}}{\nu}.$$

With respect to the applied length scale reduction it follows that the velocity in the experiments must be higher by a factor $\sqrt{2}$ provided that the same kinematic viscosity applies. In ski jumping, flight velocities between 25 m/s and 30 m/s are usual, depending on the size of the jumping hill and the flight phase. Therefore, the available velocity range up to $v_{max} \approx$

α [°]	β [°]	A [m^2]	Naked $V = 30°$		Naked $V = 35°$		Suit Red 1 $V = 35°$		Suit White $V = 35°$		Suit Red 2 $V = 35°$	
			c_D [-]	c_L [-]	c_D [-]	c_L [-]	c_D [-]	c_L [-]	c_D [-]	c_L [-]	c_D [-]	c_L [-]
	0	0.253	0.599	0.808	0.563	0.778	0.729	1.059	0.668	0.943	0.797	1.131
	5	0.269	0.612	0.792	0.575	0.769	0.792	1.095	0.699	0.958	0.843	1.145
30	10	0.283	0.644	0.781	0.614	0.770	0.882	1.129	0.740	0.960	0.915	1.156
	15	0.297	0.683	0.785	0.647	0.758	0.935	1.119	0.797	0.973	0.979	1.152
	20	0.310	0.721	0.765	0.697	0.753	1.050	1.127	0.868	0.976	1.070	1.149
	0	0.290	0.676	0.807	0.645	0.806	0.823	1.098	0.757	0.982	0.852	1.120
	5	0.305	0.694	0.792	0.665	0.791	0.888	1.098	0.780	0.975	0.919	1.128
35	10	0.318	0.718	0.784	0.701	0.785	0.985	1.125	0.828	0.975	1.003	1.145
	15	0.330	0.763	0.770	0.739	0.776	1.025	1.107	0.894	0.981	1.068	1.129
	20	0.342	0.792	0.745	0.769	0.750	1.118	1.063	0.949	0.957	1.140	1.092
	0	0.324	0.752	0.784	0.725	0.796	0.928	1.109	0.842	0.982	0.961	1.130
	5	0.337	0.774	0.779	0.754	0.791	0.994	1.102	0.874	0.978	1.018	1.120
40	10	0.349	0.807	0.771	0.790	0.783	1.059	1.098	0.930	0.975	1.073	1.108
	15	0.360	0.836	0.739	0.818	0.760	1.115	1.062	0.987	0.958	1.147	1.082
	20	0.369	0.857	0.718	0.834	0.730	1.189	1.002	1.029	0.928	1.208	1.028

Table 2: Cross sectional area A projections of the model, dimensionless coefficients c_D, c_L for selected cases; $\gamma = 160°$, $\delta = 0°$, and $\varphi = 5.3°$ during all measurements.

41 m/s is sufficient to achieve Reynolds number similarity. Compliance of similarity requirements during the measurements yields dimensionless coefficients which apply to full scale objects. The values of full scale drag and lift areas, L_{FS} and D_{FS}, can be deduced from the measurements by considering $A_{FS}/A_M = 2$ to yield:

$$D_{FS} = 2D_M, L_{FS} = 2L_M .$$

A possible dependence on the Reynolds number was tested in separate measurement series. For this purpose, the model with the normal jumping suit was held in one usual position ($\alpha = 35°$, $\beta = 10°$, $\gamma = 160°$), and the velocity was varied in the range 20 m/s $\leq v \leq$ 38 m/s. The coefficients of drag and lift were constant over the entire range, and the moment coefficient varied only slightly. Based on these results, influences on the Reynolds number, as, e. g., by the slight change of the kinematic viscosity resulting from different temperatures during competitions and the model tests, were not considered.

2.5 Measurement accuracy, test conditions

Prior to the measurements of the present work, a careful calibration of the wind tunnel balance was necessary. Subsequent measurements included a wide range of different load conditions in order to verify the achieved accuracy. Details of the calibration procedure are described by Reisenberger (2005). To illustrate the present measurements, the evaluated accuracy range is recalculated into lift and drag areas related to the naked model measurements ($V = 35°$, $\gamma = 160°$) in the complete range of investigations as given in Table 1. Uncertainties of the lift area L may range between +0.1% and −1.3% of the measured value for smaller forces, while this is changed to −0.4% to −1.1% for larger forces. Uncertainties of the drag area D amount between +0.2% and −1.75% for lower forces, and between −0.4% and −1.6% for larger forces. In general it may be stated that slightly too low values are measured, and that lift forces are determined with somewhat higher accuracy than drag forces. A graphical representation is given as Fig. 5c below for direct comparison with the absolute data.

The accuracy of the pressure transducer which measures the dynamic pressure may be specified with −0.6 Pa to 0.9 Pa throughout the effective range (1000 Pa), which was tested with the help of a Betz-type manometer. Since the present measurements were operated mainly in the upper velocity range (> 30 m/s), the maximum deviation of the dynamic pressure is better than ±0.08% of the measured value. Therefore, no corrections were applied in deriving lift and drag areas from measured forces.

The recommended blockage ratio in open test sections (< 10%) is markedly exceeded only in extreme model configurations, and we may thus expect that the measured forces are as accurate as usually accepted. Since available blockage corrections are developed mainly for very specific geometries and are not generally valid for any object shape, we did not correct data for this influence.

A further influence may exist due to the asymmetric boundary conditions resulting from the bottom plate. Indeed, a certain influence could be detected in previous test measurements (Reisenberger et al., 2004) with a flat plate at inclinations between 0° and 90° which could be compared with data from the literature (Flachsbart, 1932). At inclinations up to 50°, the measured polar agreed well with the reference data. At larger angles of attack the measured drag exceeded the reference data when the flow direction changed by the plate pointed towards the bottom plate. Repeated measurements with the plate in the reversed position yielded the desired agreement at larger angles of attack. In the present case the model inclination is set in the horizontal plane leading to a flow deflection towards the open side boundaries. Since only the tip of the left ski is near the bottom plate, while the body is far away, only a minor influence from the bottom plate was expected. However, the lack of comparable reference data prohibits a substantiated statement.

3 Numerical simulations

One of the major goals of the present research project was the evaluation of modern CFD with respect to the prediction of the flow field around ski jumpers and especially the aerodynamic forces. Since experimental investigations require large resources, even if only minor changes in posture and position should be taken into account, as e.g. for regulatory considerations or performance optimization, CFD might serve as a tool for evaluation. A further practical problem is the limited availability of highly qualified athletes for long series of wind tunnel tests. Our numerical simulation work was intended to determine current possibilities of commercial simulation tools as a reasonable alternative to wind tunnel measurements when precise results are necessary. For this reason, selected cases of posture and position measured in the wind tunnel were chosen to be simulated numerically. For providing a correct comparison, the simulation should resemble the wind tunnel testing procedure, i. e. the jumper model in its position within the 3/4-open test section should be modelled as realistically as possible. For this reason, the jumper model's hull in all possible postures (α, β, γ, $\delta = 0°$, $\varphi = 5.3°$) was constructed via CAD together with the test section floor to provide the necessary datasets for proper implementation into a mesh generator. The simulations were performed by AVL List GmbH, Graz, Austria and the Centre for Virtual Reality und Visualization (VRVis), Vienna, Austria.

In the following, we deliberately avoid the presentation and discussion of well known equations (conservation laws) and details of turbulence modelling for brevity. In the description we concentrate on numerical details (bounding geometry, mesh, boundary conditions) which are essential for the implementation in comparable commercial CFD tools.

3.1 The flow solver

The CFD code AVL SWIFT is used for the present simulations. The code employs the finite volume discretization method, which rests on the integral form of the general conservation law applied to the polyhedral control volumes (cells). All dependent variables are stored at the geometric centre of the control volume. The appropriate data structure (cell-face based connectivity) and interpolation practices for gradients and cell-face values are introduced to match an arbitrary number of cell faces. The overall solution procedure is iterative and is based on the SIMPLE-like segregated algorithm of Patankar and Spalding (1972), which ensures coupling between the velocity and pressure fields.

We assume an incompressible, Newtonian fluid flow in the absence of external forces. The flow field around the ski jumper is modelled by the Reynolds-Averaged Navier-Stokes (RANS) equations coupled with the eddy-viscosity $k-\varepsilon$ turbulence model equations. The complete framework of this model was presented by Launder and Spalding (1974), and the according standard formulation is the simplest model out of the great variety of available turbulence closure approaches, but, on the other hand, the most robust one. This linear $k-\varepsilon$ model represents the complete and computationally most efficient turbulent closure. However, its shortcomings such as the inability to predict the stress anisotropy and/or to deal properly with the effects of streamline curvature, rotation, buoyancy and streamwise pressure gradients are well known, as described, e.g., by Hanjalić (1994) and Speziale (1996).

3.2 Generation of the computational mesh

For the present task, the FAME Advanced Hybrid, AVL's automatic meshing tool, was applied, which creates an unstructured volume mesh using a surface and an edge mesh. The three-dimensional CAD data of the ski jumper are converted into a surface grid and imported in FAME, the preprocessing tool of SWIFT. The geometry of the wind tunnel's test section is used for setting the bounding geometry of the calculation mesh. However, some modifications described later were necessary because of enhanced numerical feasibility. The overall number of cells could be kept as low as approx. 1.000.000, and the whole grid is displayed in Figure 3 including the dimensions of the simulation domain and the boundary conditions.

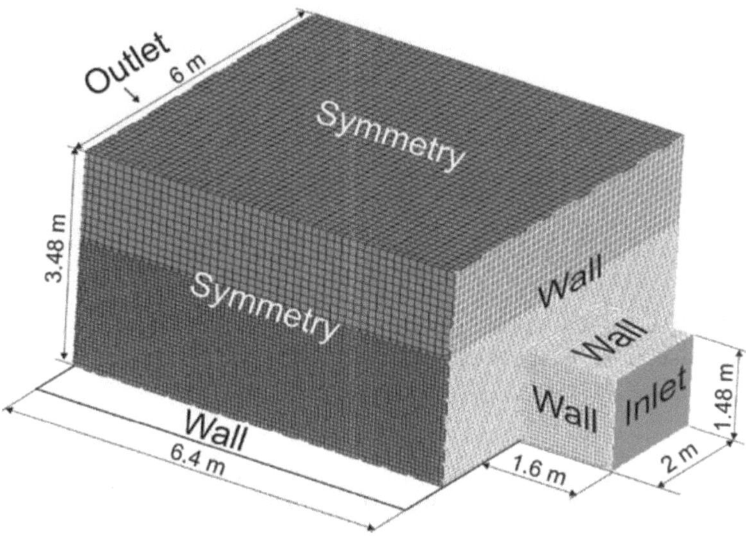

Figure 3: Numerical grid over the whole computational domain; dimensions of the domain; definition of the boundary conditions. From Meile et al. (2006), with permission.

The cell size is between 1.4 mm and 90 mm. In the region upstream of the ski jumper the cell size is kept at about 90 mm, around the model the cells are about 11 mm wide, except in regions of narrow gaps like the armpits and between the legs, where a smaller cell size prevails. In order to capture the expected separation region downstream of the jumper accurately, smaller cell sizes had to be applied. Details of the grid near the model are depicted in Figure 4.

3.3 Bounding geometry, boundary conditions, determination of forces

The inlet boundary condition is set by the normal velocity with the same value as in the measurements for each case ($v = 38$ m/s). In order to pro-

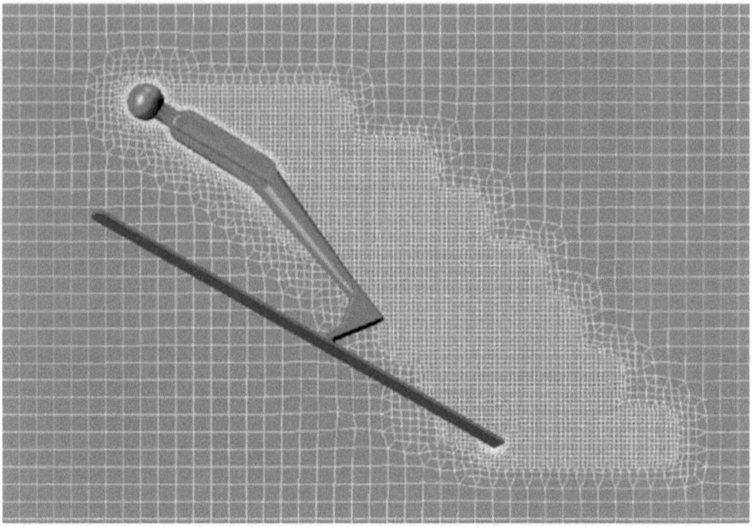

Figure 4: Refined numerical grid around the jumper surface in the wind tunnel test section. From Meile et al. (2006), with permission.

vide a somehow developed flow exhibiting a wall boundary layer of adequate thickness (as estimated for the flow through the wind tunnel nozzle), the entrance region of constant cross section was elongated to a length of 1.6 m. The lower boundary represents the test section floor and is, therefore, defined as solid wall. At the location of the nozzle exit the jet enters the hall surrounding the test section where the ambient air is at rest, which leads to intense mixing at the three free boundaries of the 3/4-open jet. For adequate representation of the free stream boundaries, the computational domain had to be enlarged as specified in Figure 3. The appropriate surfaces are defined as symmetry planes for reasons of enhanced numerical convergence. The chosen lateral extensions minimize the influence of these somehow incorrect conditions. On the side opposite to the inlet, a static pressure of 1 bar (approx. equal to the ambient pressure during the measurements) is defined as the outlet condition for all cases. The length of the "numerical" test section may be found exaggerated compared to the actual length of 3.2 m, but this method provides minimized interaction of the mixing layer or vortices from the model with the constant pressure outlet plane and, therefore, adequate numerical performance.

For determining the lift and drag forces acting on the ski jumper, the static pressure values multiplied by the corresponding model surface control areas are summed up. Separation into x-, y-, and z- components allows the calculation of the drag and lift forces, while the side force should come out as negligible.

4 Results

In this section we present the results of the measurements, partially in comparison with numerical simulations and full scale measurements. At the beginning, we discuss the findings from the experiments with the naked jumper model compared to the model clothed with jumping suits, while the influence of skis will be addressed separately. Subsequently, the influence of different arm and head positions will be considered. The angles related to head and arm variations, δ and φ, are depicted in Fig. 1. Finally, we compare the present measurements with appropriate full scale measurements and simulation results. All measurements were performed at a velocity of $v \approx 38$ m/s which corresponds to the full-scale Reynolds number Re $\approx 3.1 \times 10^5$ formed with the model's upper body thickness as the length scale. In general, the measured values of lift and drag areas are presented in the diagrams, while calculated values are given only in subsection 4.5 for comparison with full scale measurements.

4.1 Naked model, jumping suits

In the measurements with the naked model, two V-angles of the skis - the usually applied $V = 35°$ and $V = 30°$ - were compared. The measurements with jumping suits were performed only with $V = 35°$. Furthermore, a reduced number of postures with both unusual suits (white and red 2) were investigated, where we concentrated on the ranges of α and β observed during field studies (Müller et al., 1996; Schmölzer and Müller, 2002, 2005). For a clear representation avoiding overlap of individual curves, we depict drag and lift areas versus a measurement number, which combines the angles α and β, while the hip angle was set to the most usual value of $\gamma = 160°$. During all these measurements, the arm angle was adjusted to the most usual value $\varphi = 5.3°$ observed during competitions, while the head angle was set to $\delta = 0°$. The results are given in Figs. 5a and 5b. The underlying allocation of the measurement number to the various pairs of angles (α, β) is specified in Table 1. For further information, Fig. 5c provides a graphical representation of the measurement accuracy described in subsection 2.5 with regard to the actually measured values L and D. In addition, the dimensionless coefficients c_D and c_L are specified for selected cases in Table 2. The actual reference area A represents the cross sectional model area projected towards a plane perpendicular to the oncoming flow velocity vector. This area was evaluated from CAD drawings of the model hull created for implementation into the numerical mesh generator.

At the beginning we investigate the influence of the V-angle, based on comparable measurements with the naked model. As will be seen later, the posture of the jumper does not influence lift and drag of the skis alone. From this we may conclude that the skis' aerodynamic characteristics are also independent from the individual suits and, therefore, the V-angle may be assessed from the naked model measurements. From Fig. 5a it becomes evident that the drag with $V = 35°$ is generally lower than with $V = 30°$ throughout the investigated regime. Fig. 5b, in turn, exhibits lower lift up to $\alpha = 30°$, practically equal values at $\alpha = 30°$, but with increasing α this tendency is reversed and $V = 35°$ yields higher lift. This fact will be further addressed when L/D-ratios are discussed.

The influence of jumping suits on both lift and drag is significant, as could be expected. Already the skin-tight white suit (as normally used in downhill racing) leads to strongly increased lift and drag. This suit in general perfectly fits the jumper's body and does not pucker. So the fabrics may be simply regarded as surface roughness. In our previous work (Reisenberger et al., 2004), the influence of roughness was investigated exemplarily for rectangular prisms with rounded edges covered with sand paper. The corresponding results clearly exhibit the changed flow behaviour and may, therefore, explain the significant increase of aerodynamic forces with jumping suits.

The crucial influence of commonly used jumping suits (red 1) becomes evident from both figures 5a and 5b, where significantly larger values of lift and drag are recognized with this suit. Further enhancement may be detected for suit red 2, but only to a small extent. From the competitive point of view this is very essential, since a few percent of improvement in lift may lead to larger jump lengths and, therefore, decide over winning or losing in World Cup competitions. So the huge effort put into the development of appropriate suits is justified.

From the viewpoint of aerodynamics it is interesting to note that the drag curves in Fig. 5a increase very steeply, while the lift slopes in Fig. 5b generally exhibit a maximum throughout the investigated range. From $20° \leq \alpha \leq 30°$ the tendency is indicated, while at $\alpha > 30°$ the maximum occurs within the covered range of β. Closer inspection of the curves clearly exhibits that the maximum lift is achieved at $50° \leq \alpha + \beta \leq 55°$. This leads back to details of our previous investigations (Reisenberger et al., 2004) mentioned above, as follows: when all curved surfaces of the prism were

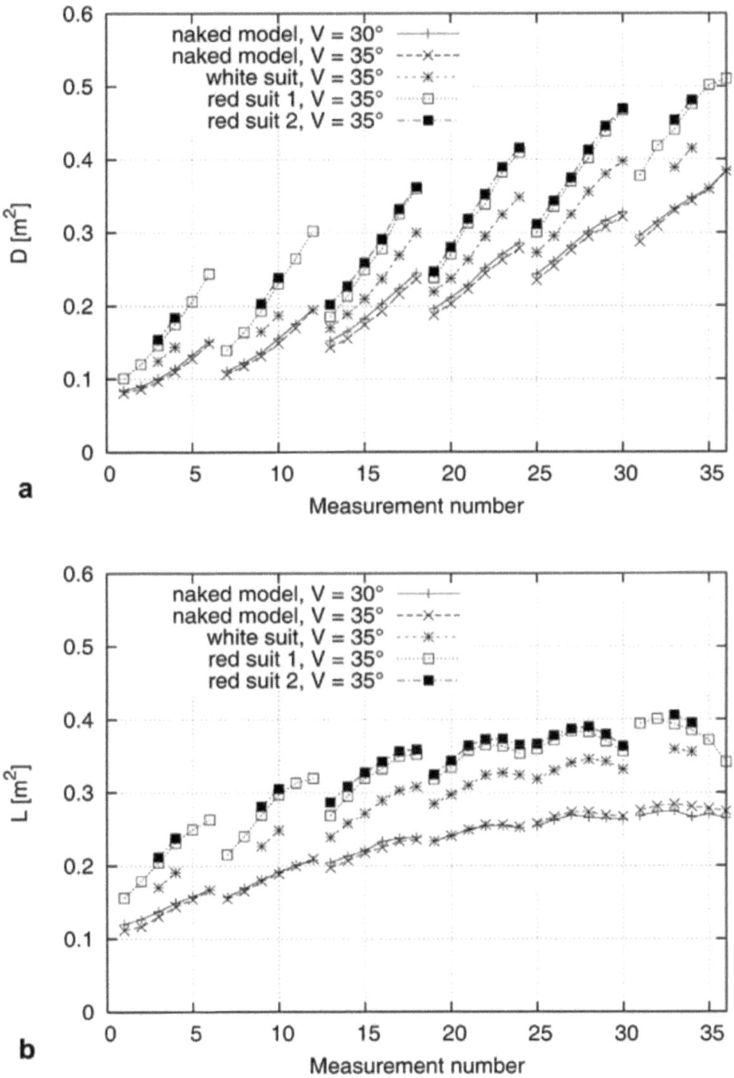

Figure 5: Drag and lift areas versus the measurement number for configurations given in Table 1; a) D, b) L.

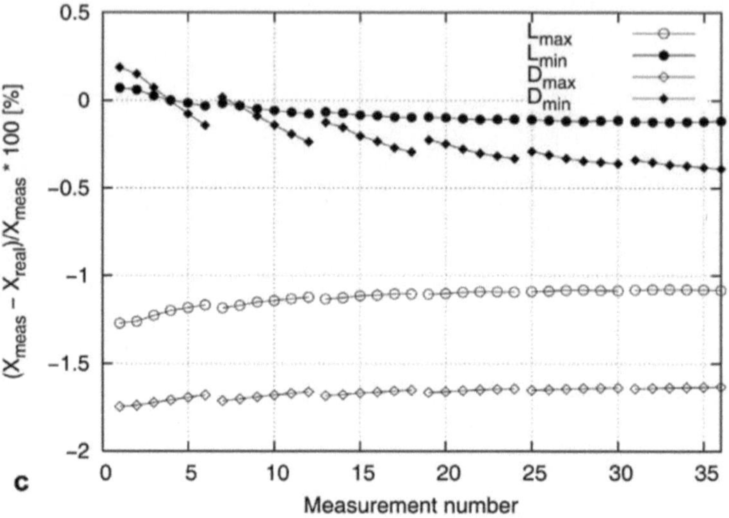

Figure 5: (continued) c) Relative measurement accuracy specifying the range of possible deviations from the measured data. From Meile et al. (2006), with permission.

covered with roughness, the flow remained attached in the range of angles between 30° and 50° producing significantly increased forces, and the point of maximum lift was shifted towards a slightly larger inclination of 50°. Similarly, in the present measurements with the complete ski-jumper model the maximum lift areas were detected in a range of angles of attack near 50°. The increase of forces due to roughness application along inclined rounded edges may be explained by the fact that the transition results in reattachment of the flow along the edges, which in turn leads to better attached flow over large parts of the bodies' backside. Therefore, the pressure is low and both lift and drag are increased.

In flight aerodynamics, the quality of lifting components is evaluated by the ratio of L/D. It was to be expected that the flight performance of ski-jumpers with $0.65 \leq L/D \leq 1.55$ is rather poor compared to streamlined bodies (wings). Within the investigated regime, the ratio L/D is largest at the lowest inclination (low values of the measurement number), and continuously decreases with increasing angles of attack - this holds for α and β as well. The corresponding graphs are depicted in Fig. 6, and a complete presentation is given by Reisenberger (2005).

With regard to the competitive performance, a comparison to standard

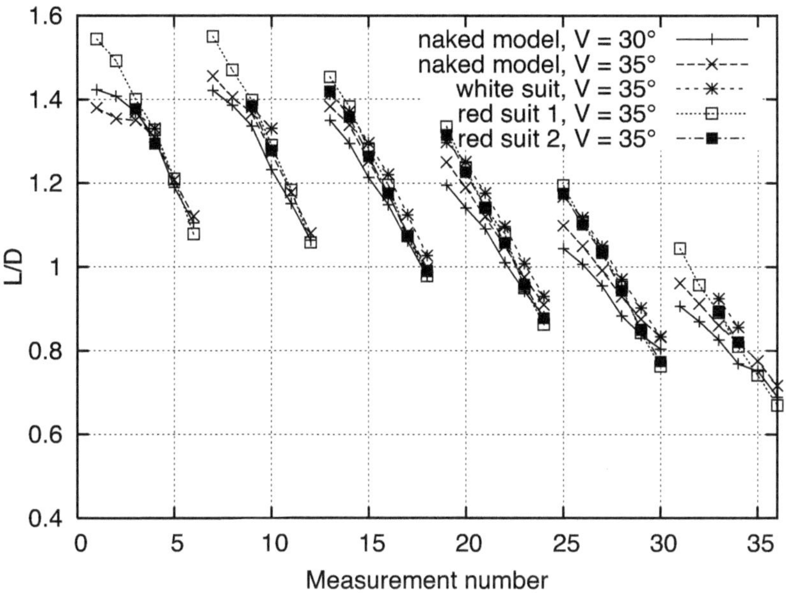

Figure 6: Lift to drag ratio L/D of jumper models with different suits. From Reisenberger (2005), with permission.

conditions seems of particular interest. Since suit red 1 is in full agreement with present regulations for material and dimensions, the according data will serve as the appropriate reference for comparison. The relative differences of L and D to the data obtained with suit red 1 are displayed in Figures 7a and 7b.

In typical flight positions (measurement number > 15), lift and drag with suit red 2 are on average both larger by about 3% than with suit red 1. At the first glance, the effect of these minor changes may seem negligible. Nevertheless, the consequences on the jump length may be evaluated by numerical simulations of the flight path. According to the simulation approach used by Müller and Schmölzer (Müller et al., 1996; Schmölzer and Müller, 2002; Müller and Schmölzer, 2002) the jump length would increase by 1.8 m when L is increased by 1% throughout the whole flight, and an increase of D by 1% would decrease the jump length by 1.2 m. The jumping hill profile in Innsbruck (K 120 m ramp) and the reference jump A with the mass set to 65 kg (as described in Schmölzer and Müller (2002)) were used for this simulation. Applying the increase of 3% in L and D between suits

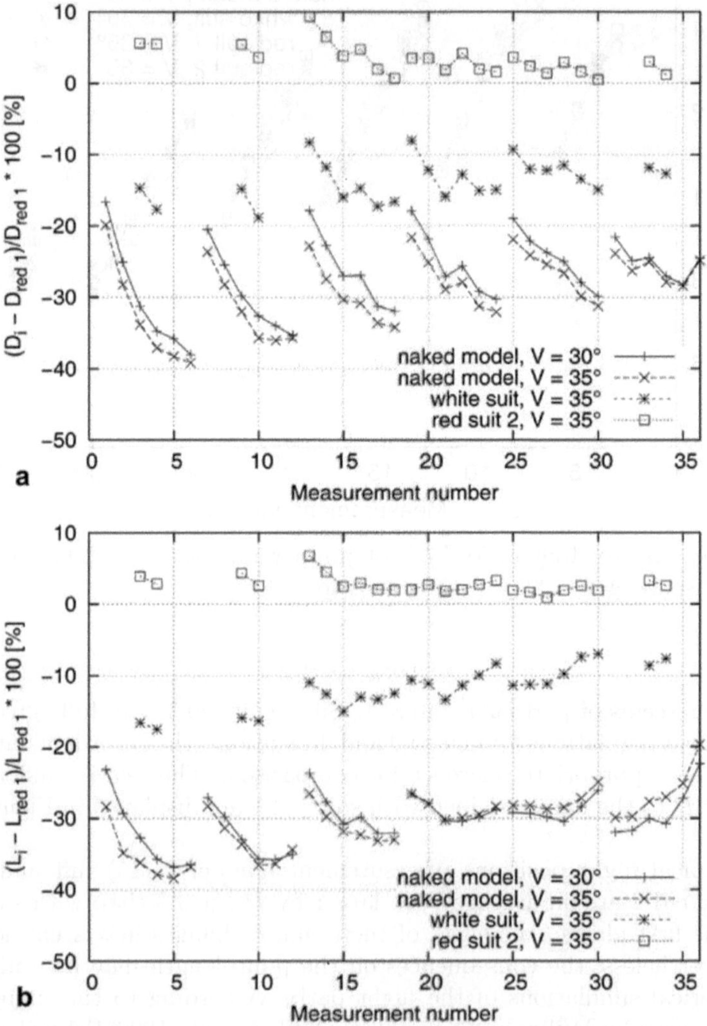

Figure 7: Difference of drag and lift areas relative to suit red 1 versus the measurement number; a) D, b) L. From Meile et al. (2006), with permission.

red 1 and 2 throughout the whole flight resulted in an increase of the jump length by about 2 m, which could be a decisive advantage. On the other hand, there exist considerations towards a general use of skin-tight suits in order to reduce the material influences and to emphasize the athletes' proficiencies. A computer simulation with L and D reduced (simultaneously) by 12% and 15%, respectively - which corresponds to typical L and D measurement values obtained with the skin-tight suit when compared to the results obtained with the suit red 1 (Figs. 7a and 7b) - resulted in a jump length reduction of 4.5 m.

4.2 Contribution of the skis to lift and drag

In former regulations, the maximum ski length was limited to body height plus 0.8m, which established a strong discrimination of tall athletes. Later, the maximum ski length regulation was changed to: 1.46 × body height. The importance of aerodynamics on ski jumping performance led to lighter ski jumpers. Many ski jumpers deliberately reduced their body weight to an alarming extent. An anthropometrical study during the Olympic Games 2002 in Salt Lake City (Müller et al., 2006) showed that 22% of Olympic-level athletes were underweight according to the World Health Organization classification criteria (body mass index *BMI* below 18.5 kg/m^2), and several cases of anorexia nervosa, which is a very dangerous disease, have come to light. One of the main goals of the present research project was the elimination of this dangerous tendency, and a reduced ski length for extremely light athletes seemed to be a possible solution to the problem (Müller et al., 1995). However, this would require appropriate knowledge about the skis' contributions to the overall aerodynamic forces. During our measurements we also determined the forces acting on the naked jumper model without skis. Basically, the correct procedure would be to measure the forces on the skis with the jumper model present as a dummy, but this could not be realized without major modifications of the set-up. Based on visualizations we assumed that the flow around the jumper model does not crucially influence the flow around the skis, and vice versa. This assumption is strongly supported by the work of Hanna (1996). For this reason, the calculation of the forces on the skis by simply subtracting the forces on the model alone from the forces acting on the complete system could be taken as a good approximation. As a result, the forces on the skis were determined from the naked model experiments for both V-angles and within the whole range of measurement numbers. It came out that the forces on the skis for constant α are indeed practically independent from the model posture (β) - deviations of max. ±2.5% from the mean seem acceptable to justify the assumption. Minor exceptions could only be detected for the lift area in the region of $\alpha \geq 40°$, where L of the skis slightly increased with increased β. However, during the important flight phase such values of α are usually not relevant. The diagrams in the following Figures 8a and 8b display the lift and drag areas of the skis, L_S and D_S, in relation to L and D of the complete system.

The most important conclusion from both figures is that the relative lift of the skis is significantly larger than the relative drag. A similar behaviour can be deduced from the work of Reisenberger et al. (2004) for the thinnest investigated prisms with sharp and rounded edges as well. At inclinations from 5° to 40°, these profiles yielded $L/D > 1$. Since skis are slender thin bodies, similar behaviour could be expected. In general, L_S/L and D_S/D are largest for the skin-tight suit and lowest for suit red 2, as could be expected. Further conclusions will be drawn for suit red 1 exemplarily as follows. For $\beta = 0°$, the relative drag D_S/D increases with α up to 47% at $\alpha = 40°$, while a slight reduction is observed in the region $40° \leq \alpha \leq 45°$. This means that in the range $20° \leq \alpha \leq 40°$ the lift slope of the skis $dD_S/d\alpha$ is larger than the lift slope of the complete system. Furthermore, the relative drag D_S/D decreases with increasing β, but remains nearly constant for $\beta \geq 20°$. In contrast, the curves of relative lift clearly show a minimum at $\alpha + \beta \approx 50°$, and this is the point where maximum lift can be observed for the jumper model without skis (see also next paragraph).

4.3 Influence of the hip-angle

First, the influence of the hip angle was investigated for both V-angles, for the naked model and with suit red 1, but only selectively for typical combinations of α and β. In the measurements with the naked model we investigated $\gamma = (150°, 160°, 170°, 180°)$, $\alpha = (30°, 35.5°, 40°)$, $\beta = (9,5°$ and $15°)$ and $V = (30°$ and $35°)$. Experiments with suit red 1 comprise $V = 35°$, $\alpha = 35.5°$, $\beta = 9.5°$, while γ was varied through all 4 angles. Based on the considerations with respect to the contribution of the skis to the overall forces, we performed similar measurements with the naked jumper model without skis in the complete range of angles for a better comparison. The positions of head and arms were the same as in the preceding paragraphs. The full list of combinations is given in Table 3.

Additionally, we included a separate row specifying a correction angle $\Delta\alpha_{UB}$, which allows the calculation of the effective inclination of the model's upper body for comparison via the following relation:

$$\alpha_{UB} = \alpha + \beta - \Delta\alpha_{UB} \, .$$

The influence of the hip angle γ is depicted in Fig. 9, where lift and

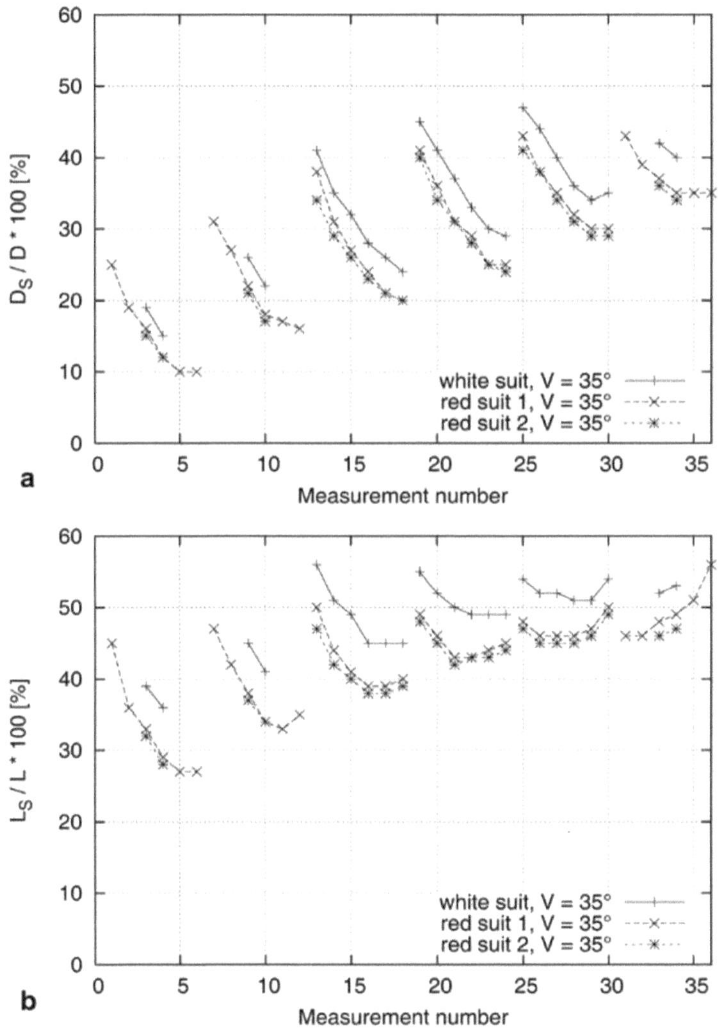

a

b

Figure 8: Relative contribution of the skis to drag and lift areas; a) D_S/D, b) L_S/L. From Meile et al. (2006), with permission.

γ [°]	$\alpha+\beta$ [°]	$\Delta\alpha_{\mathrm{UB}}$ [°]
150	39.5, 45, 49.5, 50.5, 55	19
160	20 - 70, step 5	12.6
160	39.5, 45, 49.5, 50.5, 55	12.6
170	39.5, 45, 49.5, 50.5, 55	6.3
180	39.5, 45, 49.5, 50.5, 55	0

Table 3: Measurements of the naked model without skis; $\delta = 0°$, $\varphi = 5.3°$. The combinations of $(\alpha+\beta)$ cover the full range of postures given in Table 1.

drag areas are plotted in polar form. Selected values of $(\alpha+\beta)$ are marked with arrows for clear assignment to the data points. The largest drag areas are achieved at hip angles $\gamma = 160°$ and $170°$ for a body angle $\alpha+\beta \approx 50°$ which corresponds to an inclination $\alpha_{UB} \approx 40°$ of the upper body.

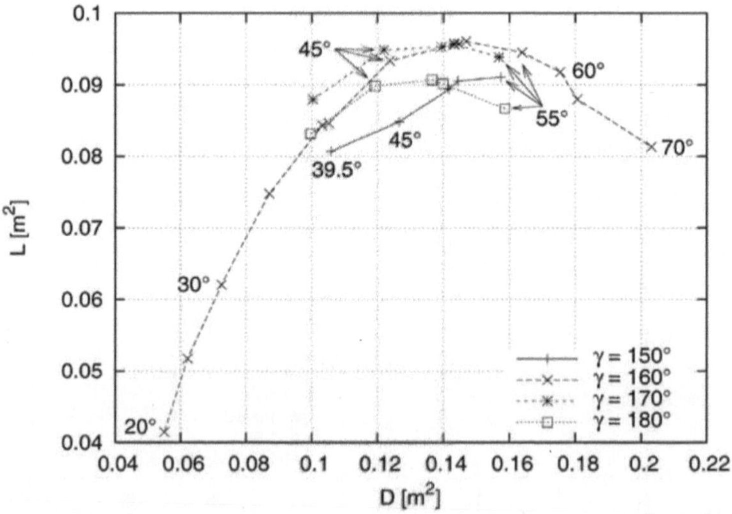

Figure 9: Influence of the hip angle; naked model without skis; drag and lift areas in polar representation. Numbers near the symbols denote the angle $\alpha+\beta$. From Meile et al. (2006), with permission.

A similar behaviour was already detected by Reisenberger et al. (2004) for prismatic bodies with rounded edges. The overall maximum of L/D is achieved at $\gamma = 170°$ and $\alpha+\beta \approx 40°$, while the slightly lower maximum

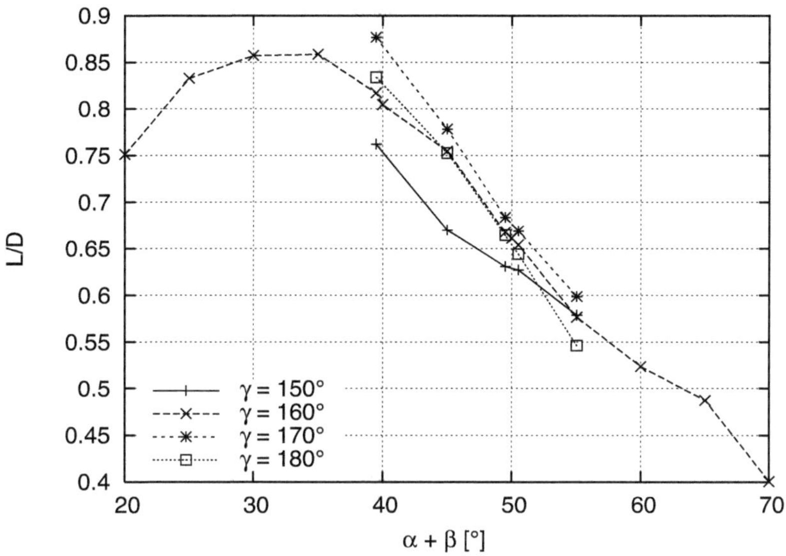

Figure 10: Lift to drag ratio L/D of the naked model at various hip angles and postures $\alpha + \beta$. From Reisenberger (2005), with permission.

L/D for $\gamma = 160°$ is achieved in a wider range of $30° \leq \alpha + \beta \leq 35°$. The corresponding graphs are depited in Fig. 10. Unfortunately, measurements with $\gamma = 170°$ are not available below $\alpha + \beta = 39.5°$. With hip angles $\gamma = 150°$ and $180°$, the maximum lift areas are lower by about 7%, and especially L/D is much smaller for $\gamma = 150°$.

A comprehensive field study during the Winter Olympic Games 2002 in Salt Lake City (Schmölzer and Müller, 2005) and several field studies during World Cup competitions (Schmölzer and Müller, 2002) recorded postures and positions of many athletes during several stages of the jump. Typical values of world class athletes during the stabilized flight phase ranged from $\alpha \approx 30°$ to $40°$, $\beta \approx 10°$ to $15°$ and $\gamma \approx 160°$. Comparing these data with the present measurements, it may be recognized that the jumpers operate in regions with maximum lift rather than with largest L/D. This is justified by the more pronounced influence of L during the later flight phases.

4.4 Influence of head and arms positions

In order to evaluate possible effects of head and arm positions, we performed measurements with the naked model and with suit red 1, but in a reduced range of postures: $V = 35°$, $\beta = 9.5°$, $\gamma = 160°$; $\alpha = 30°, 35.5°, 40°$ (naked) and $\alpha = 35.5°$ (red 1), respectively. Head variations $\delta = 0°$, $20°$ and $40°$ were tested at the usual arm position $\varphi = 5.3°$. Arm positions were varied as $\varphi = 0°, 5.3°, 10°$ at the standard head position $\delta = 0°$. The effects of different head and arm positions on lift and drag areas are depicted in Figs. 11 and 12 in polar form.

In Fig. 11 we see that the head angle causes only slight differences in lift and drag areas, the largest relative differences between comparable configurations being $\Delta D = 5.3\%$ (at $\alpha = 30°$) and $\Delta L = 1.2\%$ (at $\alpha = 35.5°$). From the present measurements a significant influence of the head angle cannot be deduced, which might be due to the spherical shape of the head without a helmet. For performance relevant statements the use of a helmet and a finer range of angles $0° \leq \delta \leq 20°$ seem necessary.

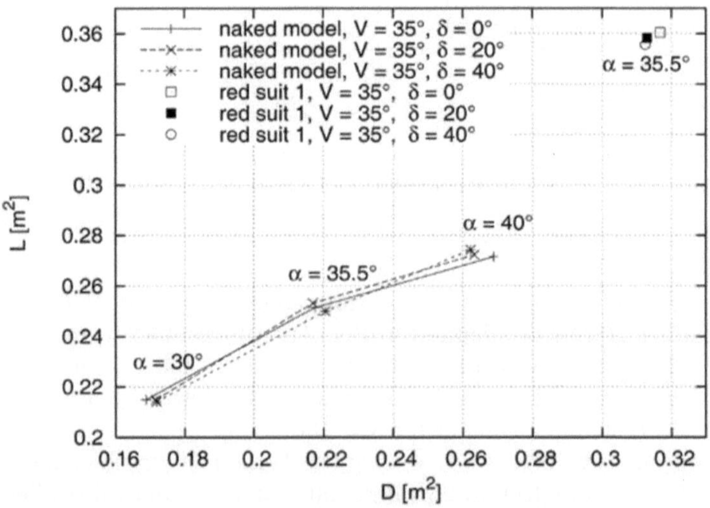

Figure 11: Influence of the head angle δ; drag and lift areas in polar representation. From Meile et al. (2006), with permission.

The arm angle $\varphi = 5.3°$ complies with the usual position during competitions, while $\varphi = 10°$ is typically applied when corrections of the flight path are necessary. The angle $\varphi = 0°$ (arms parallel to the upper body) was tested for reference. With the naked model, $\varphi = 0°$ leads to largest lift areas throughout the investigated regime, accompanied with largest drag areas, except at $\alpha = 30°$. The lowest drag areas are in general achieved with $\varphi = 5.3°$. However, the significance of these naked model results is

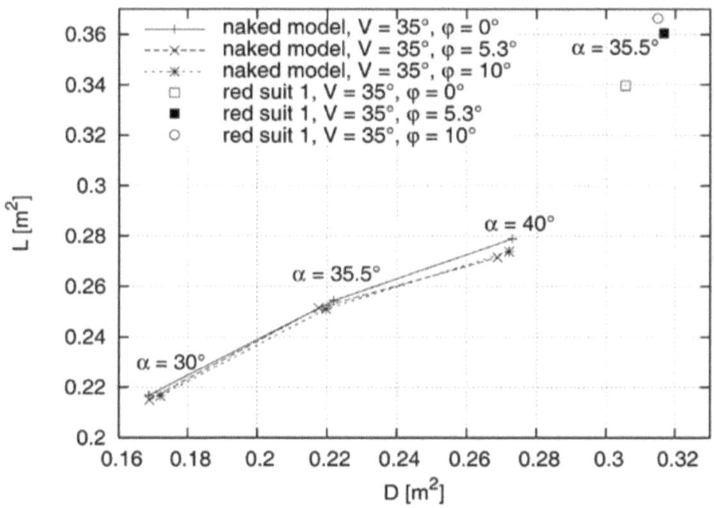

Figure 12: Influence of the arm angle φ; drag and lift areas in polar representation. From Meile et al. (2006), with permission.

restricted, as long as skin-tight suits are not used. Since the gap between the arms and the upper body is generally narrow, it will be reduced or even closed with the usual jumping suits. Therefore, the results for suit red 1 are more realistic to evaluate the arm position. The corresponding data exhibit much larger differences, especially the increase of lift with the change of the arm angle from $\varphi = 0°$ to $\varphi = 5.3°$ is remarkable. It is interesting to note that the lift further increases when the arm angle is increased, but that the drag is reduced simultaneously. At this point it should be mentioned that the differences described above were just observed from the measured data. However, the measurement parameters (α, β) defining the inclination coincide with the region of strong influences due to roughness effects (transition and possible reattachment), and thus, a closer evaluation is currently impossible without further measurements.

Within the framework of the suit regulations (suit red 1), a slight distance of the arms from the upper body may be recommended. Should skin-tight suits be prescribed in the future, the naked model results would become of greater importance, which exhibit largest lift areas with $\varphi = 0°$, while $\varphi = 5.3°$ in general yielded lowest drag areas in the investigated regime.

4.5 Comparison with full scale measurements

Finally, we compare our measurements with full scale measurements performed with 1 : 1 models of ski jumpers in the wind tunnels of Arsenal Research (Vienna, Austria) and Klotzsche (Dresden, Germany). The full scale values taken for comparison correspond to Model A presented by Schmölzer and Müller (2002). The compared data correspond to the flight phase with flight times from 0.7 s on. The present investigations did not cover postures and positions close to those investigated by Schmölzer and Müller (2002) for the initial flight phase $0 \leq t \leq 0.7$ s. The parameters (α, β, γ) and results are listed in Table 4. For better comparison, the D and L values obtained in the scaled down model measurements are recalculated according to similarity laws (see section 2.4).

In general we may notice that both lift and drag areas determined in the full scale measurements are slightly larger than the present results, even when compared to suit red 2 (typically 6% in drag and 10% in lift). Nevertheless, the agreement seems fully acceptable if we keep in mind that the reason for the obvious differences may be fourfold:

- The forces measured in open wind tunnel test sections are generally too low by a few percent. The deviations depend on the blockage ratio and the model shape.
- The present model was built from simple bodies (in view of the CFD-comparisons) and only fits the main geometric parameters, such as body and ski lengths, in the model scale.
- No helmet and ski-glasses were applied in the present measurements.
- The fit of the suits applied in the full scale tests may not be fully comparable to the scaled down case.

4.6 Numerical simulations

The results of numerical simulations are presented in Table 5, where the corresponding experimental data are also included. It should be pointed out that the numerical results were obtained without prior knowledge of the experimental data and, therefore, a possible tuning of numerical parameters or constants was avoided. A further requirement was the use of turbulence models as usually applied in standard industrial calculations which are, therefore, easily accessible. Furthermore, the investigations of Reisenberger (2005) revealed that none of the "standard" two-equation models exhibits a clear preference for accurate calculations of blunt-body high-lift configurations. On the other hand, the most advanced turbulence models available in commercial tools as, e. g., full Reynolds Stress Models (RSM) are time consuming and relatively complicated to handle. For this reason,

Scaled down model						Full-scale model			
Angles		Suit Red 1		Suit Red 2		Angles		Competition suit	
α [°]	β [°]	D [m^2]	L [m^2]	D [m^2]	L [m^2]	α [°]	β [°]	D [m^2]	L [m^2]
30	10	0.500	0.641	0.518	0.655	30	9.6	0.534	0.736
30	15	0.556	0.665	0.582	0.684	30	14.4	0.613	0.758
35	10	0.626	0.715	0.638	0.728	35.5	9.5	0.693	0.811
35	15	0.677	0.731	0.705	0.745	35.5	15.2	0.769	0.815
40	10	0.740	0.767	0.750	0.774	40	9.4	0.826	0.840
40	15	0.803	0.765	0.827	0.780	40	15.7	0.885	0.849

Table 4: Comparison of lift and drag areas between model and full scale measurements; parameters: $V = 35°$, $\gamma = 160°$, $\delta = 0°$, $\varphi = 5.3°$.

the application of such models was avoided deliberately for the present investigations.

		Measurement		Simulation		Difference [%]	
α [°]	β [°]	$D_M[m^2]$	$L_M[m^2]$	$D_S[m^2]$	$L_S[m^2]$	$\frac{(D_S-D_M)}{D_M}$	$\frac{(L_S-L_M)}{L_M}$
30	15	0.1922	0.2251	0.1445	0.1898	- 24.8	- 15.7
	5	0.2026	0.2409	0.1585	0.1585	- 21.8	- 34.2
35	10	0.2229	0.2496	0.1704	0.2185	- 23.6	- 12.5
	15	0.2441	0.2561	0.1844	0.2201	- 24.5	- 14.1
40	10	0.2759	0.2734	0.2140	0.2416	- 22.4	- 11.6

Table 5: Comparison of lift and drag areas between measurement (subscript M) and simulation (subscript S); naked model, $V = 35°$, $\gamma = 160°$, $\delta = 0°$, $\varphi = 5.3°$.

The agreement with the present measurements may be described as rather poor. Deviations of more than -20% in drag areas and between -11% and -34% in lift areas occur. It must be admitted that a ski jumper with skis is a very complicated structure with many regions of transition, separation, and vortex formation. The most likely reasons for the unrealistic results may be a too coarse grid near the jumper surface, the applied standard $k-\varepsilon$ model, and the steady-state calculation. Predicted drag areas show a similar trend as the measured data. However, one could expect an improved accuracy of transient calculations. The vortex shedding around a ski jumper may cause the change of both drag and lift areas. Such computational studies, however, require a much finer mesh resolution around the body, and especially in the wake, and more complex turbulence models as well. As depicted in Fig. 4, the computational mesh was refined around the model and in the near-wake region. The computational results indicate that the wake extends much further downstream, and vortices can be detected even at the end of the computational domain. A better treatment of separation points/lines at the model surface would require a much more advanced turbulence model allowing a near-wall resolution down to the viscous sublayer, which, in turn, would necessitate a further pronounced grid refinement. A rough estimate showed that a suitable mesh refinement would lead to more than 5 million grid cells in order to achieve acceptable numerical accuracy. Even if this could be realized, the shortcomings of the standard $k-\varepsilon$ turbulence model, as well as the transient nature of vortex shedding, remain.

Quite similar unrealistic behaviour was observed by Reisenberger (2005), who performed comprehensive simulations for a variety of geometrically simple sharp- and round-edged prisms - the corresponding experimental results were published in Reisenberger et al. (2004). For these simulations, FLUENT^{TM} was used as the tool, and the generally available two-equation turbulence models were applied (standard $k-\varepsilon$, RNG $k-\varepsilon$, $k-\varepsilon$ realizable, and $k-\omega$ SST). In the case of a flat plate with an aspect ratio of $5:1$, the simulations yielded quite realistic results with deviations lower than -10% up to inclinations of about $40°$. Beyond this angle, the results differed significantly, in positive and negative direction, depending on the particular turbulence model. While the numerical performance is roughly acceptable for sharp-edged bodies, this is not true for round-edged prismatic bodies with equal dimensions. This worse behaviour seems strongly related to the lateral curvature, where transition of the boundary layer and separation points may not be captured adequately.

With regard to the flat plate results up to an inclination of $40°$, as presented by Reisenberger et al. (2004) and Reisenberger (2005), we assume that the skis are not the reason for the observed deviations in the present simulations. For a qualitative comparison with the present simulations we pick out the rounded prism with 160 mm thickness because of the close similarity to the upper body of ski jumpers. The present simulations cover the range of $40° \leq \alpha + \beta \leq 50°$. After subtraction of the appropriate correction angle $\Delta\alpha_{UB}$ (Table 3), we obtain a range of $27°$ to $37°$ for the inclination of the upper body relative to the air stream. The corresponding simulations of Reisenberger (2005) exhibit deviations in drag areas of about -30% in this region, and deviations in lift areas are even larger (up to -40%). With regard to this fact, the present simulations closely cover the possible range of average deviations, which is - at least in direction - confirmed by numerical simulations in car aerodynamics. The weakness in numerical predictions may, therefore, be mainly ascribed to the complex flow behaviour around the upper body.

5 Summary, conclusions, outline for future work

In the present study, the aerodynamic characteristics of ski jumpers were investigated by comprehensive wind tunnel experiments. The investigations were intended to cover the range of flight positions and postures of the jumpers in a widest possible range. The major goal was to deduce the basis for changes in regulations with regard to safety and health enhancement. In addition, the possibilities and/or limits of modern commercial Computational Fluid Dynamics (CFD) tools should be evaluated with regard to predictions of aerodynamic forces in the flight.

First measurements with a "naked model" were performed in order to evaluate the fundamental aerodynamic behaviour. Based on these results, particular interest was focused on three different jumping suits - a skin-tight suit (white), one suit in compliance with current regulations (red 1) and one suit (red 2) with moderate oversize.

The investigations within the full range of postures clearly show that the drag increases practically continuously in the investigated regime with the angle $\alpha + \beta$, while the lift assumes a maximum near $\alpha + \beta \approx 50°$. The most effective hip angle turned out to be in the range $160° \leq \gamma \leq 170°$, where $\gamma = 160°$ was found in the field studies as usually applied by world class athletes. The influence of the head angle δ cannot be evaluated from the present measurements for reasons pointed out in the related paragraph. The arm angle φ should be taken into account, especially with respect to the various suits. The influence of the suit might be of competitive relevance, since the observed small differences may have decisive influence on the length of the flight path.

A major finding from the present investigations is the relative influence of the skis on the overall aerodynamic forces. In the investigated regime, the relative lift area is significantly larger than the drag area. Furthermore, in the usual range $20° \leq \alpha \leq 40°$ the lift slope of the skis was found to be larger than the lift slope of the complete system. It was particularly this result that has lead to a significant revision of the competitive regulations. Due to the efforts of Wolfram Müller, the present results enhanced the implementation of new regulations (competition season 2004/2005), where the allowed ski length is strongly coupled to the body-mass index (*BMI*) rather than simply to the body height. Observations clearly show that these new regulations were indeed able to strongly disarm the dangerous tendency to severe underweight of the athletes (Müller et al., 2006).

Concerning the numerical simulations we may state that current CFD tools do not seem to be capable of simulating the aerodynamic forces adequately, at least within the tested framework. The deviations of lift and drag areas from the experimental results are far from allowing a proper prediction of, e. g., the flight paths with the sufficient accuracy.

With regard to the numerical simulations, future work seems urgently necessary. This might include the use of a strongly refined numerical grid, the applications of much more advanced turbulence models, and/or a transient treatment with small time steps in order to capture the time-dependent separation in the wake of the jumper.

Symbols

Symbols		Greek letters	
A	cross-sectional area	α	angle of attack
c	coefficient	β	body to ski angle
D	drag area	γ	hip angle
F	force	δ, φ	head angle, arm angle
H	height	Δ	difference
L	lift area	ϱ	air density
l	reference length	μ	dynamic air viscosity
$\bar{l}$	characteristic length	ν	kinematic air viscosity
M	pitching moment	Subscripts	
q	dynamic pressure	D	drag
Re	Reynolds number	L	lift
T	thickness	FS	full scale
t	time	M	pitching moment, model
v	velocity	max	maximum
V	ski angle	S	simulation
W	width	UB	upper body

Bibliography

O. Flachsbart. Messungen an ebenen und gewölbten Platten. *Ergebnisse der Aerodynamischen Versuchsanstalt zu Göttingen*, 4:96–100, 1932.

W. Gretler and W. Meile. Eine Sechs-Komponenten-Plattform-Windkanalwaage. *ÖIAZ*, 136/9:403–408, 1991.

W. Gretler and W. Meile. Der 2m-Windkanal am Institut für Strömungslehre und Gasdynamik der Technischen Universität Graz. *ÖIAZ*, 138/3: 90–96, 1993.

K. Hanjalić. Advanced turbulence closure models: A view on the current status and future prospects. *Int. J. Heat & Fluid Flow*, 15:178–203, 1994.

R. K. Hanna. Going faster, higher and longer in sports with CFD. In S. Haake, editor, *The engineering of sport*, pages 3–10. Balkema, Rotterdam, 1996.

B. E. Launder and D. B. Spalding. The numerical computation of turbulent flows. *Comp. Meth. in Appl. Mech. and Engng.*, 3:269–289, 1974.

W. Meile, E. Reisenberger, M. Mayer, B. Schmölzer, W. Müller, and G. Brenn. Aerodynamics of ski jumping: experiments and CFD simulations. *Exp. Fluids*, 41:949–964, 2006.

W. Müller and B. Schmölzer. Computer simulated ski jumping: The tightrope walk to high performance. In *Proceedings of the 4th World Congress on Biomechanics, Calgary*, 2002.

W. Müller, D. Platzer, and B. Schmölzer. Scientific approach to ski safety. *Nature*, 375:455, 1995.

W. Müller, D. Platzer, and B. Schmölzer. Dynamics of human flight on skis: improvements on safety and fairness in ski jumping. *J. Biomech.*, 29(8):1061–1068, 1996.

W. Müller, W. Gröschl, B. Schmölzer, and K. Sudi. Body weight and performance in ski jumping: the low weight problem and a possible way to solve it. In *Proceedings of the 7th IOC World Congress on Sport Sciences, Athens*, page 43D, 2003.

W. Müller, W. Gröschl, R. Müller, and K. Sudi. Underweight in ski jumping: the solution of the problem. *Int. J. Sports Med.*, 27 (11):926–934, 2006.

S. V. Patankar and D. B. Spalding. A Calculation Procedure for Heat, Mass and Momentum Transfer in Three-Dimensional Parabolic Flows. *Int. J. Heat Mass Transfer*, 15:1787–1806, 1972.

E. Reisenberger. *Untersuchungen zum aerodynamischen Verhalten von Schispringern, PhD Thesis.* Institute of Fluid Mechanics and Heat Transfer, Graz University of Technology, 2005.

E. Reisenberger, W. Meile, G. Brenn, and W. Müller. Aerodynamic behaviour of prismatic bodies with sharp and rounded edges. *Exp. Fluids*, 37:547–558, 2004.

L. P. Remizov. Biomechanics of optimal flight in ski jumping. *J. Biomech.*, 17:167–171, 1984.

B. Schmölzer and W. Müller. The importance of being light: Aerodynamic forces and weight in ski jumping. *J. Biomech.*, 35:1059–1069, 2002.

B. Schmölzer and W. Müller. Individual flight styles in ski jumping: Results obtained during Olympic Games competitions. *J. Biomech.*, 38:1055–1065, 2005.

B. Schmölzer and W. Müller. The influence of lift and drag on the jump length in ski jumping. In *Book of Abstracts of the 1st International Congress on Science and Skiing*, page 274. Austrian Association of Sports Sciences (ÖSG), 1996.

C. G. Speziale. Modelling of turbulent transport equations. In T. B. Gatski, M. Y. Hussaini, and J. L. Lumley, editors, *Simulation and Modeling of Turbulent Flows, Ch. 5*, ICASE/LaRC Series in Comp. Sci. Eng. Oxford University Press, Oxford, 1996.

Some Aspects of Ski Jumping

Helge Nørstrud
Department of Energy and Process Engineering,
Norwegian University of Science and Technology,
Trondheim, Norway

1 Introduction

Ski jumping is very facinating in particular to the athletes necessary correct action when leaving the jump table and the ability to fast adopt to the flying position, see Figure 1.

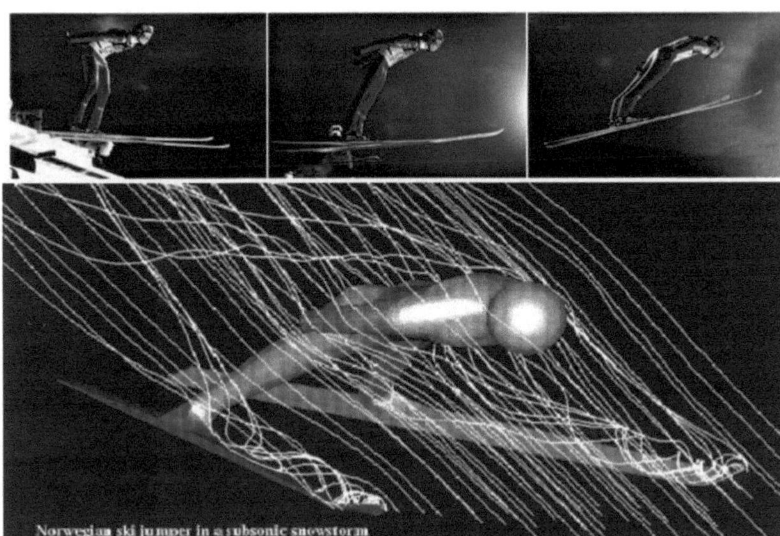

Figure 1. The two important stages (take-off and flying) of a modern ski jump. (Photo: Richard Sagen, Adresseavisen. Simulation: CFD norway).

The lower part of the figure shows the present V-style jumping which was introduced by Jan Bokløw (SWE) in 1988. As the numerical simulation shows, vortical flow is present over the upper surface of the skis which produces high velocities and, hence, low pressure to give extra lift. In contrast, Figure 2 shows the upright way of jumping in 1892 and the aerodynamics was dominated by body drag. Another advantage of the V-style is related to the goose flying formation as the show a distance apart between the wings. This can be compared with the ski jumper lifting surfaces (skiers body and a pair of skis) which are separated in order to yield a positive influence between each other, i.e. an uwash contribution to the neighbour. See Figure 7 in the chapter "Basic Aerodynamics" and imagine that you have three cars side by side.

Figure 2. Ski activity in the old days. (Illustration: Sverresborg Museum, Trondheim).

Between the early days of ski jumping and the introduction of the V-style lies a century of improvements of ski bindings, head gear and jumping dress fabric. The attitude of the jumpers body relative to the skis also underwent changes which are illustrated in Figure 3.

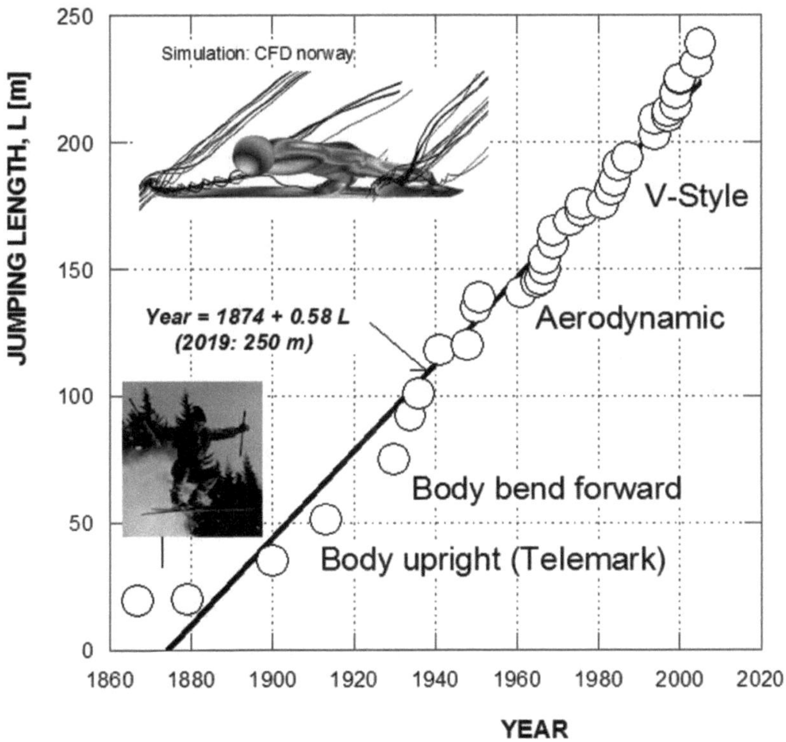

Figure 3. World ski jumping records over the last century.

2 Evaluation of ski jumping results with emphasis on jumper's weight

During the FIS Summer Grand Prix ski jump competition on August 18, 1996 in Trondheim, Norway the body weight and height of 50 competitors were obtained prior to the first round, see Figure 4. From linear interpolation of the data a reference jumper is defined as a weight $G = 65$ kg and height $H = 1.77$ m.

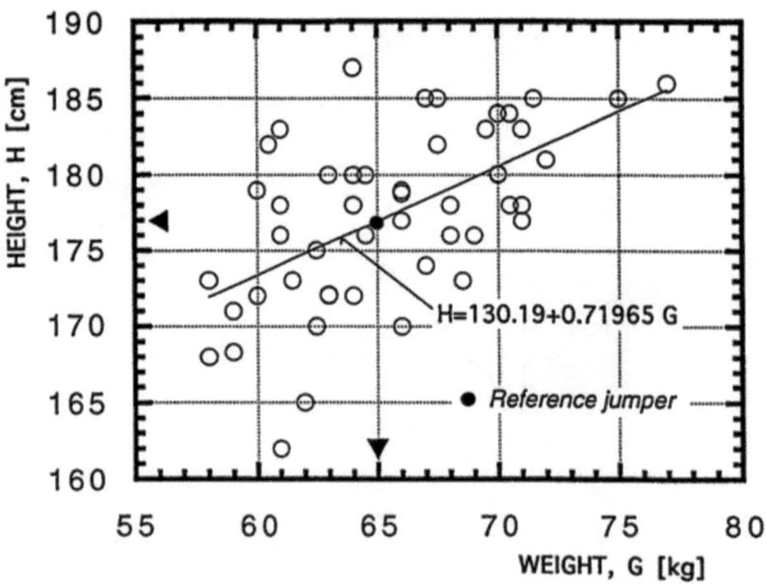

Figure 4. Data for 50 ski jumpers.

The final jumping results (Figure 5) indicates a clear influence of weight on the ranking number, i.e. 8 jumpers from the best 10 are lightweigthers.

In order to formulate a new rule for the ski length we adopt the DuBois formulae which relate the human surface area S [m^2] to the weight G [kg] of the person and to the height H [m], i.e.

$$S = 0.2029 G^{0.425} H^{0.725} \tag{1}$$

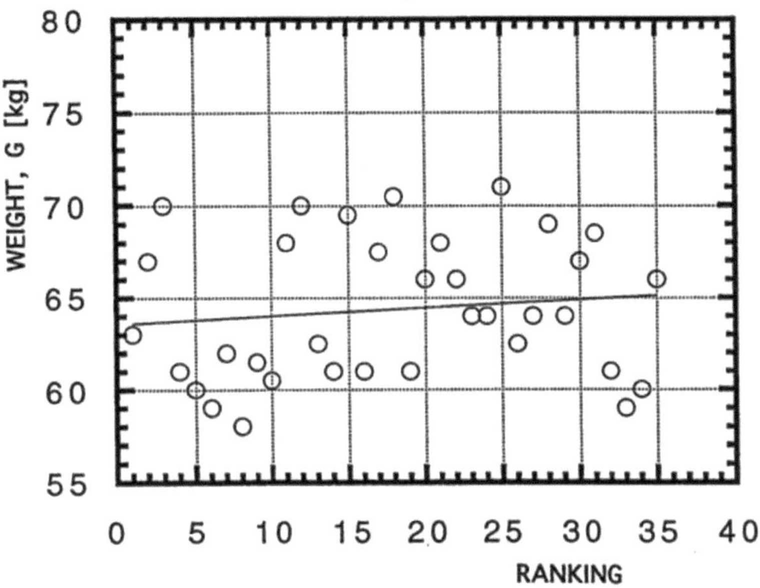

Figure 5. Weight influence on ranking.

Since the frontal area $A_{frontal}$ [m^2] of a person can be estimated as $A_{frontal} =$ 0.35 S (see Figure 6) and numerical simulations of a V-style ski jumper shows that the lifting surface of the two skis is equal to the lifting surface of the ski jumper ($= A_{frontal}$) the DuBois formulae can be used to postulate the relation

$$A_{skis}[m^2] = 0.0714G^{0.4}H^{0.7} \tag{2}$$

Based on a ski width of 0.11 m the length of the ski can then be evaluated from the above relation as $\ell_{new} = 0.325G^{0.4}H^{0.7}$.

Table 1 gives some values for the ski length ℓ_{new} as calculated from the new formulae compared to the old FIS ruling of $\ell_{FIS,old} = H + 0.8$ m. The present FIS ruling for allowed ski length is based on the body mass index (BMI) which is the ratio between the weight G [kg] and the square of the length H [m], i.e. $BMI = G/H^2$.

As Table 1 shows the two realistic jumpers have the same height, but jumper H is heavier than the jumper L (light). The heavier jumper can increase

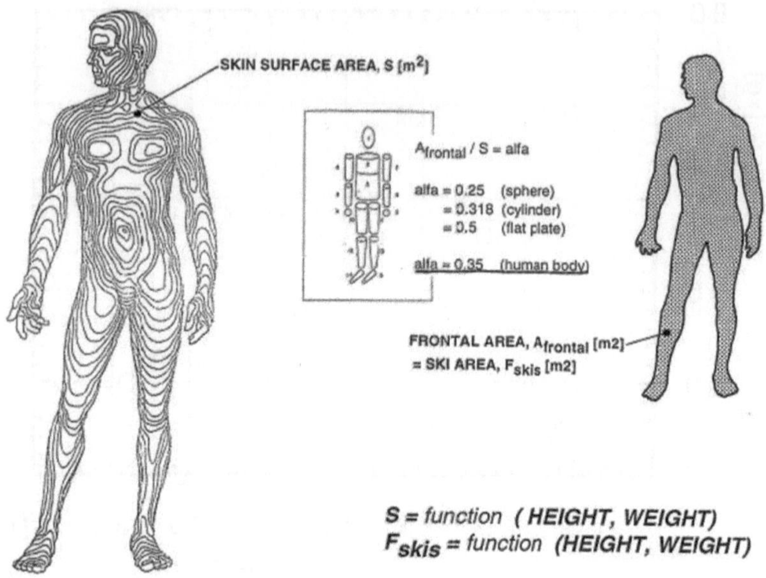

Figure 6. Illustration of the use of DuBois formulae to define the ski area.

Table 1. Ski lengths for three selected ski jumpers.

	G [kg]	H [m]	S [m^2]	A_{skis} [m^2]	ℓ_{new} [cm]	$\ell_{lFIS,old}$ [m^2]	$\Delta\ell$ [cm]
"SUMO"	**190**	**1.9**	**3.00**	**0.94**	**427**	**270**	**+ 157**
Jumper H	68.5	1.73	1.82	0.57	259	253	+ 6
Jumper L	58.0	1.73	1.70	0.53	241	253	- 12

the ski length by 6 cm (relative to the old FIS rule) whereas the lighter jumper have to reduce the ski length by 12 cm.

Figure 7. The ski length has varied in the various ski activities over time, and the "SUMO" ski jumper (with a BMI index of 52.6) as an extreme case (Photo: Skiforeningen).

3 Static stability of a ski jumper in pitch

If we regard the ski jumper as a body of mass moving steadely in a two-dimensional frame of space, an unbalance between the total aerodynamic force and the force of gravity can be written in a vectorial form as

$$\vec{g} \cdot \vec{F}_{aero} + mg^2 < 0 \tag{3}$$

where m [kg] is the total mass of the skijumper, see Figure 8. Hence the skijumper has an attitude corresponding to a position reached after passing through his maximum flight path level. Furthermore, if we select the aerodynamic centre (ac) as the point of reference, located at a distance from the center of gravity (cg), we can represent the action of the aerodynamic force by the components L and D. The aerodynamic centre is defined as the unique point where the aerodynamic force act in such a way that the following condition condition for the associated pitching moment $M_{ac} \neq 0$ can be stated as

$$M_{ac}[Nm] = C_{M,ac} A \ell_{ref} q = const. \tag{4}$$

or

$$C_{M,ac}(\alpha_{abs}) = C_{M,ac}(C_L) = const. \tag{5}$$

Here C_M [-] designates the pitching moment coefficient, A is a reference area conveniently taken as the total ski area ($\sim 0.52m^2$) and ℓ_{ref} is a reference length. The dynamic pressure is represented as q.

The absolute angle of attack α_{abs} [degree] is measured from the zero-lift line as shown in Figure 9. Hence, the moment around the aerodynamic centre is per definition independent (within a certain range) of the lift force L or the angle of attack α_{abs}.

The conditions for static stability in pitch can now be written as

$$\text{i) } M_{cg} = C_{M,cg} A \ell_{ref} q > 0 \text{ for } L = 0 \tag{6}$$

$$\text{ii) } \partial M_{cg} / \partial \alpha_{abs} < 0 \tag{7}$$

and these two conditions for static stability in longitudinal motion are illustrated in Figure 10. It should be noted that e.g. a pitch up of the skier due to a sudden wind gust will increase the absolute angle of attack α_{abs} from the equilibrium position $C_{M,cg} = 0$ and produce a negative and restoring

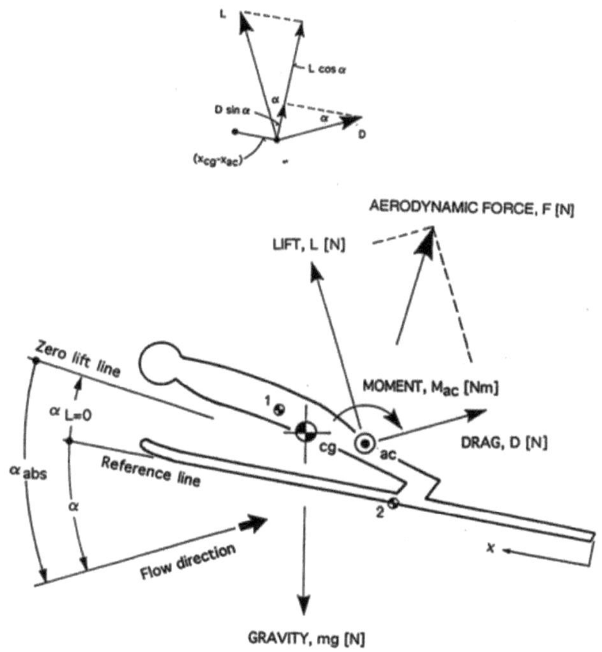

Figure 8. A ski jumper in steady flight.

moment. A pitch down disturbance will reduce the angle of attack and, hence, induce a positive moment which again will bring the jumper back to the equilibrium position.

The moment about the centre of gravity (cg) can be written as

$$M_{cg} = M_{ac} - (L \cos \alpha + D \sin \alpha)(x_{cg} - x_{ac}) \tag{8}$$

where the x-axis is defined in Figure 8 and where the factor $(x_{cg} - x_{ac})$ represents the projection of the distance between the centre of gravity and the aerodynamic centre normal on that axis. With the approximation:

$$L \approx L \cos \alpha + D \sin \alpha \tag{9}$$

Eq. (8) can be rewritten into

$$M_{cg} = M_{ac} - L(x_{cg} - x_{ac}) \tag{10}$$

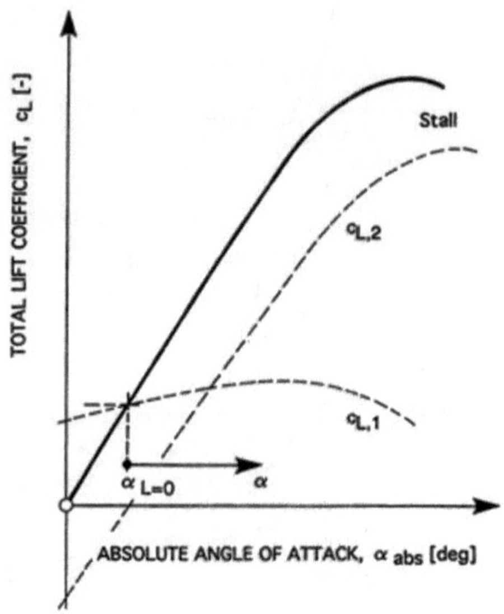

Figure 9. Aerodynamic lift vs. angle of attack.

or with the aid of Eqs. (4) and (6) we have

$$C_{M,cg} = C_{M,ac} - C_L(x_{cg} - x_{ac})/\ell_{ref} \tag{11}$$

Differentiating Eq. (11) with respect to α_{abs} (or C_L) will yield

$$\frac{\partial C_{M,cg}}{\partial \alpha_{abs}} = -(x_{cg} - x_{ac})/\ell_{ref} \tag{12}$$

since $\partial C_{M,ac}/\partial \alpha_{abs} = \partial C_{M,ac}/\partial C_L = 0$. Hence, comparing Eqs. (7) and (12) we can observe that the stability requirement (ii) is fulfilled when the centre of gravity lies ahead of the aerodynamic centre, see again Figure 8.

As a natural closure of the present section we present a picture of a ski jumper which faces the often problem of strong winds in jumping hills. One method of reducing this unfavorable effect is to mount wind sails along the flight path. This is also seen in Figure 11.

Finally, it should be emphasized that a ski jumper has also to fly with high loading on his/hers moderate sized lifting surfaces, see Figure 12. *This requires a sharp pilot.*

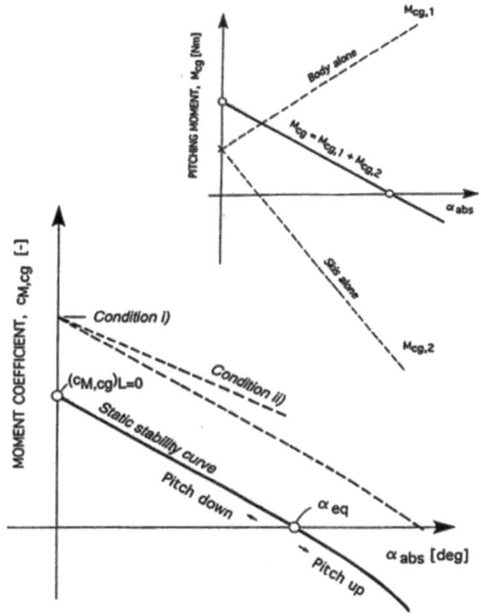

Figure 10. Conditions for static stability in pitch.

Figure 11. Adam Malysz taken by the wind in the Holmenkollen ski jump in Oslo, Norway on March 19, 2007. (Photo: Aftenposten).

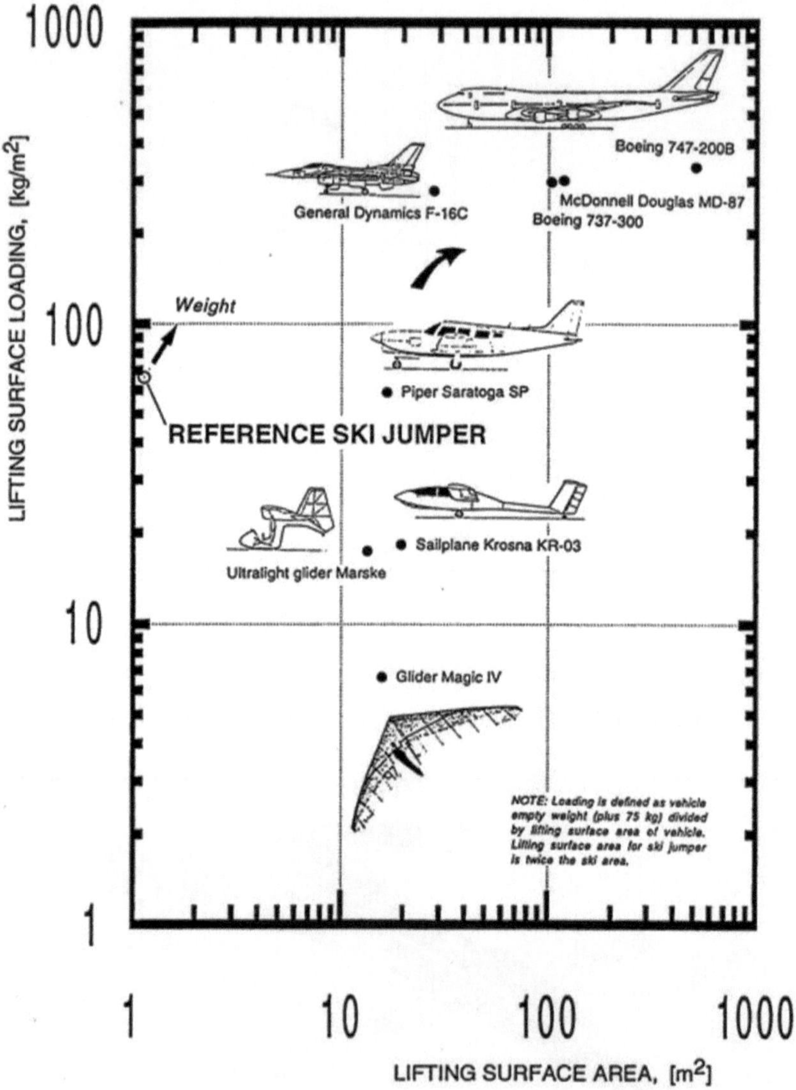

Figure 12. Comparison of lifting surface loading for various flying objects.

Sports Ball Aerodynamics

Rabindra D. Mehta
Sports Aerodynamics Consultant
Mountain View, California, USA

1 Introduction

Aerodynamics plays a prominent role in defining the flight of a ball that is struck or thrown through the air in almost all ball sports. The main interest is in the fact that the ball can often deviate from its initial straight path, resulting in a curved, or sometimes an unpredictable, flight path. It is particularly fascinating that not all the parameters that affect the flight of a ball are always under human influence. Lateral deflection in flight, commonly known as swing, swerve or curve, is well recognized in cricket, tennis, golf, soccer, volleyball and baseball. In most of these sports, the lateral deflection is produced by spinning the ball about an axis perpendicular to the line of flight, which gives rise to what is commonly known as the *Magnus effect*, named after the German chemist/physicist, Gustav Magnus.

It was this very effect that first inspired scientists to comment on the flight of sports balls. In 1666, Sir Isaac Newton at the very ripe age of 23, had noted how the flight of a tennis ball was affected by spin, and he gave this profound explanation: "For, a circular as well as a progressive motion..., its parts on that side, where the motions conspire, must press and beat the contiguous air more violently than on the other, and there excite a reluctancy and reaction of the air proportionably greater" (Newton, 1672). Some 70 years later, in 1742, an English scientist and engineer, Benjamin Robins showed that a transverse aerodynamic force could be detected on a rotating sphere. However, the Swiss mathematician and physicist, Leonhard Euler completely rejected this possibility in 1777. The association of this effect with the name of Magnus was due to Lord Rayleigh (1877), who, in his paper on the irregular flight of a tennis ball, credited Magnus with the first "true explanation" of the effect. Magnus had found that a rotating cylinder moved sideways when mounted perpendicular to the airflow. Rayleigh also gave a simple analysis for a "frictionless fluid," which showed that the side force was proportional to the free-stream velocity and the rotational speed of the cylinder. A Scottish mathematical physicist, Peter Tait (1890, 1891, 1893) used these results to try and explain the forces on a golf ball in flight by observing the trajectory and time of flight. This was all before the introduction of the boundary layer concept by the German physicist, Ludwig Prandtl. In 1904, he delivered a groundbreaking paper, *Fluid Flow in Very Little Friction*, in which he described the boundary layer and its importance for drag and streamlining. The paper also described flow separation as a result of the boundary layer, clearly explaining the concept of wing stall for the first time. Wing stall occurs when the boundary layer over the upper part separates from the surface and the wing loses all the lift force. Since then, the Magnus effect has been correctly attributed to asymmetric boundary layer separation. In most situations, the effect of spin is to delay boundary layer separation on the retreating side and to enhance it on the advancing side. In certain instances, when a rotating ball is released close to the "critical" Reyolds number, a *negative* Magnus force can be generated when boundary layer transition occurs first on the advancing side, thus delaying separation. The Reynolds number represents a ratio between the inertial and viscous forces and was named after the British fluid dynamics engineer, Osborne Reynolds. The Reynolds number is defined as, $Re = Ud/\nu$, where U is the ball velocity, d is the ball diameter and ν is the air kinematic viscosity. The kinematic viscosity is defined as $\nu = \mu/\rho$, where μ is the dynamic viscosity of air and ρ is its density.

It is now well recognized that the aerodynamics of sports balls is strongly dependent on the detailed development and behavior of the boundary layer on the ball's surface. A side force, which makes a ball curve or swing through the air, can also be generated in the absence of the Magnus effect. In one of the cricket deliveries, the ball is released with the seam angled, which creates the boundary layer asymmetry necessary to produce swing. In baseball, volleyball and soccer there is an interesting variation whereby the ball is released without any spin imparted to it. In this case, depending on the seam or stitch orientation, an asymmetric, and sometimes timevarying, flow field can be generated, thus resulting in an unpredictable flight path. Almost all ball games are played in the *Re* range of between about 40,000 to 400,000. It is particularly fascinating that, purely through historical accidents, small disturbances on the ball surface, such as the stitching on cricket balls and baseballs, the felt cover on tennis balls and patch-seams on volleyballs and soccer balls, are all about the right size to affect boundary layer transition and development in this *Re* range.

There has been a lot of research on sports ball aerodynamics since the first review article on the topic was published (Mehta 1985). Some of the more relevant data from that earlier article on the three balls that were covered (cricket ball, baseball and golf ball) are presented here, together with some new data and analyses. In addition, the aerodynamics of tennis balls, volleyballs and soccer balls are also included here. Two recent articles which reviewed the aerodynamics of all these balls include Mehta and Pallis (2001a) and Pallis and Mehta (2003); a more in depth review is presented here. The flow regimes are presented and discussed using recent flow visualization data and wind tunnel measurements of the aerodynamic forces that are generated on the balls. Photographs of the six sports balls together with typical operating conditions are given in Figs. 1-6. They are presented in the order of discussion in this chapter.

Figure 1. A 4-piece (Red) Cricket Ball. Diameter = 7.19 cm, Mass = 0.15 kg, Typical $Re = 1 \times 10^5$.

Figure 2. Wimbledon Tennis Ball. Diameter = 6.7 cm, Mass = 0.058 kg, Typical $Re = 2.2 \times 10^5$.

Apart from round balls, aerodynamics can also play a significant role in the flight of other (non circular) balls. However, research on these balls has been somewhat limited and their discussion is therefore not included here. A couple of notable references include that due to Alam et al. (2005b) who conducted studies on the aerodynamics of rugby balls and Australian

Figure 3. A Modern Day Golf Ball. Diameter = 4.27 cm, Mass = 0.046 kg, Typical $Re = 2 \times 10^5$.

Figure 4. Modern Day Soccer Ball. Diameter = 22.3 cm, Mass = 0.44 kg, Typical $Re = 3.9 \times 10^5$.

Figure 5. A Modern Day Volleyball. Diameter = 21 cm, Mass = 0.27 kg, Typical $Re = 2.8 \times 10^5$.

Figure 6. An Official (National League) Baseball. Diameter = 7.23 cm. Mass = 0.156 kg, Typical Re = 1.5 $\times 10^5$.

rules footballs and the one by Rae and Streit (2002) who investigated the aerodynamics of American footballs.

2 Basic Fluid Mechanics Principles

Let us first consider the flight of a smooth sphere through an ideal (non-viscous or inviscid fluid). To make the example even simpler, consider a fixed sphere in a moving stream of such an ideal fluid. In aerodynamic testing terms, the flow over a sphere flying through the air is equivalent to that of moving flow over a stationary sphere (as in typical wind tunnel testing). The flow in this ideal case remains parallel and "attached" over the whole surface of the sphere, as shown in Fig. 7a. The inviscid fluid strikes the sphere at point A and then as it follows the surface round to point B, the fluid near the surface is "squashed." The reason for the squashing effect is that a given amount of fluid has a narrower "passage" through which to travel (this is similar to the velocity of a jet of water increasing when the nozzle of a hose is closed). As this happens, the fluid velocity near the surface must increase over the front part of the sphere, reaching a maximum at point B, and then decrease over the back part as the quashing is relaxed. The flow velocity returns to that of the mainstream fluid behind the sphere (downstream of point C).

Now as the velocity of the fluid changes, so does the pressure at the surface of the sphere, as shown in Fig. 7b. The pressure is lowest when the velocity is highest, at point B. This inverse relation comes from a basic concept in aerodynamics known as Bernoullis principle or theorem, named after the Dutch/Swiss mathematician/physicist, Daniel Bernoulli. The Bernoulli theorem states that:

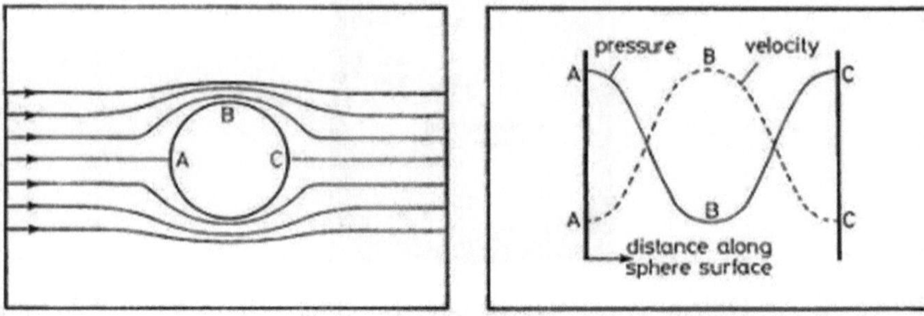

Figure 7. (a) Inviscid flow over a sphere. (b) Pressure and velocity distributions for inviscid flow over a sphere (Mehta & Wood 1980).

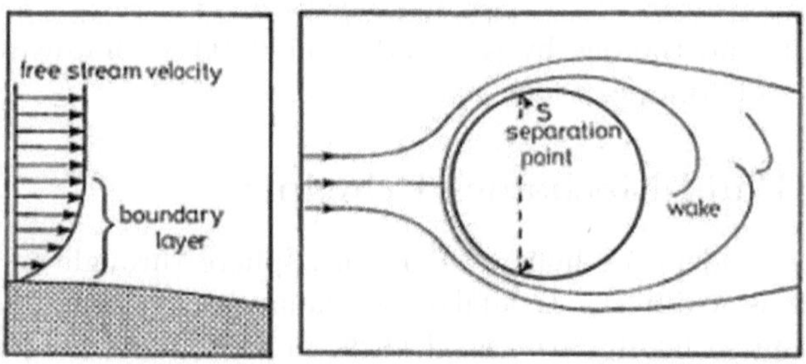

Figure 8. Boundary layer profile and development over a sphere for a viscous fluid (Mehta & Wood 1980).

$$p + \tfrac{1}{2}\rho U^2 + \rho g h = \mathit{constant},$$

where p is the static pressure, ρ is the density, U is the velocity, h is the elevation and g is the gravitational acceleration. So clearly, for this equation to hold, the pressure must decrease as the flow velocity increases. A good illustration of this effect is obtained by blowing between two sheets of paper; the sheets tend to cling together (because of the slightly reduced pressure) rather than flying apart. The theorem generally applies along a *streamline* to inviscid, steady, and *incompressible* flow. The surface pressure can be thought of as a force acting at right angles to the surface over a given area. So on integrating the surface pressures, one ends up with the aerodynamic force acting on the sphere. For example, in Fig. 7b when the pressure over the front of the sphere is integrated and compared to that over the back, the net sum is zero and this implies a zero *drag* force acting on the sphere in this inviscid flow. Similarly, when the pressure distribution over the top part of the sphere is compared to that over the bottom part, they are symmetric and so there would be no *lift* force exerted on the sphere either. So under these obviously unreal conditions, there are no aerodynamic forces acting on the sphere and so it would neither slow down nor deviate sideways in flight.

In real fluids, such as air, the picture is more complex. Real fluids are in fact viscous or sticky. Water is more viscous than air and oil is more viscous than water. This viscosity or stickiness itself leads to drag because of the friction between the sticky air and the sphere of the surface. This is often known as the viscous or "*skin friction*" drag. Another, perhaps more important, consequence of this real fluid is that it gives rise to a layer of relatively slow-moving fluid near the sphere's surface. At the surface, there can be no relative motion between the sphere and the air because of the so-called "*no-slip*" condition, a basic principle in fluid mechanics. So the flow velocity is zero at the surface and then increases as one moves away from the surface until it reaches the same value as that of the "*free stream.*" This layer is known as the *boundary layer* and its thickness is defined as that distance in which the relative air velocity increases from zero at the surface to that of the free stream (Fig. 8).

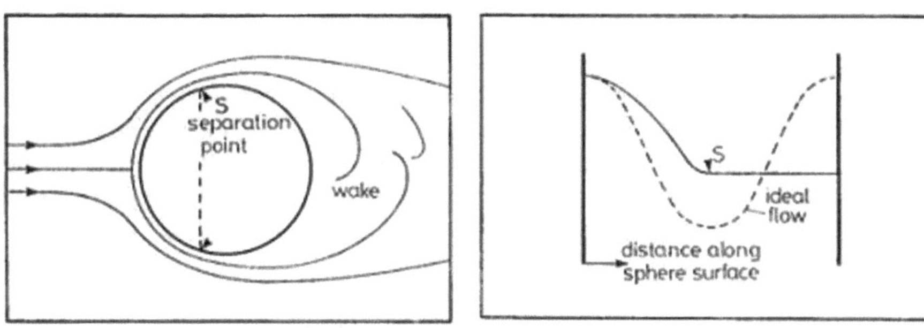

Figure 9. (a) Laminar boundary layer separation on a sphere. (b) Pressure and velocity distributions for laminar flow over a sphere (Mehta & Wood 1980).

The aerodynamics of a sports ball under real conditions depends on the behavior of the two possible states of the boundary layer: *laminar* or *turbulent*. In a laminar boundary layer, the flow is regular, steady, smooth and nearly parallel to the surface. In a turbulent boundary layer, the general mean motion is still roughly parallel to the surface, but in addition there are rapid, random fluctuations in velocity magnitude and direction, giving a rather chaotic appearance. The two types of flow are easily demonstrated by slowly turning on a water tap. At first the water emerges in a clear, steady and smooth stream which may persist until it hits the sink. This is laminar flow. As the tap is opened further and the speed of flow increases, the bottom part of the stream appears to take on some random motions which make it appear splashy. This is turbulent flow. As the speed is increased still further, the random motions spread further upstream, towards the tap so the flow becomes turbulent earlier. The change or "*transition*" from laminar to turbulent flow depends on the Reynolds number ($Re = Ud/\nu$). In the above water tap example, the appropriate length dimension would be the width of the water jet (for a sphere or sports ball the diameter would be the relevant length dimension). The transition in a given flow occurs if the flow Reynolds number exceeds a critical value, which happens sooner if the sphere is bigger, moves faster or the fluid viscosity is lower. In the above water tap example, by increasing the speed of the water, this critical Re is reached sooner and so the flow becomes turbulent closer to the tap.

Now a boundary layer cannot typically negotiate the increasing pressure (*adverse pressure gradient*) over the back part of the sphere and it will tend to peel away or "*separate*" from the surface (Smits 2000). Basically, it is like pushing the fluid uphill whereby the flow close to the surface is continually decelerated until it finally leaves the surface. The boundary layer thickness increases as the boundary layer travels "uphill." As a result, the skin friction coefficient, which is proportional to the velocity gradient close to surface, decreases continually and becomes zero once the boundary layer separates. The pressure becomes constant once the boundary layer has separated giving a pressure distribution like that shown in Fig. 9. Now the pressure over the front part of the sphere is not the same as that over the back. This pressure difference between the front and back of the sphere results in a drag force that slows down the sphere, as one would expect. A laminar boundary layer tends to separate as soon as it encounters the increasing pressure (around the location of the apex, point B in Fig. 7a). However, a turbulent boundary layer has higher momentum near the wall, compared to the laminar layer, and it is continually replenished by turbulent mixing and transport. It is therefore better able to withstand the adverse pressure gradient over the back part of the sphere and, as a result, separates relatively late (downstream of the apex) compared to a laminar boundary layer. This necessarily means that the pressure behind a sphere with turbulent boundary layer separation is higher than that behind one with laminar separation, as shown in Fig. 10. Hence, the drag force on a sphere with turbulent boundary layer separation is lower than that on one with laminar separation. It is also worth noting that the separated region or "wake" behind the ball is smaller on a sphere with turbulent separation; the size of the wake gives an excellent qualitative indication of the pressure drag on a model.

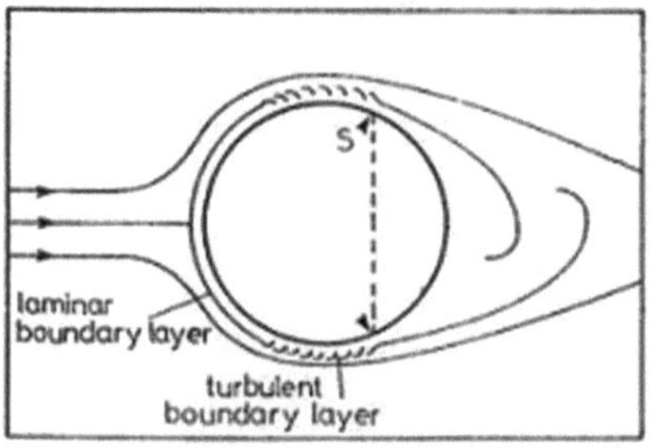

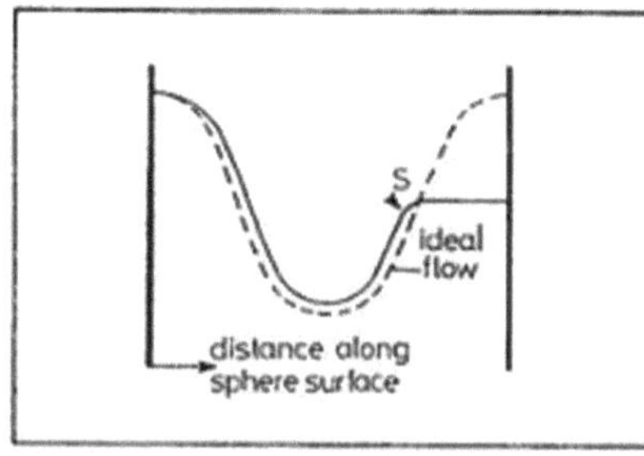

Figure 10. (a) Turbulent boundary layer separation on a sphere. (b) Pressure and velocity distributions for turbulent flow over a sphere (Mehta & Wood 1980).

The interesting point about the critical Reynolds number is that it can be affected by changing the surface roughness on the sphere. The laminar boundary layer can be encouraged or "*tripped*" into a turbulent state by adding a protuberance or roughness onto the surface. This effect is nicely illustrated in Figs. 11 and 12. In Fig. 11, the laminar boundary layer separates relatively early (around the region of the sphere apex). However, with a wire fixed on the sphere surface (Fig. 12), the boundary layer is tripped into a turbulent state and separation is delayed, thus resulting in a smaller wake region and lower aerodynamic drag in this case.

The flow over a sphere can be divided into four distinct regimes (Achenbach 1972). These regimes are illustrated in Fig. 13, in which the drag coefficient (C_D) is plotted against the Reynolds number (Re). The drag coefficient is defined as:

$$C_D = D/(\tfrac{1}{2}\rho U^2 A),$$

where D is the drag force, ρ is the air density and A is the cross-sectional area of the sphere. Similarly, for the lift force (L) the coefficient (C_L) is defined as:

$$C_L = L/(\tfrac{1}{2}\rho U^2 A)$$

In the subcritical regime, laminar boundary layer separation occurs at an angle from the front stagnation point (θ_s) of about 80° and the C_D is

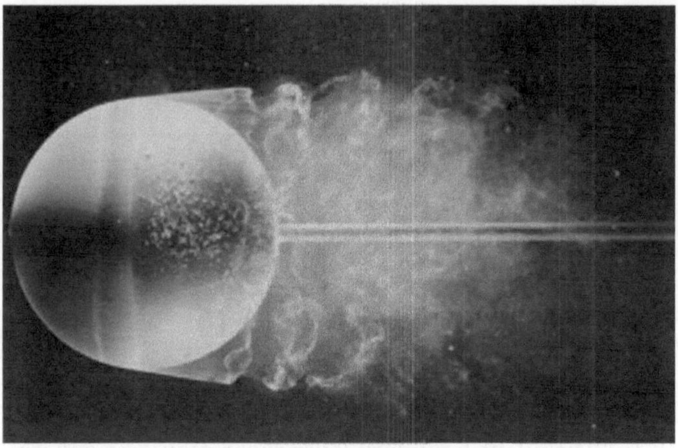

Figure 11. Laminar boundary layer separation on a smooth sphere; boundary layer separates near the sphere apex. Flow is from left to right and note that a wide wake implies high drag (from Van Dyke 1982, photograph by H. Werlé 1980, copyright ONERA, the French Aerospace Laboratory).

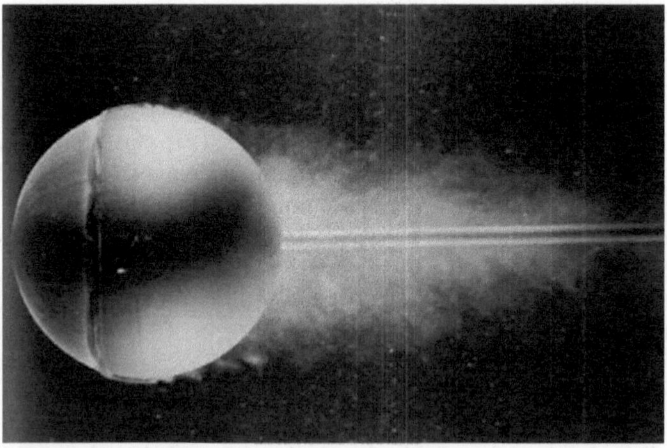

Figure 12. Turbulent boundary layer separation; boundary layer is tripped into a turbulent state by a thin wire attached to the front of the sphere and separation is delayed. Flow is from left to right and note the narrower wake compared to that in Fig. 11, implying lower drag (from Van Dyke 1982, photograph by H. Werlé 1980, copyright ONERA, the French Aerospace Laboratory).

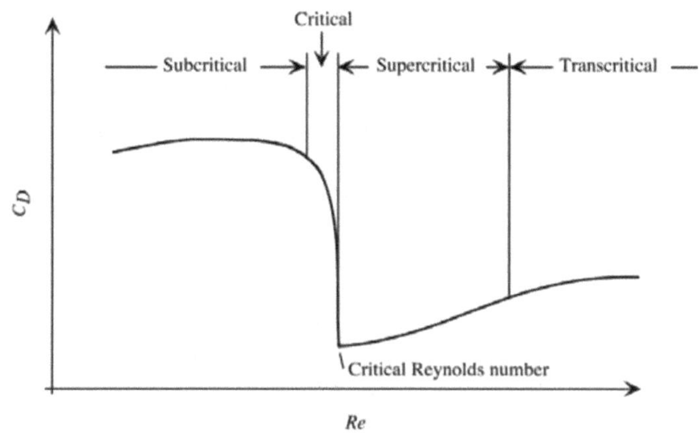

Figure 13. Flow regimes on a sphere (based on Achenbach 1972).

nearly independent of Re. In the critical regime, the C_D drops rapidly and reaches a minimum at the critical Re. The initial drop in C_D is due to the laminar boundary layer separation location moving downstream ($\theta_s \approx 95°$). At the critical Re, a separation bubble is established at this location whereby the laminar boundary layer separates, transition occurs in the free-shear layer and the layer reattaches to the sphere surface in a turbulent state. Achenbach (1972) proposed that the presence of a separation bubble was consistent with the relatively low value of the local minimum in the measured skin friction coefficient in this region. The turbulent boundary layer is better able to withstand the adverse pressure gradient over the back part of the ball and separation is delayed to $\theta_s \approx 120°$. In the supercritical regime, transition occurs in the attached laminar boundary layer and the C_D increases gradually as the transition and the separation locations creep upstream with increasing Re. A limit is reached in the transcritical regime when the transition location moves all the way upstream, very close to the front stagnation point. The turbulent boundary layer development and separation is then determined solely by the sphere surface roughness, and the C_D becomes independent of Re since the transition location cannot be further affected by increasing Re.

As discussed above, earlier transition of the boundary layer can be induced by tripping the laminar boundary layer using a protuberance (e.g. seam on a cricket ball or baseball) or surface roughness (e.g. dimples on a golf ball or fabric cover on a tennis ball). The C_D versus Re plot shown in Fig. 14 contains data for a variety of sports balls together with Achenbach's

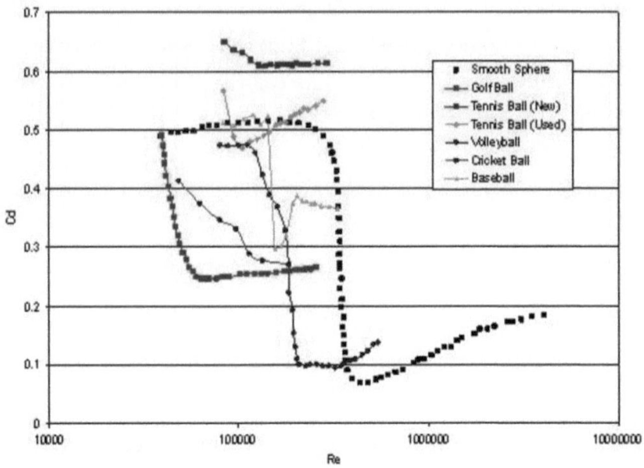

Figure 14. Drag coefficient versus Reynolds number for different nonspinning sports balls (Mehta & Pallis 2001a).

(1972) curve for a smooth sphere. All these data are for non-spinning test articles, and all except the cricket ball, were held stationary in wind tunnels for the drag measurements. The cricket ball was projected through a wind tunnel test section and the drag determined from the measured deflection. For the smooth sphere, the C_D in the subcritical regime is about 0.5 and at the critical *Re* of about 400,000 the C_D drops to a minimum of about 0.07, before increasing again in the supercritical regime. The critical *Re*, and amount by which the C_D drops, both decrease as the surface roughness increases on the sports balls. The specific trends for each of the sports balls are discussed below in the individual ball sections. The appropriate non-dimensional roughness parameter (ϵ) for sports balls is defined as:

$$\epsilon = k/d,$$

where d is the ball diameter and k is a representative roughness length scale, for example the height of the stitching on a cricket ball or baseball. Of course, representing the roughness by just a single parameter which describes just the height does not seem appropriate. The type and distribution of the roughness must be expected to play a role as well. And indeed, in some recent studies by Haake et al. (2007), a "skewness" parameter has been proposed which takes into account the shape of the surface profiles, in addition to the roughness height. A negative skewness indicates a profile where wide peaks are separated by narrow valleys, whereas a positive skewness indicates a profile with wide valleys separated by narrow peaks.

In a viscous flow such as air, a sphere that is spinning at a relatively high rate can generate a flow field that is very similar to that of a sphere in an inviscid flow with added circulation. That is because the boundary layer is forced to spin with the ball due to viscous friction, which produces a circulation around the ball, and hence a side force. At more nominal spin rates, such as those encountered on sports balls, the boundary layers cannot negotiate the adverse pressure gradient on the back part of the ball and they tend to separate, somewhere in the vicinity of the sphere apex. The extra momentum applied to the boundary layer on the retreating side of the ball allows it to negotiate a higher pressure rise before separating and so the separation point moves downstream. The reverse occurs on the advancing side and so the separation point moves upstream, thus generating an asymmetric separation and an upward deflected wake, as shown in Fig. 15.

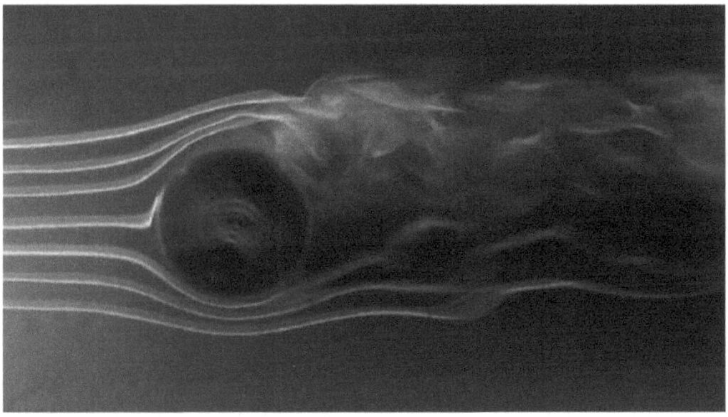

Figure 15. Dye flow visualization over a spinning ball in a water channel. Flow is from left to right and the ball is rotating in a counter-clockwise direction (photograph courtesy of NASA Ames Research Center).

Following Newtons 3rd Law of Motion, the upward deflected wake implies a downward (Magnus) force acting on the ball. Now the dependence of the boundary layer transition and separation locations on Re can either enhance or oppose (even overwhelm) this effect. Since the effective Re on the advancing side of the ball is higher than that on the retreating side, in the subcritical or (especially) supercritical regimes, the separation location on the advancing side will tend to be more upstream compared to that on the retreating side. This is because the C_D increases with Re in these regions, thus indicating an upstream moving separation location. However, in the region of the critical Re, a situation can develop whereby the advancing side winds up in the supercritical regime with turbulent boundary layer separation, whereas the retreating side is still in the subcritical regime with laminar separation. This would result in a negative Magnus force, since the turbulent boundary layer on the advancing side will now separate later compared to the laminar layer on the retreating side. Therefore, a sphere with topspin for example, would experience an *upward* aerodynamic force. So in order to maximize the amount of (positive) Magnus force, it helps to be in the supercritical regime and this can be ensured by lowering the critical Re by adding surface roughness (e.g. dimples on a golf ball). For spinning balls, the appropriate non-dimensional parameter is the spin parameter (S), which is defined as:

$$S = V/U,$$

where U is the ball translation velocity and V is the equatorial velocity at the edge of the ball.

3 CRICKET BALL AERODYNAMICS

3.1 Introduction

The origins of cricket are obscured and the source of much speculation, but it was probably played in as early as the 1300s in England. It was certainly well established by the time of the Tudor monarchs (1485-1603). The first reference to cricket was contained in a document dated December 1478 and it referred to 'criquet' near St. Olmer, in what is now north-eastern France. The first recorded cricket match took place at Coxheath in Kent, England in 1646 and the first 'Test' (international) Match took place between England and Australia in Melbourne, Australia in 1877. The infamous 'Ashes' match was played in London in 1882 when an English newspaper printed a mock obituary notice after Australia had defeated England.

Aficionados know cricket as a game of infinite subtlety, not only in strategy and tactics, but also in its most basic mechanics. On each delivery, the ball can have a different trajectory, varied by changing the pace (speed), length, line or, most subtly of all, by moving or 'swinging' the ball through the air so that it drifts sideways. The actual construction of a cricket ball and the principle by which the faster bowlers swing the ball is unique to cricket. The outer cover of a cricket ball consists of four or two pieces of leather, which are stitched together (by hand on the better quality four-piece balls). Six rows of prominent stitching along its equator make up the primary seam, with typically 60 to 80 stitches in each row. On the four-piece balls, each hemisphere also has a line of internal stitching forming the 'quarter' or 'secondary' seam.

The earliest cricket 'balls' consisted of stones, pieces of wood and lumps of hide. The first 'manufactured' ball was made around 1658 by interlacing narrow strips of hide. The first 'six-seamed' ball was made in 1775 by Dukes, a family firm in Kent, England. The earlier cricket balls had a cork center with layers of cork and wool yarn around it. The wool was wound wet under tension so that it compressed the center as it dried. Today's cricket balls have a composite cork/rubber or a cork/rubber laminate center. The weight of an approved (official) ball is between 0.15 kg (5.5 oz) and 0.16 kg (5.75 oz) and the circumference is between 22 cm (8.81 inches) and 23 cm (9 inches).

It was the curved flight of tennis balls that first inspired scientists to comment on the subject (Newton 1672, Rayleigh 1877). This type of "spin-swing" also occurs in cricket where it is usually employed by the spin bowlers to generate the Magnus effect (Mehta and Wood 1980). However, there is another type of swing that is unique to cricket and one that is perhaps more intriguing. The actual construction of a cricket ball and the principle by which the fast bowlers swing the ball is unique to cricket. The primary and quarter seams play a critical role in the aerodynamics of a swinging cricket ball. It is said that this latter type of swing originated around the turn of the century, but there is evidence that it was in existence well before that time. Dr. W.G. Grace, often acknowledged as the "father" of modern day cricket, and who played in the late 19th century was apparently an exponent of swing bowling. Bowlers from that era had realised that a perfectly new ball favoured the "peculiar flight," so there is not much doubt that it was the traditional cricket ball swing and not spin-swing that the bowlers were referring to.

The first published scientific account of cricket ball swing was by Cooke (1955), who gave an explanation of why it was possible for fast bowlers to make a new cricket ball "swerve" and why it became more difficult to do this when the shine on the ball had worn off. Since then, several articles have been published on the theories of cricket ball swing (Lyttleton 1957, Horlock 1973, Mehta and Wood 1980). More recently, Barton (1982), Bentley et al. (1982) and Mehta et al. (1983) described detailed experimental investigations where the magnitude of the side force that produces swing and the factors that affect it were determined; see Mehta (1985) for a detailed review of the earlier work. The relatively new concept of "reverse swing," which first became popular in the late 1980s and 1990s, was first explained and discussed by Bown and Mehta (1993). A preliminary analysis of cricket ball swing using computational fluid dynamics was described by Penrose et al. (1996). More recently, the flow field around a cricket ball was measured and described by Grant et al. (1998) and Sayers and Hill (1999) published some measurements of the aerodynamic forces on a spinning cricket ball. Some of the myths and misconceptions surrounding cricket ball aerodynamics were presented in Mehta (2000) and a recent overview of cricket ball aerodynamics is given in Mehta (2005). All the measurements shown in this article are taken from the author's own research described by Bentley et al. (1982).

3.2 Aerodynamics of Conventional Swing

Fast bowlers in cricket make the ball swing by a judicious use of the primary seam. The ball is released with the seam at an angle to the initial line of flight. Over a certain Reynolds number (*Re*) range, the seam trips the laminar boundary layer into turbulence on one side of the ball whereas that on the other (nonseam) side remains laminar (Fig. 16). By virtue of its increased energy, the turbulent boundary layer, separates later compared to the laminar layer and so a pressure differential, which results in a side force, is generated on the ball as shown in Fig. 16. In order to show that such an asymmetric boundary layer separation can indeed occur on a cricket ball, a ball was mounted in a wind tunnel and smoke was injected into the separated region (wake) behind the ball where it was entrained right up to the separation points (Fig. 17). The seam has tripped the boundary layer on the lower surface into turbulence, evidenced by the chaotic nature of the smoke edge just downstream of the separation point. On the upper surface, a smooth, clean edge confirms that the separating boundary layer was in a laminar state. Note how the laminar boundary layer on the upper surface has separated relatively early compared to the turbulent layer on the lower surface. The asymmetric separation of the boundary layers is further confirmed by the upward deflected wake, which implies that a downward force is acting on the ball.

In order to confirm that an asymmetric boundary layer separation on a cricket ball leads to a pressure differential across it, 24 pressure taps were installed on a ball along its equator, in a plane perpendicular to that of the seam (Fig. 18). This pressure model was produced by cutting open a normally manufactured cricket ball, removing its core and gluing the two halves together after the pressure taps and support tube had been installed. Figure 19 shows the measured pressures on the ball mounted in a wind tunnel with the seam angled at 20° to the oncoming flow. At low values

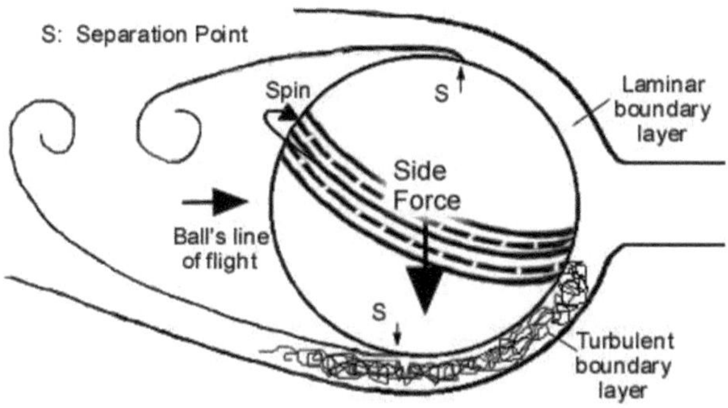

Figure 16. Schematic of flow over a cricket ball for conventional swing (Mehta 2005).

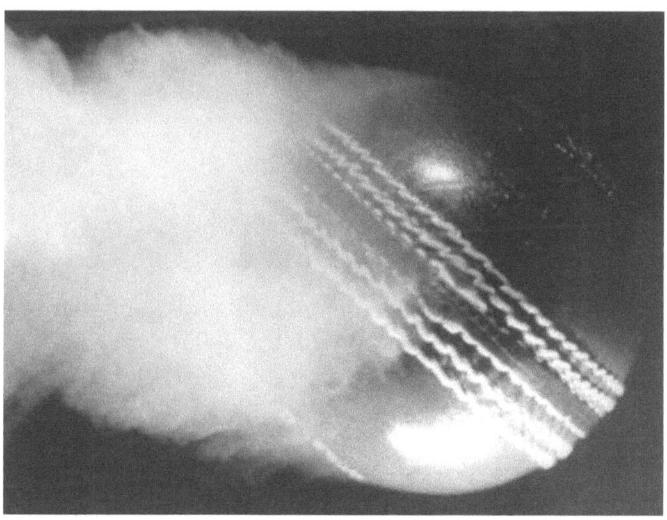

Figure 17. Smoke flow visualization of flow over a cricket ball. Flow is from right to left. Seam angle = 40°, flow speed = 17 m/s, Re = 850,000 (Mehta 2005).

of Re or U, the pressure distributions on the two hemispheres are equal and symmetric, so there would be no side force. At $U = 25$ m/s, the pressure dip on the right-hand (seam-side) face of the ball is clearly lower than that on the left-hand (nonseam-side) face, which would result in the ball swinging towards the seam side. The maximum pressure difference between the two sides occurs at $U = 29$ m/s (65 mph), when the boundary layer on the seam side is fully turbulent while that on the nonseam side is still laminar. Even at the highest velocity achieved in this test ($U = 37$ m/s), the asymmetry in pressure distributions is still clearly exhibited, although the pressure difference is reduced. The actual (critical) velocities or Re at which the asymmetry appears or disappears were found to be a function of the seam angle, surface roughness, and free-stream turbulence; in practice it also depends on the spin rate of the ball, as shown and discussed below.

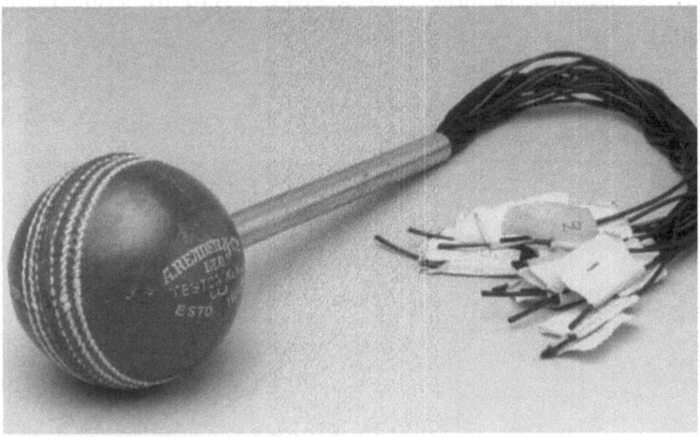

Figure 18. Cricket ball with the core removed and 24 pressure taps (1 mm diameter) installed along the equator (Mehta 2005).

When a cricket ball is bowled, with a round arm action as the laws insist, there will always be some backspin imparted to it. In simple terms, the ball rolls-off the fingers as it is released. In scientific terms, the spin is necessarily imparted to conserve angular momentum. The ball is usually held along the seam so that the backspin is also imparted along the seam (the ball spins about an axis perpendicular to the seam plane). At least this is what should be attempted, since a "wobbling" seam will not be very efficient at producing the necessary asymmetric orientation, and hence asymmetric boundary layer separation. This type of release is obviously not very easy to master, which is the main reason why not every bowler

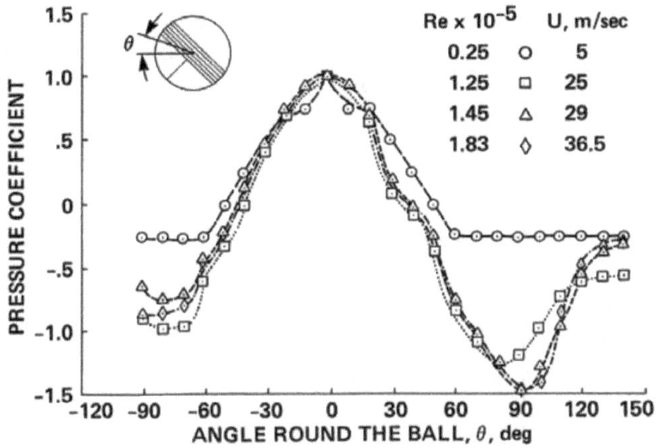

Figure 19. Pressure distributions on a cricket ball held at a seam angle of 20° (Mehta 1985, 2005).

can swing a cricket ball effectively, even a brand new one which is more conducive to swing.

In order to measure the forces on spinning cricket balls, balls were rolled along their primary seams down a ramp and projected into a wind tunnel test section through a small opening in the ceiling (Bentley et al. 1982). The spin rate was varied by changing the starting point along the ramp, and the seam angle was varied by adjusting the alignment of the ramp with the airflow. Once the conditions at the entry to the wind tunnel and the deflection from the datum are known, the forces due to the airflow can be easily evaluated. The spin rate and velocity of the ball at the end of the ramp were measured using strobe photography. Figure 20 shows the measured side force (F), normalised by the weight of the ball (mg), and plotted against the ball's velocity; the side force is averaged over five cricket balls that were tested extensively. At nominally zero seam angle (seam straight up, facing the batsman) there is no significant side force, except at high velocities when local roughness, such as an embossment mark, starts to have an effect by inducing transition on one side of the ball. However, when the seam is set at an incidence to the oncoming flow, the side force starts to increase at about U = 15 m/s (34 mph). The normalised side force increases with ball velocity, reaching a maximum of about 0.3 before declining rapidly. The critical velocity at which the side force starts to decrease is about 30 m/s. This is the velocity at which the laminar boundary layer on the nonseam side also undergoes transition and becomes turbulent. As a result, the asymmetry between the two sides (difference in the locations of the boundary layer separation points) is reduced and the side force starts to decrease.

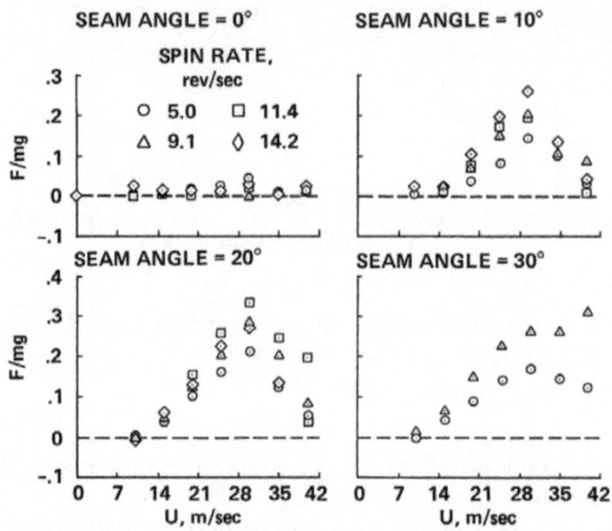

Figure 20. Variation of normalized side force with flow speed; averaged over five balls (Mehta 1985, 2005).

The maximum side force is obtained at a bowling speed of about 30 m/s (67 mph) with the seam angled at 20° and the ball spinning backwards at a rate of 11.4 revs/s. At a seam angle of 20°, the *Re* based on seam height is about optimal for effective tripping of the laminar boundary layer. At lower speeds, a bowler should select a larger seam angle so that by the time the flow accelerates around to the seam location, the critical speed for efficient tripping has been reached. Of course, releasing a ball spinning along the seam (without much wobble) becomes more difficult as the seam angle is increased. Spin on the ball helps to stabilize the seam orientation. Basically, for stability, the angular momentum associated with the spin should be greater than that caused by the torque about the vertical axis due to the flow asymmetry. Too much spin is also detrimental, since the effect of the balls surface roughness is increased and the critical *Re* is achieved sooner on the nonseam side. In order to maximize the amount of conventional swing, the ball surface on the nonseam side should be kept as smooth as possible so that a laminar boundary layer can be maintained.

The actual trajectory of a cricket ball can be computed using the measured forces. Figure 21 shows the computed trajectories at five bowling speeds for the ball exhibiting the best swing properties ($F/mg = 0.4$ at $U = 32$ m/s, seam angle = 20°, spin rate = 14 revs/s). The results illustrate that the flight path is almost independent of speed in the range $24 < U < 32$ m/s ($54 < U < 72$ mph). The trajectories were computed using a simple relation, which assumes that the side force is constant and acts perpendicular to the initial trajectory. This gives a lateral deflection that is proportional to the square of the elapsed time (t^2) and hence a parabolic flight path. In some photographic studies of a swing bowler (Gary Gilmour, who played for Australia in the 1970s), it was confirmed that the trajectories were indeed parabolic (Imbrosciano 1981). Those studies also confirmed that the final deflections of over 0.8 m predicted here are not unreasonable. One of the photographed sequences was analyzed and the actual flight path is also plotted in Fig. 21. The agreement is rather remarkable considering the simplicity of the image processing and analytical techniques. The data in Fig. 21 also have a bearing on the phenomenon of the so-called "late swing."

One of the popular theories for late swing suggests that a ball released at a speed just above the critical (with the boundary layers on both sides or hemispheres turbulent) may slow down enough during flight so that the boundary layer on the nonseam side reverts to a laminar state, thus creating a late movement of the ball. However, it turns out that a ball released at postcritical *Re* slows down by less than 5% in flight, and from the shapes of the curves in Fig. 20, it does not seem likely that this effect would occur in practice. Another theory relies on a change in the ball orientation (through the gyroscopic precession effect), but test results indicate that this is not a significant effect (Bentley et al. 1982). The suggestion that sudden changes in wind direction can lead to late swing (through a change in the seam angle) is also not very likely to occur in practice (Wilkins 1991). In fact, the data in Fig. 21 offer the best explanation for the phenomenon of late swing. Since the flight paths are parabolic, late swing is in fact "built-in," whereby 75% of the lateral deflection occurs over the second half of the flight from the bowler to the batsman. So late swing is most likely a natural, built-in part of cricket ball swing, rather than an artifact of some new, unknown phenomenon.

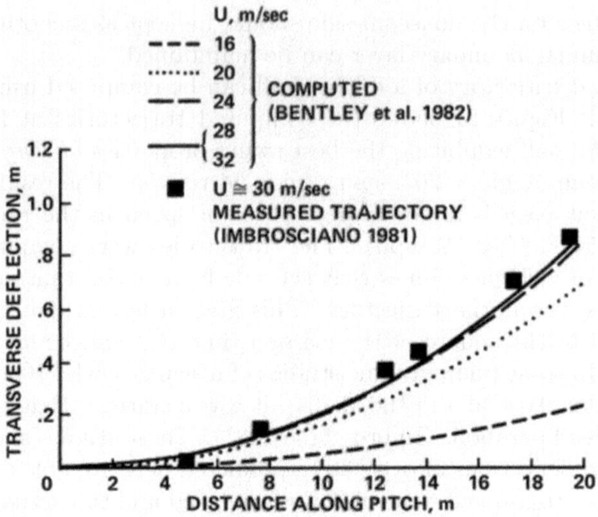

Figure 21. Comparison of computed flight paths using measured forces for the cricket ball with the best swing properties. Seam angle = 20°, spin rate = 14 revs/sec (Mehta 1985, 2005).

3.3 Aerodynamics of Reverse Swing

Since the mid-1980s, there has been a lot of talk in the cricketing world of a supposedly new bowling concept employed by swing bowlers. The new concept or phenomenon is popularly known as "reverse swing" since the ball swings in a direction opposite (or reversed) to that expected based on conventional cricketing wisdom and previously accepted aerodynamic principles. This new form of swing bowling was first demonstrated (with astonishing success) by the Pakistani bowlers, in particular Imran Khan and Sarfraz Nawaz in the early years, followed by Wasim Akram and Waqar Younis. They produced reverse swing very effectively, and generally using older cricket balls, which obviously added to the intrigue.

Ironically, I first heard about the phenomenon of reverse swing in the summer of 1980 from an old school mate of mine, none other than Imran Khan himself, who is often considered to be the first exponent of reverse swing. In talking about some issues regarding cricket ball swing, Imran told me about a curious effect he had observed when bowling. He was predominantly an inswing bowler, but he remarked that with the same grip and bowling action, the ball would swing away (outswinger) on the odd occasion. At the time, I honestly did not believe that such a phenomenon could occur since I could not explain it using cricket ball aerodynamics as they were understood at the time. However, in the following year when we started conducting experiments on cricket ball swing, the whole "mystery" was revealed (Bentley et al. 1982). As discussed above, for conventional swing it is essential to have a smooth polished surface on the nonseam side facing the batsman so that a laminar boundary layer is maintained. At the critical *Re*, the laminar boundary layer on the nonseam side undergoes transition and the flow asymmetry, and hence side force, starts to decrease. A further increase in *Re* results in the transition and separation points moving upstream, towards the front of the ball. A zero side force is obtained when the flow fields (boundary layer separation locations) on the two sides of the ball become completely symmetric. In terms of reverse swing, the really interesting flow events start to occur when the *Re* is increased beyond that for zero side force. As mentioned above, the transition point will continue to move upstream (on both sides now) setting up the flow field shown in Fig. 22. The transition points on the two sides are symmetrically located, but the turbulent boundary layer on the seam side still has to encounter the seam. In this case, the seam has a "detrimental" effect whereby the boundary layer is thickened and weakened, making it more susceptible to separation compared to the thinner turbulent boundary layer on the nonseam side. The turbulent boundary layer on the seam side separates relatively early and an asymmetric flow is set up once again, only now the orientation of the asymmetry is reversed such that the side force, and hence swing, occurs towards the nonseam side, as shown in Fig. 22; *this is reverse swing.*

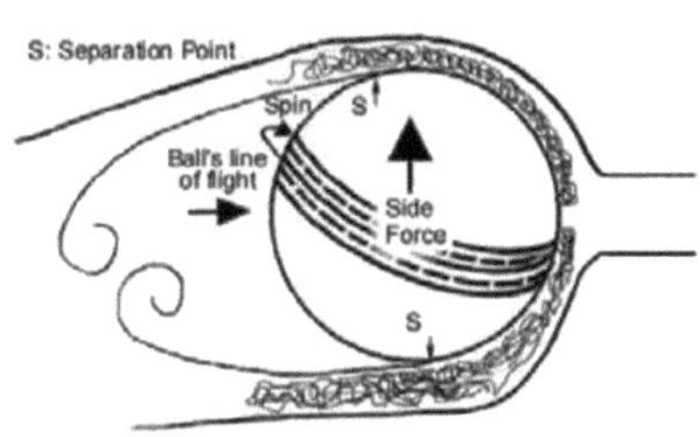

Figure 22. Schematic of flow over a cricket ball for reverse swing (Mehta 2005).

Amongst other factors, transition is strongly dependent on the condition (roughness) of the ball's surface. This is demonstrated in the side force results for three cricket balls with contrasting surface conditions (Fig. 23). The new two-piece ball (without the quarter seams) exhibits a higher maximum (positive) side force than the other two balls and the side force does not start to decline until $U = 36$ m/s. This ball will only produce reverse swing for velocities above 45 m/s (100 mph), which is not very useful in practice, although it is worth noting that two-piece cricket balls are generally not used in competitive cricket matches. However, the side force measurements for a new four-piece ball (with quarter seams) show that it achieves significant negative side force or reverse swing at velocities above about 36 m/s (80 mph). Note how the magnitude of the negative side force at 40 m/s is not much less than the positive force at 30 m/s. So it seems as though reverse swing can be obtained at realistic, albeit relatively high, bowling velocities. In particular, reverse swing can be clearly obtained *even on a new ball, without any tampering of the surface.*

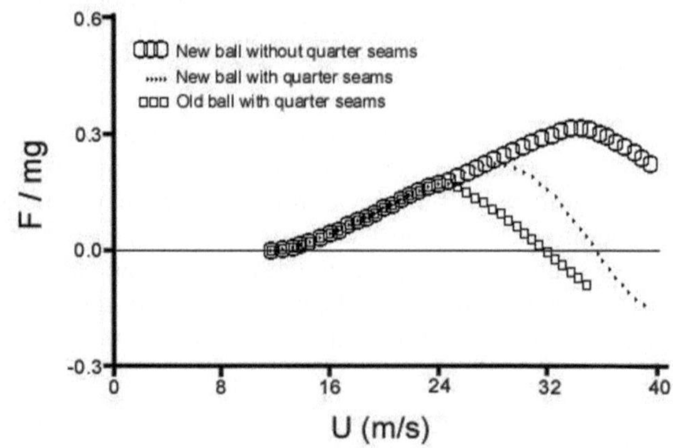

Figure 23. Normalized side force versus ball speed showing reverse swing (Mehta 2005).

Some of the fastest bowlers, such as Jeff Thomson (Australia), Michael Holding (West Indies) and Imran Khan (Pakistan) from prior years and Shoaib Akhtar (Pakistan) and Brett Lee (Australia) from recent times, have been measured bowling in the 40+ m/s (90+ mph) range and so in theory reverse swing would certainly be achievable by them. Alas, not every bowler can bowl at 40 m/s, so what about the mere mortals who would still like to employ this new art? Well, there is hope as shown in Fig. 23. The "old" ball, with an estimated use of about 100 overs, gives less positive side force compared to the new balls, but it also produces reverse swing at a lower velocity of about 30 m/s (67 mph). The contrasting results for the three balls are directly attributable to the effects of surface roughness on the critical Reynolds number. Due to the absence of the quarter seams, the new two-piece ball has a smoother surface compared to the new four-piece ball and the critical Reynolds number at which transition occurs on the nonseam side is therefore higher. Conversely, the critical Reynolds number on the used ball is lower because of the rougher surface. The key to reverse swing is early transition of the boundary layers on the ball's surface and the exact velocity beyond which reverse swing is obtained in practice will decrease with increasing roughness.

Contrary to popular belief, based largely on comments initially made by the Pakistani bowlers, Imran Khan and Sarfraz Nawaz, the cricket ball does not have to be wet on one side to produce reverse swing. The notion that this makes the ball heavier on this side and it would therefore swing in that direction has no aerodynamic basis to it whatsoever. However, there are some possible advantages to wetting the ball (Wilkins 1991). For one thing, it makes it easier to gouge the softer leather with the fingernails. Also, it is possible that the quarter seam may produce more roughness by absorbing water underneath the exposed stitches, thus making the ridges more pronounced. As discussed above, additional roughness reduces the bowling speed at which reverse swing can be obtained. The disadvantage of a wet ball is that it is very likely to become heavier since the (bare) leather tends to absorb water and it will therefore swing less.

3.4 Contrast Swing, Effects of Ball Condition and Tampering

For conventional swing, a prominent primary seam obviously helps the transition process, whereas a smooth polished surface on the nonseam side helps to maintain a laminar boundary layer. Historically, bowlers have always paid a lot of attention to these two features of the ball, although the scientific reasons for doing so may not be totally obvious to them. As all true gentlemen cricketers know, only natural substances such as sweat or saliva can be legally used as a polishing agent, although the odd use of *Vaseline*, *Brylcreem* or sunscreen lotion was often at the centre of a ball tampering controversy, especially in the 1970s. Nowadays, there is lot of controversy over "special mints" that bowlers suck; this supposedly affects the saliva in a positive way for polishing. "Picking" of the primary seam (lifting the stitching so that the primary seam height is retained) on aging balls is also technically illegal, but bowlers can be often seen running their fingernails over the stitching.

Regardless of the chosen "procedure" for polishing, in order to continue obtaining conventional swing from a new ball, it is wise to polish the new ball right from the start, *but not on both sides.* At the outset, the opening bowler should pick the side on the ball with the smaller or lighter (less rough) embossment and continue to polish only that side during the course of the innings. The other (seam) side of the ball should be allowed to roughen during the course of play to aid the production of reverse swing. As shown above, the exact velocity at which reverse swing occurs, and how much negative side force is generated at a given speed above the critical, is a strong function of the ball's surface roughness. Once the seam side has roughened enough, reverse swing is simply obtained by *turning the ball over* so that the rough side faces the batsman. In general, the production of conventional and reverse swing will not be affected significantly by having a contrasting surface condition on the side facing away from the batsman. So a bowler bowling outswingers will still have the seam pointed towards the slips, but with the rough side facing the batsman, instead of the smooth for conventional swing, and the ball will now behave like an inswinger and swing into the batsman. The whole beauty (and success) of this phenomenon is that a bowler who could only bowl outswingers at the onset (with the new ball) can now bowl inswingers *without any change in the grip or bowling action.* Similarly, a predominantly inswing bowler can now bowl outswingers. Of course, if the contrast in surface roughness on the two sides of a ball is successfully created and maintained, the bowler becomes even more lethal since he can now bowl outswingers and inswingers at will by simply changing the ball orientation. Needless to say, this would make for a highly successful ability since there are not many bowlers who can make the new ball swing both ways using conventional bowling techniques. Moreover, the few that can bowl inswingers and outswingers are generally not equally effective with both types of swing and, of course, cannot do it with the same grip and bowling action. So the key to conventional swing bowling is keeping the nonseam side as smooth as possible, whereas for reverse swing the nonseam side needs to be as rough as possible.

One of the reasons why reverse swing has gained such notoriety is its constant link to accusations of ball tampering (Mehta 2006d, 2007a). The fact that bowlers started to illegally roughen the ball surface since the early 1980s is now well documented. Oslear & Bannister (1996) quote and show several examples and I have also personally examined several balls that were confiscated by umpires due to suspicions of ball tampering. The most popular forms of tampering consisted of gouging the surface and attempting to open up the quarter seam by using either fingernails or foreign objects such as bottle tops. It is rather ironic that a law prohibiting the rubbing of the ball on the ground was introduced in the same year that I first heard about reverse swing (1980). Also, in the following year (1981), the Test and County Cricket Board (TCCB) standardized the balls so that they now had smaller seams. Of course, the implications of these changes as they relate to reverse swing had yet to be realised.

There is another distinct advantage in maintaining a sharp contrast in surface roughness on the two sides or hemispheres of the ball. The primary seam plays a crucial role in both types of swing. It trips the laminar boundary layer into a turbulent state for conventional swing and thickens and weakens the turbulent boundary layer for reverse swing. During the course of play, the primary seam becomes worn and less pronounced and not much can be done about it unless illegal procedures are invoked to restore it, as discussed above. However, a ball with a worn seam can still be swung, as long as a sharp contrast in surface roughness exists between the two sides. In this case, the difference in roughness, rather than the seam, can be used to produce the asymmetric flow. The seam is oriented facing the batsman (straight down the pitch) at zero degrees incidence. The critical Re is lower for the rough side and so, in a certain Re range, the boundary layer on the rough side will become turbulent, while that on the smooth side remains laminar. The laminar boundary layer separates early compared to the turbulent boundary layer, in the same way as for conventional swing, and an asymmetric flow, and hence side force, is produced. The ball in this case will swing towards the rough side. At very high bowling speeds, the boundary layers on both surfaces will be turbulent and the ball will swing towards the smooth side, much like in the case of reverse swing. This type of swing, which tends to occur when the ball is older and a contrast in surface roughness has been established, is often erroneously referred to as reverse swing. In order to avoid this confusion and distinguish this type of swing from conventional and reverse swing, I gave it the name, "*contrast swing*" (Mehta 2006a).

The most exciting feature about contrast swing is that just about any bowler can implement it in practice. As most cricketers are aware, it is much easier to release the ball (spinning backwards along the seam) with the seam straight up, rather than angled towards the slips or fine leg. Thus, even mere mortals should be able to swing such a ball, and in either direction, since the bowling action is the same for both types of swing, the only difference being the orientation of the ball with regards to the rough and smooth sides. In fact, the medium pace "seam" or "stock" bowlers usually bowl with the seam in this orientation in an attempt to make the ball bounce on its seam so that it may gain sideways movement off the ground. With a contrast in surface roughness, these bowlers could suddenly turn into effective swing bowlers, without any additional effort, thus confusing not only the batsman, but perhaps themselves as well.

During the last three World Cups, there was a lot of discussion about the swing properties of the white ball used in the tournaments. The white ball was introduced since it was apparently easier to see, both for the players on the field and for television viewers. The main contention was that the white ball swung significantly more than the conventional red ball. Some players also contend that the white ball swings more after some use. According to the manufacturer of the white ball used in the 1999 tournament (British Cricket Balls, Ltd.), the only difference between the two balls is in the coating. With the conventional red ball, the leather is dyed red, greased and polished with a shellac topcoat. This final polish disappears very quickly during play and it is the grease in the leather that produces the shine when polished by the bowler. The finish applied to the white ball is slightly different. The leather is sprayed with a polyurethane white paint-like fluid and then heat-treated so that it bonds to the leather like a hard skin. As a final treatment, one coat of clear polyurethane-based topcoat is applied to further protect the white surface so that it does not get dirty easily. On inspecting the "Dukes" white ball used in the 1999 World Cup and the "Kookaburra" ball used in 2003 and 2007, it is quite apparent that the surfaces over the quarter seams are much smoother compared to those on a conventional red ball where the ridges created by the internal stitching can be clearly seen and felt. As a consequence, a new white ball should swing more, especially at the higher bowling speeds since a laminar boundary layer is more readily maintained on the smoother surface. Another consequence of the smoother surface is that reverse swing will occur at higher bowling speeds with a new white ball and later in the innings at more reasonable bowling speeds. It was apparent during all three World Cup tournaments that the ball became rough and dirty during the later stages of an innings and reverse swing was readily observed on several occasions. A more detailed discussion of the differences between the red and white balls is given in Mehta (2006d, 2007b).

3.5 Effects of Weather Conditions on Swing

The effect of weather on swing is by far the most discussed and most controversial topic in cricket, both on and off the field. It is quite fascinating that this topic was discussed in the very first scientific paper on cricket ball swing (Cooke 1955). The one bit of advice that cricket "Gurus" have consistently passed down over the years is that a humid or damp day is conducive to swing bowling. However, the correlation between humid conditions and swing has not always been obvious and most of the scientific explanations put forward have also been somewhat far-fetched. Of course, on a day when the ground is soft with green wet grass, the new ball will retain its shine for a longer time, thus helping to maintain a laminar boundary layer on the nonseam side. However, the real question is whether a given ball will swing *more* on a humid or damp day.

As shown in the previous sections, the flow regime over a cricket ball depends only on the properties of the air and the ball itself. The only properties of the air that may conceivably be influenced by a change in meteorological conditions are the dynamic viscosity and density. The dynamic viscosity and density both appear in the definition of Re, but small changes in Re are unlikely to affect the side force coefficient significantly [the side force coefficient (C_F) is defined as, $C_F = F/(\frac{1}{2}\rho U^2 A)$]. However, changes in air density can affect the side force directly since for a given side force coefficient, the side force is proportional to the density. However, it is rather ironic that humid or damp air is often referred to as constituting a heavy atmosphere by cricket commentators, when in fact, humid air is less dense than dry air.

A popular theory that has circulated for years, especially amongst the scientific community, is that the primary seam swells by absorbing moisture, thus making it a more efficient boundary layer trip. Bentley et al. (1982) investigated this possibility in detail. Profiles were measured across the primary seam on a new ball before and after a few minutes soaking in water. Even in this extreme example, there was no sign of any change in the seam dimensions (Fig. 24). A similar test on a used ball (where the varnish on the seam had worn-off) also showed no swelling of the seam. Rather than soaking the ball in water, a more controlled test was also conducted whereby a ball was left in a humidity chamber (relative humidity of 75%) for 48 hours. The projection test was performed on these balls with the surface dry, humid and wet and no increase in side force was noted for the humid or wet balls, as shown in Fig. 25.

Several investigators (Horlock 1973, Barton 1982, Sherwin & Sproston 1982, Wilkins 1992) have confirmed that no change was observed in the pressures or forces when the relative humidity of the air changed by up to 40%. It has been suggested that humid days are perhaps associated with general calmness in the air and thus less atmospheric turbulence (Sherwin & Sproston 1982, Wilkins 1992). On the other hand, Lyttleton (1957) and Horlock (1973) conjectured that humid conditions might result in increased atmospheric turbulence. However, there is no real evidence or basis for either of these scenarios, and even if it were the case, the turbulence scales (size of the turbulent eddies) would generally be too large to have any

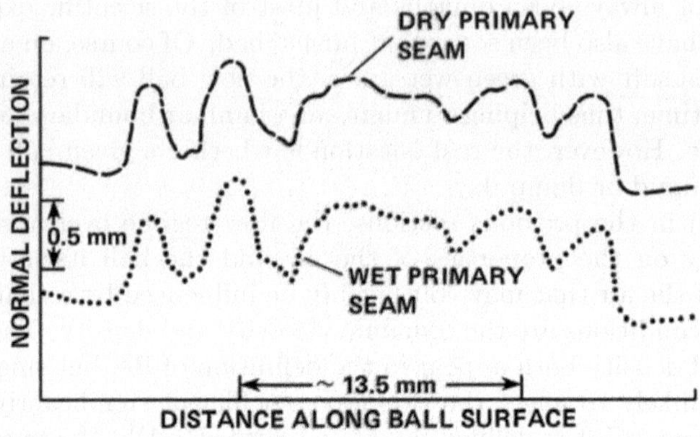

Figure 24. Surface contour plots of the primary seam on a cricket ball to investigate the effects of humidity (Mehta 1985, 2005).

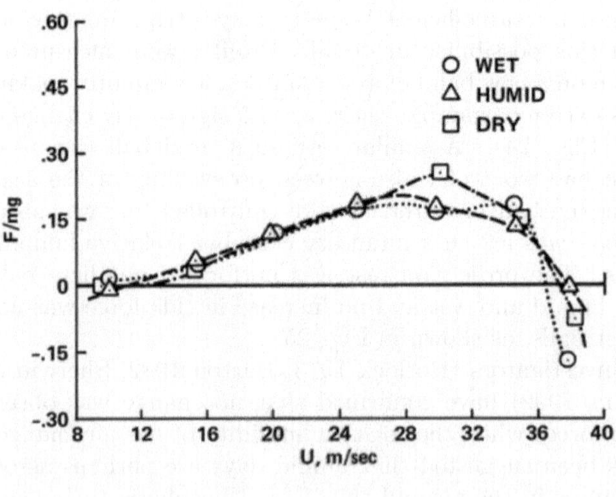

Figure 25. Effect of humidity on the measured side forces on a spinning cricket ball. Seam angle = 20°, spin rate = 5 revs/sec (Mehta 1985, 2005).

significant effect on the flow regime over the ball. Binnie (1976) suggested that the observed increase in swing under conditions of high humidity is caused by "condensation shock" which helps to cause transition. However, his calculations showed that this effect could only occur when the relative humidity was nearly 100%. Also, as shown by Bentley et al. (1982), the primary seam on almost all new cricket balls is already adequate in tripping the boundary layer in the Reynolds number range of interest.

There is only one published paper which claims that the positive effects of increased humidity on swing were observed in a wind tunnel test (Bowen 1995). Only two data points were presented which showed that the side force coefficient was higher and the drag coefficient lower for a relative humidity of 54% compared to those at 36%. However, the shapes of the curves and the proposed explanation, that humidity increases the surface roughness on the ball are both hard to believe. One would need to see a lot more evidence and better explanations before such an important result can be accepted for the first time. So there seems to be no (positive) scientific evidence which supports the view that humid conditions are more conducive to swing. The only viable explanation, which was first proposed by Bentley et al. (1982), is that humidity must affect the initial flight conditions of the ball. There is a possibility that the amount of spin imparted to the ball may be affected. The varnish painted on all new balls reacts with moisture to produce a somewhat tacky surface. The tacky surface would ensure a better grip and thus result in more spin as the ball rolls-off the fingers, and as shown above in Fig. 20, an increase in spin rate (at least up to about 11 revs/sec) certainly increases the side force. So, perhaps without actually realising it, the bowler may just be imparting more spin on a humid or damp day. There is one other possibility. Could it be that on a day which is supposedly conducive to swing bowling, the bowlers concentrate more on the optimum release for swing (seam angled and the ball spinning steadily along the seam without wobble) rather than trying to bowl too fast or trying to extract that extra bounce?

3.6 Conclusions

The basic flow physics of conventional swing and the parameters that affect it are now well established and understood. However, some confusion still remains over what reverse swing is, and how it can be achieved on a cricket field. A popular misconception is that when an old ball swings, it must be reverse swing. It is only reverse swing if the ball swings in a direction that is opposed to the one the seam is pointing in so that, for example, a ball released with the seam pointed towards the slip fielders swings *into* the batsman. While it is generally believed (with some justification) that tampering with the ball's surface helps in achieving reverse swing, the exact form of the advantage is still not generally understood. It is shown here that the critical bowling speed at which reverse swing can be achieved is lowered as the ball's surface roughness increases. One of the more important points to note is that ball tampering is not essential in order to achieve reverse swing. Reverse swing can be obtained with a brand new (red) four-piece ball, but only at bowling speeds of more than 36 m/s (80 mph). The whole beauty of reverse swing is that by simply changing the ball orientation, and nothing else, the ball will swing the "wrong" way. With a sharp contrast in surface roughness between the two sides of a cricket ball, "contrast swing" can be obtained with the seam oriented vertically and pointed straight down the pitch.

It is shown here how late swing is actually built into the flight path of a swinging cricket ball and it is this, rather than some special phenomenon, that is often observed on the cricket field. The question of the effects of humidity on cricket ball swing is still not totally resolved. While the effect is often observed on the cricket ground, there is not enough laboratory evidence to explain how the amount of swing may be increased in humid conditions. The introduction of the new white ball with its unique outer cover finish has started a new controversy on how its swing properties may differ from those of a conventional red ball. So while most of the mysteries surrounding cricket ball aerodynamics have now been resolved, there are still a few intriguing effects that will keep researchers busy and motivated for some time ahead.

4 TENNIS BALL AERODYNAMICS

4.1 Introduction

The game of tennis originated in France some time in the 12th century and was referred to as 'je de paume,' the game of the palm played with the bare hand. As early as the 12th century, a glove was used to protect the hand. Starting in the 16th century and continuing until the middle of the 18th century, rackets of various shapes and sizes were introduced. Around 1750, the present configuration of a lopsided head, thick gut and a longer handle emerged. The original game known as Real Tennis, was played on a stone surface surrounded by four high walls and covered by a sloping roof. The shape of the new racket enabled players to scoop balls out of the corners and also to put 'cut' or 'spin' on the ball. The rackets were usually made from hickory or ash and heavy sheep gut was used for the strings. The old way of stringing a racket consisted of looping the side strings round the main strings. This produced a rough and smooth effect in the strings and hence came the practice of calling 'rough' or 'smooth' to win the toss at the start of a tennis match. Only Royalty and the very wealthy played the game and the oldest surviving Real Tennis court, located at Hampton Court Palace, was built by King Henry VIII in about 1530. The present day game of Lawn Tennis was derived from Real Tennis in 1873 by a Welsh army officer, Major Walter Wingfield.

Balls used in the early days of Real Tennis were made of leather stuffed with wool or hair. They were hard enough to cause injury and even death. Starting in the 18th century, strips of wool were wound tightly around a nucleus made by rolling a number of strips into a small ball. Then string was tied in many directions around the ball and a white cloth covering sewn around the ball. The original Lawn tennis ball was made out of India rubber, made from a vulcanisation process invented by Charles Goodyear in the 1850s.

Balls approved for play under the Rules of Tennis must comply with International Tennis Federation (ITF, the world governing body for tennis) regulations covering size, bounce, deformation and colour. Ball performance characteristics are based on varying dynamic and aerodynamic properties. Tennis balls are classified as Type 1 (fast speed), Type 2 (medium speed), Type 3 (slow speed) and high altitude. Type 1 balls are intended for slow pace court surfaces such as clay. Type 2 balls, the traditional standard tennis ball, is meant for medium paced courts such as a hard court. Type 3 balls are intended for fast courts such as grass. High altitude balls are designed for play above 1219 meters (4000 feet). Tennis balls may be pressurised or pressureless. Today's pressurised ball design consists of a hollow rubber-compound core, with a slightly pressurized gas within and covered with a felt fabric cover. The hourglass 'seam' on the ball is a result of the adhesive drying during the curing process. Once removed from its pressurised container, the gases within the pressurised balls begin to leak through the core and fabric and the balls eventually lose bounce. Pressureless balls are filled with micro-cellular material. Subsequently, pressureless balls wear from play, but do not lose bounce through gas leakage. As a cost saving measure, pressureless balls are often recommended for individuals who play infrequently. The tennis ball must have a uniform outer surface consisting of a fabric cover and be white or yellow in colour. Any ball seams must be stitchless. All balls must weigh more than 56.0 grams (1.975 ounces) and less than 59.4 grams (2.095 ounces). Type 1 and Type 2 ball diameters must be between 6.541 cm (2.575 inches) and 6.858 cm (2.700 inches); Type 3 balls must be between 6.985cm (2.750 inches) and 7.302 cm (2.875 inches) in diameter.

It was in fact the flight of a tennis ball that first inspired scientists to think and write about sports ball aerodynamics (Newton 1676, Rayleigh 1877). Despite all this early attention, when the first review article on sports ball aerodynamics was published (Mehta 1985), no scientific studies on tennis balls could be identified and so tennis ball aerodynamics were not discussed. The first published study of tennis ball aerodynamics was that due to Stepanek (1988) who measured the lift and drag coefficients on a spinning tennis ball simulating the topspin lob. The aerodynamic forces were determined by projecting spinning tennis balls into a wind tunnel test section. Empirical correlations for the lift and drag coefficients (C_L and C_D) were derived in terms of the spin parameter (S) only; it was concluded that a Reynolds number dependence could be neglected. Stepanek measured values of between 0.55 and 0.75 for C_D, and between 0.075 and 0.275 for C_L, depending on the spin parameter (S), which was varied between about 0.05 and 0.6. The extrapolated C_D for the non-spinning case was found to be about 0.51. Some work on the aeromechanical and aerodynamic behavior of tennis balls was conducted at Cambridge University in the late 1990s (Cooke 2000, Brown and Cooke 2000). One of the more significant conclusions of these investigations was that the tennis ball would reach a quasi-steady aerodynamic state very soon after leaving the racket, in approximately 10 ball diameters, which is equivalent to only 3% of its trajectory (Cooke 2000). So the initial transient stage, when the ball is still deformed and the flow around it is still developing, will not generally make a significant contribution to the overall flight path. Based on comparisons with Achenbach's (1974) drag measurements on rough spheres, it was estimated that the critical Re for a tennis ball would be about 85,000, based on a "nap" or "fuzz" height of about 1 mm. It was therefore deduced that for Reynolds numbers normally encountered during a serve, $100{,}000 < Re < 200{,}000$ (corresponding to a serving velocity range of $26 < U < 46$ m/s or $57 < U < 104$ mph), the ball would be in the supercritical regime giving a drag coefficient of about 0.3 to 0.4. However, some recent measurements on non-spinning tennis balls (discussed below) showed that the drag coefficient was higher and appeared to be independent of Re.

Most of the recent research work on tennis ball aerodynamics was initially inspired by a decision made by the International Tennis Federation (ITF) in the 1990s to start field testing of a slightly larger oversized tennis ball, roughly 6.5% larger diameter (Fig. 26). This decision was instigated by a concern that the serving speed in (men's) tennis had increased to the point where the serve dominates the game. The fastest recorded serve was produced by Greg Ruzedski in March 1998 and it was measured at 66.6 m/s or 149 mph (Guinness 2000). The main evidence for the domination

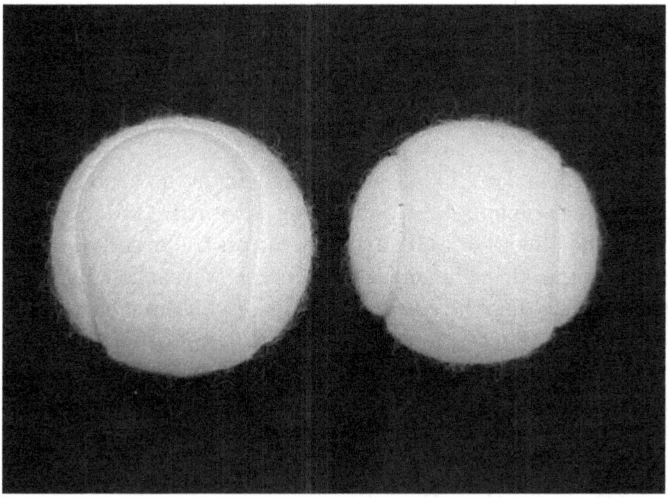

Figure 26. Comparison of a normal sized tennis ball (right) to the oversized ball (left) with a 6.5% larger diameter (photograph courtesy of NASA Ames Research Center and Cislunar Aerospace Inc.).

of the serve in men's tennis has been the increase in the number of sets decided by tie breaks at the major tournaments (Haake et al. 2000). This is particularly noticeable on the faster grass courts such as those used at Wimbledon.

Chadwick and Haake (2000a) obtained tennis ball C_D measurements using a force balance mounted in a wind tunnel. The initial measurements gave a C_D of about 0.52 for a standard tennis ball as well as the larger ball and it was found to be independent of Re over the range, $200,000 < \mathrm{Re} < 270,000$. Later, Chadwick and Haake (2000b) and Haake et al. (2000) reported that $C_D \approx 0.55$ over the same Re range for a standard tennis ball, a pressureless ball and a larger ball. The difference between the two reported levels is attributed to the technique used to measure the ball diameter (Cooke 2000). Chadwick and Haake (2000a) used an outer (projected) diameter, which included the nap or fuzz height. Their results also showed that the tennis ball C_D could be increased (by raising the fuzz) or decreased (by shaving off the fuzz) by up to 10% (Chadwick and Haake 2000a,b; Haake et al. 2000, Chadwick 2003).

Mehta and Pallis (2001b) conducted a detailed experimental study of tennis ball aerodynamics. Apart from flow visualization studies on non-spinning and spinning tennis balls, the drag coefficient on a variety of tennis balls, including oversized and worn balls, was measured over a wide range of *Re*. The critical role of the fuzz elements was identified and discussed in detail. A lot of the discussion of tennis ball aerodynamics in this section is based on that article and also on Pallis and Mehta (2003) and Mehta and Pallis (2004).

Goodwill and Haake (2004) and Goodwill et al (2004) measured the aerodynamic forces on spinning balls using a rotating balance mounted in a wind tunnel. The effects of ball surface wear on the aerodynamic properties were also investigated. The measured forces were then used to compute the ball trajectories for the various conditions. Alam et al. (2003, 2004, 2005a and 2007) conducted a series of similar experimental studies using a 6 component force sensor in a wind tunnel. The lift and drag forces were measured as a function of *Re* and S. The results from these two studies are presented below in Section 4.5.

4.2 Flow Visualization Results

The first tests were conducted with the 28 cm diameter tennis ball stationary (not spinning). The first observation at the lower test velocities was that the boundary layer over the top and bottom of the ball separated relatively early, near the apex at about 80 to 90 degrees from the front stagnation point (Fig. 27). This normally implies that the boundary layer is laminar (transition to a turbulent state has yet to occur). However, on increasing the wind tunnel velocity, even up to the maximum, resulting in *Re* = 284,000 and standard-sized ball velocity of 66 m/s (148 mph), no significant changes were observed. This was somewhat surprising since, at some point, the assumed laminar boundary layer was expected to undergo transition, which would be evidenced by a sudden rearward movement of the separation point, thus leading to a smaller wake and less drag. So a new conclusion was reached which presumed that transition had already occurred and that a turbulent boundary layer separation was obtained over the whole *Re* range tested. Although the felt was expected to affect the critical *Re* at which transition occurs, it seemed as though the felt was a more effective boundary layer trip than had been anticipated. Since the flow behavior, in particular the boundary layer separation location, in the flow visualization studies appeared to be independent of *Re*, the tennis ball was believed to be in this transcritical flow regime. The fact that the boundary layer separation over the top and bottom of the nonspinning ball was symmetric leading to a horizontal wake was, of course, anticipated since a side force (upward or downward) is not expected in this case. In one series, the ball orientation was altered to check for seam effects, in particular to see if the seam could generate an asymmetric boundary layer separation, and hence side force, on the ball. However, the boundary layer separation remained symmetric and unaffected by the seam orientation, even when the seam was in the vicinity of the separation location. So the flow visualization data suggested that the seam probably does not play a critical role in determining the aerodynamics of a tennis ball. The seam orientation was also investigated during the drag measurement phase of the investigation, as discussed below.

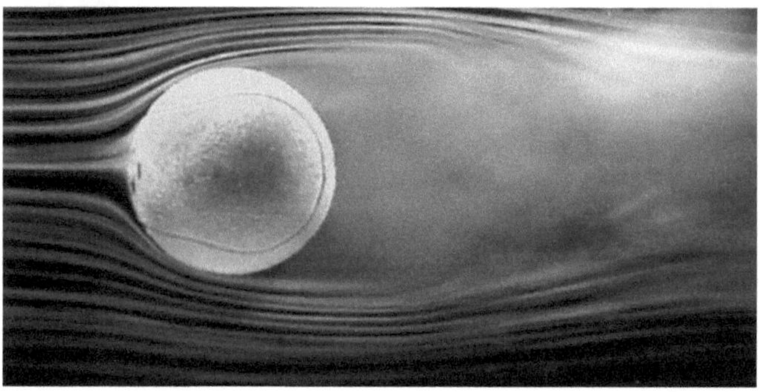

Figure 27. Flow visualization of 28 cm diameter tennis ball model with no spin (Re = 167,000). Flow is from left to right (Mehta and Pallis 2001b).

In the second round of testing, spin was imparted to the ball by rotating the support rod. In Fig. 28, the ball is spun in a counter-clockwise direction to simulate a ball with topspin while in Fig. 29 the ball is spun in a clockwise direction (underspin or backspin). In Figs. 28 and 29, the wind tunnel flow speed was 9 m/s and the model was spun at 4 revs/sec, which corresponds to a standard tennis ball velocity of 39 m/s (87 mph) and spin rate of 72 revs/sec (4320 rpm); this would represent a typical second serve in men's professional tennis. In Fig. 28, the boundary layer separates earlier at the top of the ball compared to the bottom. This results in an upward deflection of the wake behind the ball, and following Newton's 3rd Law of Motion, implies a downward (Magnus) force acting on it which would make it drop faster than a non-spinning ball. On the other hand, in Fig. 29 the wake is deflected downwards which means that the ball has an upward Magnus force acting on it that would make it travel further than a non-spinning ball. The fact that the Magnus force occurred in a direction that was expected (positive Magnus force) over the whole speed and spin ranges further suggested that transition had occurred and that the boundary layer over both sides (top and bottom) of the ball separated in a turbulent state.

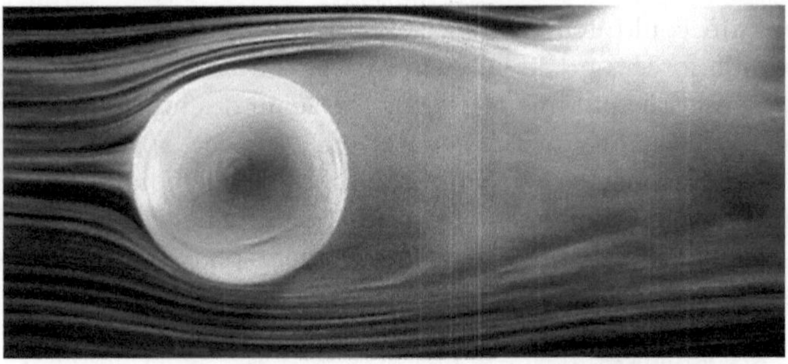

Figure 28. Flow visualization on ball with topspin (counter-clockwise rotation at 4 revs/sec, Re = 167,000). Flow is from left to right (Mehta and Pallis 2001b).

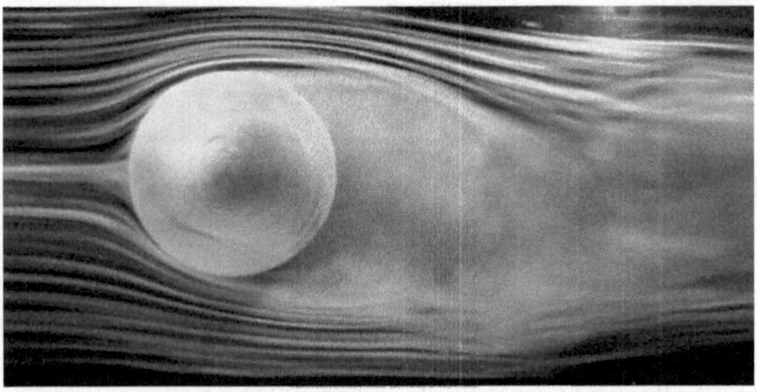

Figure 29. Flow visualization on ball with underspin (clockwise rotation at 4 revs/sec, Re = 167,000). Flow is from left to right (Mehta and Pallis 2001b).

4.3 Drag Measurements on Nonspinning Balls

In order to verify the accuracy of the whole measurement system, the first test was conducted on a smooth Plexiglas sphere with about the same diameter (d = 6.35 cm) as a standard tennis ball. The results (Fig. 30) compare very well with the classic data of Achenbach (1972, 1974b). A C_D of about 0.5 with a slight increase with Re initially, followed by a decrease at the higher Re as boundary layer transition is about to occur, are all evident in these measurements. Transition, evidenced by a sudden drop in C_D, occurs at a Re of about 400,000 for a smooth sphere, as seen in this figure. Premature transition can occur (at a lower Re) if the wind tunnel flow quality is inadequate (high turbulence levels) or if the model vibrates significantly; the fact that it does not occur here is again evidence of a suitable experimental set up. Achenbach's data for rough spheres show that the critical Re, and the amount by which the C_D drops, both decrease with increasing surface roughness. At the same time, the (constant) C_D level in the transcritical region increases with roughness for the roughness levels shown.

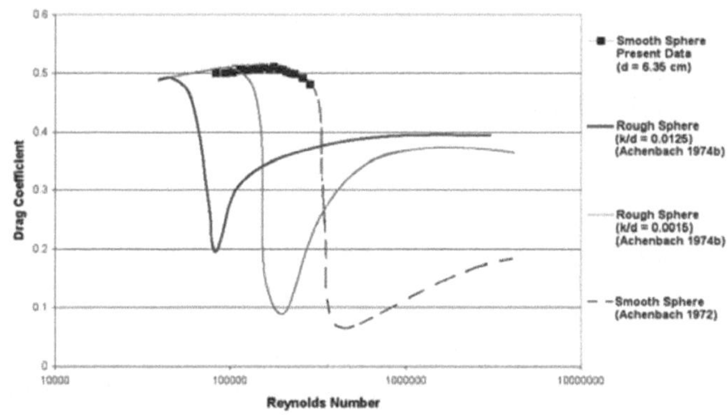

Figure 30. Drag Coefficient versus Reynolds Number for smooth and rough spheres (Mehta and Pallis 2001b).

The new (unused) tennis ball C_D data are presented in Fig. 31, together with the smooth sphere data and those for a "bald" ball (the inner rubber core of a tennis ball) over a Re range of about 80,000 to 300,000. In Fig. 31, the ball velocity quoted along the upper horizontal axis is that for a standard-sized tennis ball (d = 6.6 cm). The bald ball surface was quite smooth except for a small ridge along the equator where the two halves of the ball are joined together. This ball was mounted with the ridge perpendicular to the flow. Transition occurs at a Re of about 140,000, considerably lower than that for a smooth sphere since the ridge acts as a boundary layer trip. The C_D in the supercritical regime remains more or less constant up to about Re = 200,000. In this Re range, transition is still triggered by the ridge on the ball. However, beyond Re = 200,000 the C_D starts to increase again as transition moves upstream of the ridge and the separation

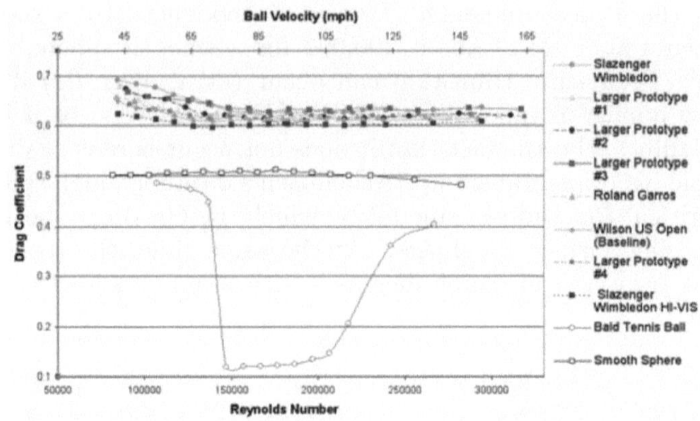

Figure 31. Drag Coefficient versus Reynolds Number for new tennis balls (Mehta and Pallis 2001b).

also moves towards the front of the ball. The rise in C_D is rather abrupt because the turbulent boundary layer is thickened by the ridge, thus making it more susceptible to separation. Beyond $Re \approx 280{,}000$, the ball would be in the transcritical regime and the C_D would be expected to level off at a constant value, perhaps slightly above 0.4, similar to Achenbach's (1974b) observations

The results for the new (unused) tennis balls show an approximately constant C_D of about 0.6 to 0.65 at the higher Re ($> 150,000$). In the lower Re range, $(85,000 < \mathrm{Re} < 150,000)$, it was rather surprising to see higher values of C_D, achieving values of up to 0.7. A Reynolds number dependence also develops in this range with the C_D decreasing with increasing Re. These data include measurements for some of the new oversized tennis balls. At the higher Re, the tennis ball is clearly in the transcritical regime, only at relatively low Re compared to the smooth and rough sphere data of Achenbach (1972, 1974b). So it appears as though the fabric cover on the ball is very effective at causing early transition and also in thickening the turbulent boundary layer. The C_D data for each ball presented in this figure were averaged over several (between 3 and 10) separate runs. The run to run repeatability was better on some balls than others. For example, the C_D repeatability on the baseline Wilson US Open ball was about $\pm$ 0.01. Although some of the variation was no doubt due to the unsteady nature of the flow field, because of vortex shedding in the wake, this ball to ball difference was somewhat of a mystery at first; this aspect of the results is further discussed below in the next section.

In one phase of the drag measurement testing, the effects of tennis ball seam orientation were investigated to verify the observations and conclusions drawn from the flow visualization studies. For most of the drag measurement tests, the tennis ball was mounted with the manufacturers marking in the horizontal plane and facing the oncoming flow. Two additional Wilson US Open balls were also tested with the marking facing to the side on one ball and facing downstream on the other. The drag measurements repeated to within the ± 0.01 tolerance quoted above, thus further indicating a lack of seam effects. Although ball seam orientation can affect the flight and trajectory of other sports balls, these effects were not observed on the tennis ball. Unlike cricket balls and baseballs (Mehta 1985), the seam on a tennis ball is indented and the cover surface is very rough, thus obscuring or overwhelming any seam effects.

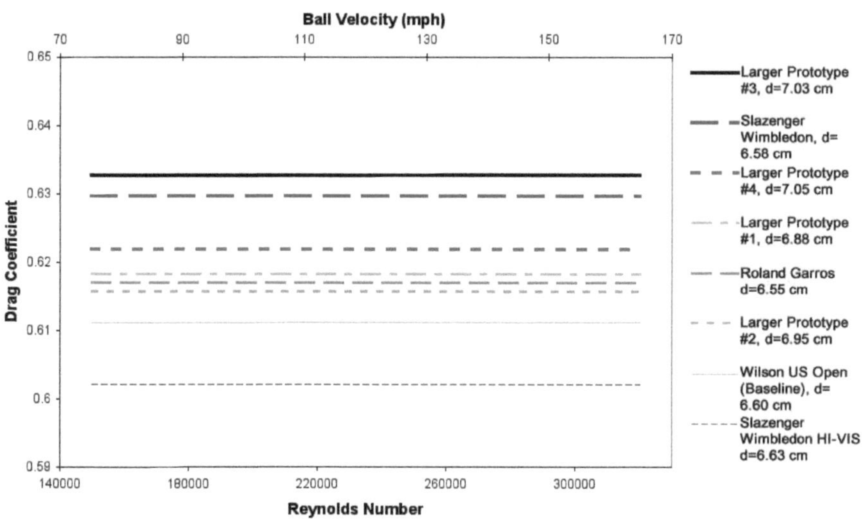

Figure 32. Drag Coefficient versus Reynolds Number for new tennis balls in transcritical regime (Mehta and Pallis 2001b).

For $Re > 150{,}000$, the data in the transcritical regime for each ball are averaged to give a single value for the C_D and these data are presented in Fig. 32. The lowest C_D is that for the Slazenger Wimbledon Hi-Vis at just above 0.6 while the highest C_D is that for one of the prototype oversized balls at just above 0.63, with all the other balls tested scattered in between. So basically the C_D for all the new balls tested, including the larger balls, falls within about 5% of each other. Furthermore, given the repeatability quoted above, not too much should be made of the differences in C_D between the balls tested. The important point to note is that the C_D values for the larger balls are comparable to those for the regular balls. Of course, this is not all that surprising since a simple scaling of the size should not affect the C_D, as long as other parameters, such as the surface characteristics (e.g. the fuzz), are not altered significantly. As a result, the drag on the oversized ball will increase by an amount proportional to the cross-sectional area, and the desired effect of increasing the flight time for a given serve velocity will be attained.

4.4 Drag Coefficient: Further Discussion

Right from the time the very first flow visualization tests were conducted, the indications were that the fabric cover on the tennis ball was very effective at triggering relatively early transition of the laminar boundary layer and also at rapid thickening of the turbulent boundary layer beyond transition. The fact that the flow pattern, in particular the separation location, was not affected by increasing *Re*, was the first indication that the ball was in the transcritical regime with turbulent boundary layer separation. In the transcritical regime, boundary layer transition occurs right next to the stagnation point and so a limit is reached whereby the transition and separation points cannot move further upstream with increasing *Re*.

The main point of concern was the relatively high values of the measured C_D for the tennis balls. The turbulent boundary layer separation location for the tennis ball appeared to be comparable to that seen for laminar separation at relatively low *Re*. Now, the total drag on a bluff body, such as a sphere or tennis ball, consists mainly of pressure drag (due to the pressure difference between the front and back of the ball) and a very small contribution due to viscous or skin friction drag (due to the no slip condition). Achenbach (1972) showed that the viscous drag for a smooth sphere approaching the transcritical regime was about 2% of the total drag. So the bulk of the total drag is accounted for by pressure drag, which in turn is determined solely by the boundary layer separation location on the ball. Since the separation location for the tennis ball is comparable to that for laminar separation at low *Re*, one would expect the C_D for the tennis ball to be around 0.5. However, the present measurements give a C_D of about 0.62 for the tested (new or unused) tennis balls. The previously measured C_D values of 0.51 (Stepanek 1988) and 0.55 (Chadwick and Haake, 2000b and Haake et al., 2000) are also above the 0.5 level. Some of the difference in the measured C_D values between the different investigations can be attributed to the different techniques used to measure the ball diameter. The main task in the present investigation was to determine why the tennis ball C_D was higher than 0.5 and where the additional drag contribution was coming from.

As explained above, in the supercritical regime, once transition has occurred, the transition and separation locations start to creep upstream and so the C_D starts to increase. At some point the transition location moves all the way up to the stagnation location and the separation location is then totally determined by the development of the turbulent boundary layer. With increasing roughness, the boundary layer growth rate is increased, thus resulting in earlier separation and higher C_D. The constant level achieved by the C_D in the transcritical regime is also expected to increase with increasing roughness, as evidenced in Achenbach's (1974b) measurements (data for two roughness levels are shown in Fig. 30). However, Achenbach's data show an upper limit of $C_D \approx 0.4$ on spheres with increasing roughness (Figs. 2 and 4 in Achenbach 1974b show this limit for a k/d range of 0.0025 to 0.0125). The measured separation location for this value of C_D was about $\theta_s \approx 100°$. This is still in the region of the adverse pressure gradient and so one would expect the boundary layer separation location to continue moving upstream with increasing surface roughness. However, the point to note is that while the boundary layer growth (rate of thickening) increases with increasing roughness, so does its skin friction coefficient and the behavior of the separation location is then determined by the behavior of these competing effects. The increasing skin friction coefficient makes the boundary layer more resilient to separation, thus opposing the tendency of a boundary layer to separate as it thickens. So it is entirely possible that for certain types of roughness, such as the round glass beads investigated by Achenbach for example, a limit is reached for the C_D level in the transcritical regime because the effects of the boundary layer thickening are offset by those due to the increasing skin friction coefficient.

In principle, though, there is no reason why the separation location cannot continue to creep forward for other types of roughness elements, which may be more effective at thickening the boundary layer than increasing the skin friction coefficient. It is proposed here that the absolute limit for the turbulent boundary layer separation location in the transcritical regime is the same as that for laminar boundary layer separation in the subcritical regime ($\theta_s \approx 80°$). Laminar boundary layer separation occurs upstream of the sphere apex because of the presence of an adverse pressure gradient in this region (see Fig. 7b in Achenbach 1972). The adverse pressure gradient is generated in this region due to an upstream influence of the separated near wake. One effect which occurs is that initially, when the flow is first turned on, the laminar boundary layer separates at the apex and immediately a pressure minimum is generated upstream of it due to streamline curvature effects, much in the same way as that generated near the exit region of a contraction (Bell and Mehta 1988). Once this adverse pressure gradient is generated, the laminar boundary layer separation will tend to move to that location. This is probably the most upstream location that the adverse pressure gradient can move up to. Assuming that a very thick (weak) turbulent boundary layer can become as prone to separation as a laminar layer, then it will separate as soon as it encounters an adverse pressure gradient (at about $\theta_s \approx 80°$), just like the laminar layer. So if the location for turbulent separation in the transcritical regime is similar to that of laminar separation in the subcritical regime, then the pressure drag should also be comparable, thus giving a total drag of $C_D \approx 0.5$.

On examining the tennis ball, the relatively rough surface on the felt is readily apparent. The roughness actually results from the junctions of the fuzz elements, where they are embedded within the fabric covering on the ball. However, in addition, the fuzz elements have a finite thickness and length and this forms an additional porous coating on the ball through which air can still flow. So the tennis ball can be thought of as a very rough sphere with a porous coating. Subsequently, each fuzz element will also experience pressure drag and when this is summed up for all the fuzz elements on the ball's surface, one obtains the additional drag contribution and this is herein termed the "fuzz drag." So the present data suggest that the contribution of the fuzz drag to the total drag on the tennis ball is between 20% and 40%, depending on the *Re*.

The other trend in the tennis ball C_D measurements, which was initially puzzling, was the higher values of C_D at the lower *Re* (Fig. 31). At first it was tempting to discard the trend by attributing it to experimental error since both, the tunnel reference pressure and drag force (drag count), become harder to measure accurately as the wind tunnel flow speed is reduced (the percentage error increases as the mean values are lower). However, in this particular case, the smooth sphere data in the same *Re* range look consistent and agree very well with previous measurements, as shown in Fig. 30. Moreover, compared to the smooth sphere, the overall drag-count error for the tennis balls would be lower since the drag is higher. All past measurements for rough spheres had shown that the approach to a constant C_D level in the transcritical regime was from below, from a lower, critical value (C_D increases with *Re* in the supercritical and early part of the transcritical regimes until a constant level is attained). Once again the fuzz elements and their behavior with varying *Re* were identified as the only likely causes for the observed effect.

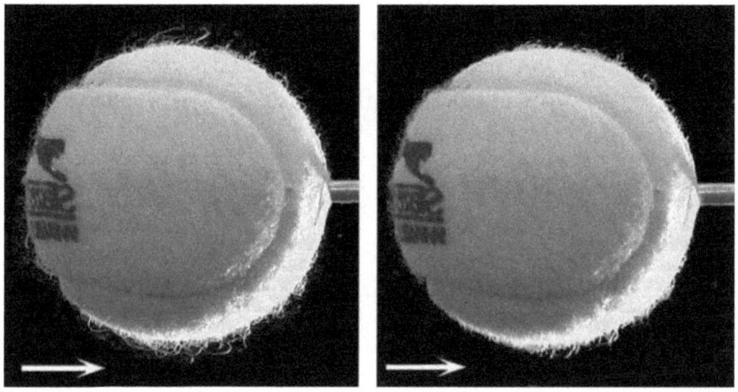

Figure 33. Effect of flow velocity on fuzz element orientation. Left-hand photo, $U = 20$ m/s (45 mph, $Re = 100{,}000$). Right-hand photo, $U = 60$ m/s (135 mph, $Re = 260{,}000$). Flow is from left to right (Mehta and Pallis 2001b).

The first effect, which is perhaps not too surprising, is the change in orientation of some of the filaments. As the flow velocity is increased, many of the filaments that are initially standing almost perpendicular to the ball's surface are forced to lay down due to aerodynamic drag effects. Note how in Fig. 33, the fuzz filaments, particularly over the front face of the ball, and up to the apex region, tend to lay down at the higher flow speed. Hence the contribution of the fuzz drag is reduced at the higher flow speeds or *Re*. Also, the fuzz element *Re* (based on filament diameter) is estimated to be of order 20, and this puts it in a range where the C_D (for a circular cylinder) is much higher ($C_D \approx 3$) and a strong function of the *Re*, with the C_D decreasing with increasing *Re* (see Fig. 10.12 in Smits 2000). So the higher C_D level at the lower ball *Re* investigated is attributed to the combined effect of fuzz filament orientation and *Re* effects on the individual filaments.

All the new (unused) tennis balls exhibited this trend of higher C_D at the lower tested *Re* (Fig. 31), but not to the same extent. The data for the two balls, which showed the maximum and minimum rise in C_D at the lower *Re*, are shown in Fig. 34. Ironically, they are both Slazenger Wimbledon balls, but it was noted while testing that the fuzz filaments on the Hi-Vis ball were not affected as much (did not lay down as much) at the higher *Re* as the other Slazenger ball. Also, the drag measurement repeatability on the Hi-Vis ball was much better than on any other ball. Apparently, the only difference between the two Slazenger balls was in the felt dyeing process and it is possible that this resulted in stiffer fuzz filaments on the Hi-Vis ball. The stiffer filaments led to a lower dependence on Re, since the filament orientation is not affected as much by Re, and also to better repeatability since the filaments are affected less by operational procedures, such as handling and storing.

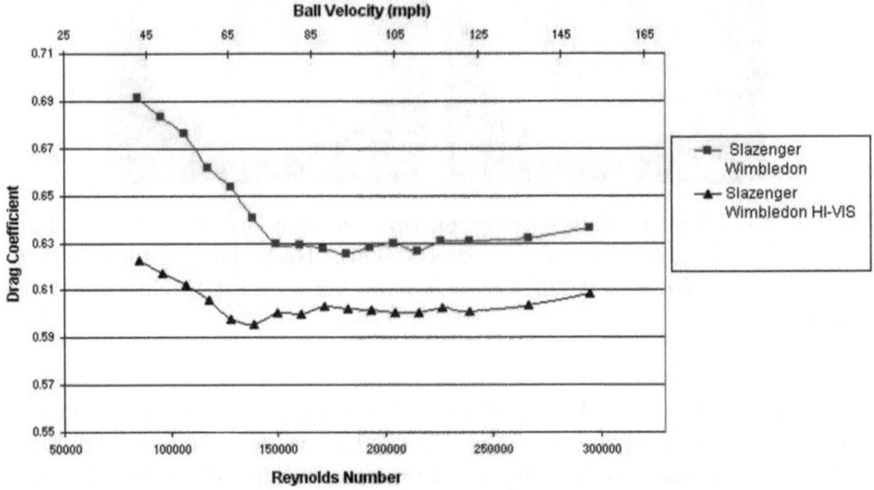

Figure 34. Drag Coefficient versus Reynolds Number for new Slazenger balls (Mehta and Pallis 2001b).

The critical role of the fuzz in determining the tennis ball drag was borne out most succinctly in the results for the used balls (Fig. 35). One of the balls was used in the 1997 US Open for nine games and the other used balls were played with by recreational players for the noted number of games, using only two balls at a time. The baseline data for the new Wilson US Open ball are also shown for reference. The 1997 US Open ball seems to indicate a supercritical behavior with the critical Re of about 100,000 and then a gradual approach towards the transcritical regime. For the balls used by the recreational players, after three games the C_D behavior is comparable to that of the new ball. However, after six games the C_D is clearly higher and to confirm this increase, both the used balls were tested and they show a very consistent trend with excellent repeatability. This initial increase in C_D is supposedly known to tennis players who often refer to the felt as having "fluffed-up" (fuzz is raised), which would obviously account for the higher drag. The fluffing-up was not apparent when these two balls were examined

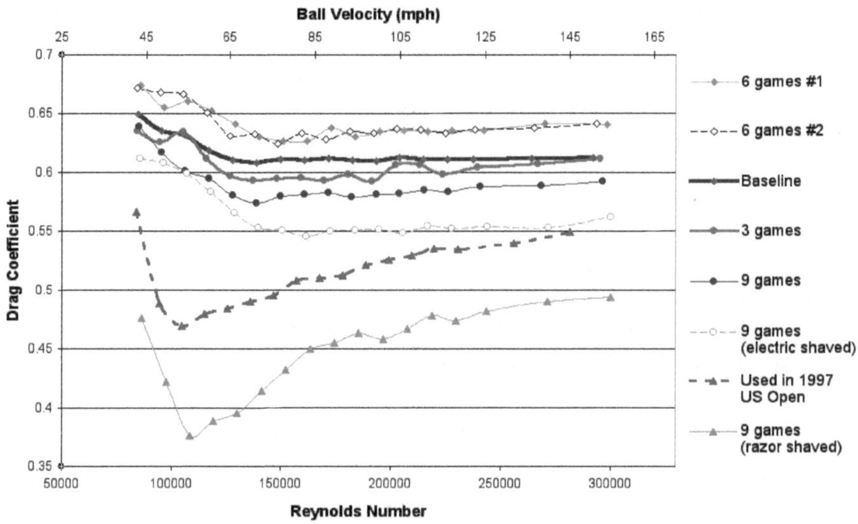

Figure 35. Drag Coefficient versus Reynolds Number for used Wilson US Open balls (Mehta and Pallis 2001b).

visually, but the present C_D measurements strongly suggest that the felt texture, perhaps the internal structure, must have been affected. The nine game ball clearly exhibited a worn felt and so it was not too surprising to see a lower measured C_D. The fact that a fluffed-up felt results in an increased C_D and a worn felt in a reduced C_D was not too surprising since the effect had been previously observed by Chadwick (Chadwick and Haake 2000a, Chadwick and Haake 2000b, Haake et al. 2000).

The C_D for the nine game ball was still a lot higher and the behavior with *Re* quite different compared to that of the 1997 US Open ball. This is not too surprising since the recreational players are obviously not expected to hit the ball as hard or have as many rallies as the professionals at the US Open. In order to try and replicate the 1997 US Open ball data, it was decided to accelerate the wear on the nine game ball by shaving-off the remaining fuzz. An electric razor was used at first, which removed some of the fuzz and the C_D in the transcritical regime was reduced as a result. However, the electric razor also tended to fluff-up the felt, since the filaments often caught in the razor when it was lifted off the surface, and the higher rise in C_D at the lower *Re* is attributed to this effect. As a final effort, a razor blade was used to shaveoff as much of the fuzz as possible. Note that the even this level of shaving does not remove the fuzz and associated roughness completely; the surface is similar to that on a man's face with a about a two day stubble. This ball clearly shows a C_D behavior that is comparable to that of the 1997 US Open ball, with about the same critical Re and supercritical behavior. However, the C_D levels are lower throughout the Re range and this is because there is no contribution due to the fuzz drag. The effective surface roughness on this shaved ball and the 1997 US Open ball is probably comparable, which explains the similar value of the critical Re. Also, as is hypothesized above, the maximum C_D in the transcritical regime on a very rough sphere, with no contribution from the roughness element drag (fuzz drag), should be 0.5 and that is exactly the level this shaved ball C_D seems to be asymptoting towards.

4.5 Aerodynamic Force Measurements on Spinning Tennis Balls

In tennis, apart from the flat serve where there is zero or very little spin imparted to the ball, almost all other shots involve the ball rotating about some axis. In addition to Stepanek's (1988) earlier work, the aerodynamics of spinning tennis balls were recently studied by several research groups. In this case, apart from the drag and gravitational forces, the lift (or side) force also come into play since a Magnus force is generated due to the spin.

Goodwill and Haake (2004) and Goodwill et al. (2004) supported the balls on a rotating force balance mounted in a 305 × 355 mm wind tunnel with a slotted test section. The ball was supported using two horizontal stings and rotated at a constant rate while measuring the drag and lift forces. In addition to new balls, the effects of wear on the aerodynamic properties of the ball were also investigated. The wear was simulated using a "wear rig" located at the International Tennis Federation (ITF) laboratory in London. They arbitrarily chose for the balls to undergo 0, 60, 500, 1000 and 1500 impacts, which corresponds to between 2 and 50 games if only one ball is used in the game.

In order to verify their overall measurement system, the drag coefficient on a nonspinning smooth sphere with a diameter of 66 mm (about the same diameter as a standard tennis ball) was measured first. As shown in Fig. 36, they obtained an average C_D of about 0.54, about 5% higher than that measured by Achenbach (1972) and Mehta and Pallis (2001b); see Fig. 30. The C_D results for the new tennis ball are comparable to those shown in Fig. 31. The data in Fig. 36 show the C_D decreasing with increasing wear on the balls. The C_D drops by about 0.04 for the ball that has undergone 1500 impacts (50 games). Mehta and Pallis (2001b) measured a much larger drop for the ball used in the US Open, as shown in Fig. 35, but this is not

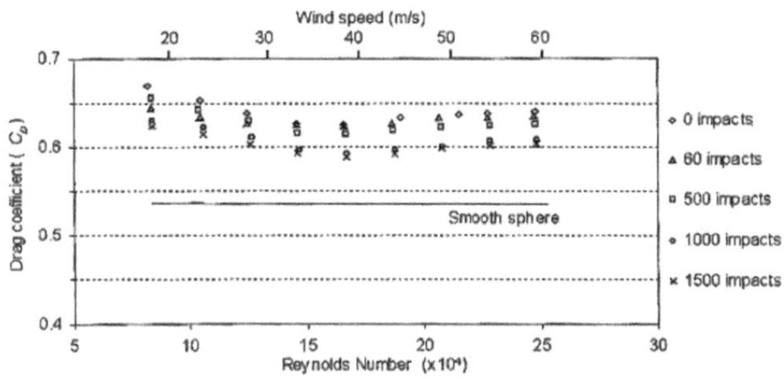

Figure 36. Drag Coefficient of ten worn balls (two of each category) versus Reynolds number (Goodwill and Haake 2004).

too surprising since the contact surface in the two cases and the amount of wear are not likely to be very similar. The really interesting point to note in the data presented in Fig. 36 is that the trend of higher C_D at the lower Re, seen by Mehta and Pallis (2001b) is repeated here (compare Figs. 31 and 36). The additional interest here is that the trend is maintained for the worn balls.

The C_D for the spinning balls are shown in Figs. 37a and b as a function of the spin parameter (S) for Re = 105,000 and 210,000, respectively. For the balls subjected to 0 and 60 impacts, the C_D increases with S, presumably due to the fuzz elements "standing up" when the ball is rotated (Chadwick, 2003). Also, note that with lift generated on spinning balls, there will also be a contribution of induced drag. The lower C_D on the worn balls is still evident with the maximum difference apparent at S = 0.15 with the new ball C_D = 0.67 versus 0.61 for one with 1500 impacts. For the higher Re of 210,000, the C_D for the new ball is about 0.03-0.04 higher than that of the heavily worn ball for all values of S. The data for the lift coefficient, C_L, are shown Figs. 38a and b, again for Re = 105,000 and 210,000, respectively. In general, the C_L increases with S for all the balls, as would be expected, with almost linear relations at both Re. For the lower Re, there is some effect of wear on the C_L, especially at the lower values of S, but this trend is not repeated at the higher Re. In general, there does not appear to be any strong effect of wear on the ball lift coefficient.

Goodwill and Haake (2004) used the measured aerodynamic forces to compute the ball trajectories for the various balls. They assumed that the ball left the racket at a speed of 50 m/s, 4.25° below the horizontal and 2.7 m above the ground. The spin was assumed to be constant at 32

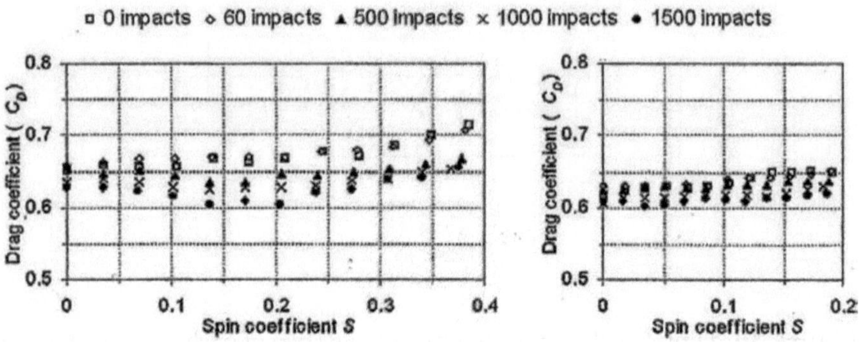

Figure 37. Drag Coefficient for spinning balls. (a) U = 25 m/s (Re= 105,000); (b) U = 50 m/s (Re=210,000) (Goodwill and Haake 2004).

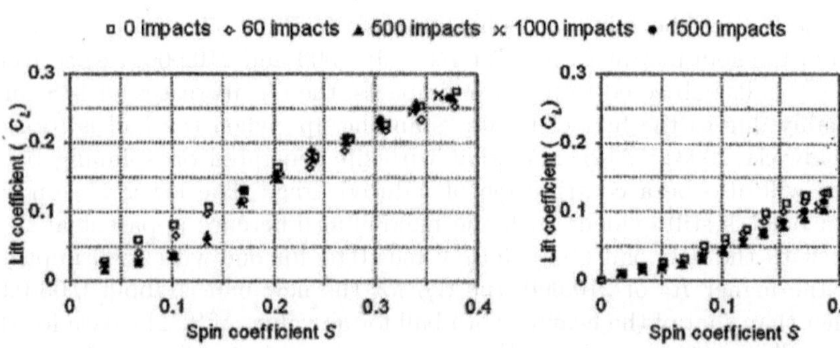

Figure 38. Lift Coefficient for spinning balls. (a) U = 25 m/s (Re= 105,000); (b) U = 50 m/s (Re=210,000) (Goodwill and Haake 2004).

revs/sec (1920 rpm). Since the ball speed, and hence Re, decrease during flight, the C_L and C_D were linearly interpolated using their measured data. The computed trajectories are shown in Fig. 39 which also includes the trajectory for an oversized ball (used the same C_L and C_D values as for a standard ball). It is clearly seen that for a given elapsed time (0.7 seconds), the worn ball has traveled 400 mm further than the standard new ball and the oversized ball is about 700 mm behind the standard ball. This amounts to about 2.5% more reaction time for the oversized ball, while the worn ball gives about 1.5% less time. It appears that the players obsession with new balls needs to be revisited, especially if they happen to have a relatively fast serve.

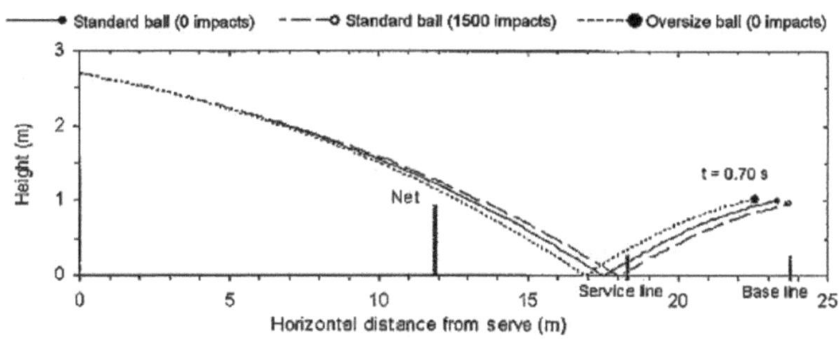

Figure 39. Predicted trajectory for new and worn standard size balls and an oversize ball (Goodwill and Haake 2004).

Alam et al. (2003, 2004, 2005a and 2007) conducted a series of experimental studies using a 6 component force sensor in a wind tunnel with a test section 3 m wide, 2 m high and 9 m long. They reported that the average drag coefficient for non-spinning new tennis balls varied between 0.55 and 0.65 (Fig. 40). These values were slightly higher compared to some previous studies (Stepanek 1988, Haake et al. 2000, Chadwick and Haake et al. 2000a). However, they agreed extremely well with measurements reported by Mehta and Pallis (2001b). For the spinning tests, twelve balls were used over a range of Reynolds numbers (46,000 and 161,000; speeds of 11 m/s to 39 m/s) at spin rates of 500 rpm to 3000 rpm (8.33 rev/s to 50 rev/s). The results indicated that with an increase of spin rate, the lift coefficient or down force coefficient depending on top spin or back spin increased (Fig. 41). However, with an increase of Reynolds numbers, the lift force coefficient decreased. The reduction of lift force coefficients at high Reynolds numbers (over 33 m/s) was minimal. The studies also found that with an increase of

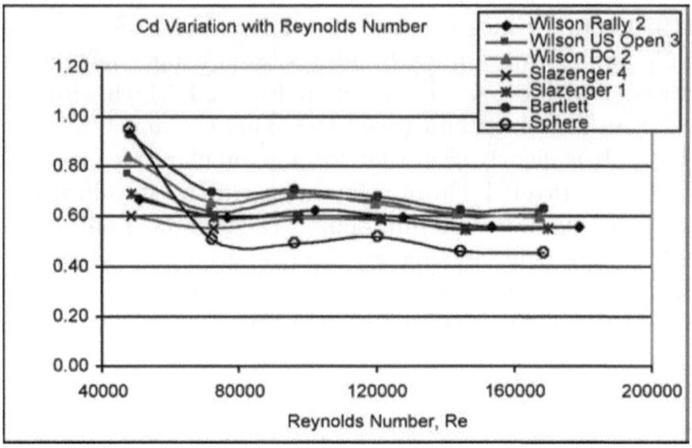

Figure 40. Drag coefficients as function of Reynolds number for a series of new tennis balls (Alam et al. 2004).

spin rate, the drag coefficient also increased. The average C_D value varied between 0.6 and 0.8 which was somewhat higher compared to non-spinning balls. One of the reasons for higher drag coefficients of a tennis ball when spun is believed to be the effect of the fuzz elements. A close visual inspection of each ball after the spin revealed that the hairy stuff (fuzz) comes outward from the surface (but not uprooted) and the surface becomes very rough. As a result, it is believed that the fuzz element generates additional drag. However, as Reynolds number increases, the rough surface (fuzz elements) becomes streamlined and reduces the drag. Of course, it is worth noting that a spinning ball will develop a lift or side force and so there will be a contribution from induced drag to the overall drag force as well.

4.6 Conclusions

The fact that the flow over a new tennis ball was essentially in the transcritical regime, where the separation location does not move significantly with *Re*, was clearly suggested by the flow visualization results. This in turn implies that the C_D is independent of *Re* since the total drag on a bluff body, such as a round ball, is almost totally accounted for by the pressure drag and this was confirmed by the drag measurements. The fuzz on a tennis ball is very effective at causing very early transition of the laminar boundary layer and rapid thickening of the turbulent boundary layer. This results in the separation location moving up to the apex region, comparable

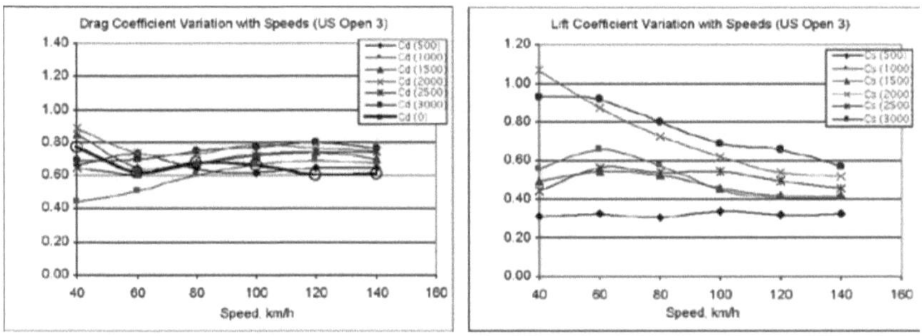

Figure 41. Drag and lift coefficients as a function of speed and spin rate for Wilson US Open 3 (Alam et al. 2007).

to that for laminar boundary layer separation at subcritical Re. In this case, the total drag for the tennis ball would be expected to be about the same as that for a ball in the subcritical regime, thus giving $C_D \approx 0.5$. However, all the previous and present measurements indicate that the C_D for a new (unused) tennis ball is higher than 0.5, the exact value depending to some extent on the measurement and definition of the ball diameter. C_D values of between 0.6 and 0.7 were measured in the present investigation for new tennis balls in the Re range, $80{,}000 < Re < 300{,}000$. So there must be another drag contribution and it is proposed here that the additional drag is due to the pressure drag on the individual fuzz elements. This additional contribution is herein defined as the "fuzz drag" and the present measurements show that, depending on the ball Re, it can account for 20% to 40% of the total drag on new tennis balls. The higher levels of C_D at the lower Re ($80{,}000 < Re < 150{,}000$) are attributed to the dependence of fuzz element orientation on flow (or ball) velocity and the stronger dependence of C_D on Re at the very low fuzz element Re. The recently approved oversized tennis ball C_D is comparable to that for the standard-sized balls. However, the drag on the oversized balls is higher by virtue of the larger cross-sectional area and so the desired effect of "slowing down the game" (increased tennis ball flight time) will be achieved.

For the spinning tennis balls, the lift coefficient increases with the spin parameter (S), as expected and the drag coefficient also increases due to the contribution from the induced drag. Some dependence on Re has been noted, especially at the relatively low Re.

The most revealing results from this investigation are those for the used balls. The C_D increases at first, presumably due to the ball fuzz "fluffing-up," but it decreases with further use as the felt becomes worn and the fuzz starts coming off. In addition, the C_D of the tennis ball with worn felt becomes a function of Re and the critical Re is identifiable. It is interesting to note that a ball used in a major tournament (1997 US Open) can achieve this state of lower C_D, which should make it more attractive for the faster servers in tennis. Of course, tennis players are also concerned about other attributes of the ball, such as "liveliness" and "controllability" and they therefore generally prefer to serve with new balls. For the ball with the razor-shaved fuzz, the critical Re is about 100,000 and the maximum C_D level in the transcritical regime approaches 0.5, as predicted by the present analysis. In view of the present findings, it is particularly intriguing to note that the current rules of tennis do not stipulate any specifications regarding the fuzz. The rule states: "The ball shall have a uniform outer surface consisting of a fabric cover and shall be white or yellow in colour."

5 GOLF BALL AERODYNAMICS

5.1 Introduction

The early golf ball in the 1400s, known as a "featherie," was simply a leather pouch filled with goose feathers. In order to obtain a hard ball, the pouch was filled while wet with wet goose feathers. Since people believed a smooth sphere would result in less drag (and thus fly farther), the pouch was stitched inside out. Once the pouch was filled, it was stitched shut, thus leaving a few stitches on the outside of the ball. The ball was then dried, oiled, and painted white. The typical drive with this type of ball was about 135 to 160 meters. Once this ball became wet, it was totally useless.

In 1845, the "gutta-percha" ball was introduced. This ball was made from the gum of the Malaysian Sapodilla tree. This gum was heated and molded into a sphere. This resulted in a very smooth surface. The typical drive with the gutta-percha ball was shorter than that obtained with the featherie. However, according to golf legend, a professor at Saint Andrews University in Scotland soon discovered that the ball flew farther if the surface was scored or marked (Chase, 1981). This led to a variety of surface designs which were chosen more or less by intuition. By 1930, the golf ball with round dimples in regular rows was accepted as the standard design. The modern golf ball generally consists of two, three or even four pieces or layers. The multi-piece ball inner core is generally made of solid or liquid-filled rubber. The inner core is wrapped with rubber thread on wound balls or nowadays usually replaced by a solid polybutadiene layer. The outer dimpled coating or skin is generally made out of an ionomer (Surlyn), natural or synthetic Balata or polyurethane. The dimples come in various shapes, sizes and patterns. The number of dimples has varied between 250 and 800 dimples with shapes varying from round, square and rectangular to hexagonal. Their arrangement on the surface has also seen many changes. Perhaps the most popular is one based on the icosahedral design, where the ball surface is divided into segments defined by 20 equilateral triangular patches. A modern day ball is likely to have about 380 dimples of up to six different sizes and covering over 75% of the balls surface. The typical drive with a modern golf ball is about 230 to 275 meters.

The rules of golf state that: "Foreign material must not be applied to a ball for the purpose of changing its playing characteristics. A ball is unfit for play if it is visibly cut, cracked or out of shape. A ball is not unfit for play solely because mud or other materials adhere to it, its surface is scratched or scraped or its paint is damaged or discolored. The weight of the ball shall not be greater than 45.93 gm avoirdupois (1.620 ounces). The diameter of the ball shall not be less than 42.67 mm (1.680 inches). This specification will be satisfied if, under its own weight, a ball falls through a 42.67 mm diameter ring gauge in fewer than 25 out of 100 randomly selected positions, the test being carried out at a temperature of 23 $\pm 1°$C. The ball must not be designed, manufactured or intentionally modified to have properties which differ from those of a spherically symmetrical ball."

5.2 Aerodynamic Forces on a Golf Ball

The two components of the total aerodynamic force that act on a golf ball flying through the air are the lift and drag. The lift force (L) acts in a direction normal to the direction of motion, whereas the drag force (D) acts along the direction of motion. The other non-dimensional groups that are relevant to golf ball aerodynamics are the Reynolds number (Re), the spin parameter (S) and the roughness parameter (ϵ).

Perhaps the most popular question in golf science is: 'Why does a golf ball have dimples?' The answer to this question can be found by looking at the aerodynamic drag on a sphere. As discussed above, most of the total drag on a bluff body such as a golf ball is due to pressure drag with a very small contribution from the skin friction drag. The pressure drag is minimized when the boundary layer on the ball is turbulent and separation is delayed, thus leading to a smaller wake and less drag. Transition of the laminar boundary layer on a ball can be achieved at relatively low Re by introducing surface roughness. This is why the professor in Scotland experienced a longer drive with the marked ball.

So, why dimples? The critical Reynolds number, Re_{cr}, holds the answer to this question. Re_{cr} is the Reynolds number at which the boundary layer transitions from a laminar to a turbulent state. For a smooth sphere, Re_{cr}

is much larger ($\approx 4 \times 10^5$) than the average Reynolds number experienced by a golf ball. For a sand roughened golf ball or a one with a bramble surface, the reduction in drag at Re_{cr} is greater than that of the dimpled golf ball. However, as the Reynolds number continues to increase beyond the critical value, the drag increases. The dimpled ball, on the other hand, has a lower Re_{cr} ($\approx 0.6 \times 10^5$), and the drag is fairly constant for Reynolds numbers greater than Re_{cr} (see Fig. 42). Therefore, the dimples cause Re_{cr} to decrease which implies that the flow becomes turbulent at a lower velocity than on a smooth sphere. This in turn causes the flow to remain attached longer on a dimpled golf ball which implies a reduction in drag. As the speed of the dimpled golf ball is increased, the drag does not change significantly. This is a good property in a sport like golf where the ball when it is driven starts off at a relatively high Re and so the lower drag is beneficial. However, as it slows down towards the end of the flight and the Re is lower, the dimples must still trigger transition and maintain the low drag condition. The main goal in golf is to maintain the ball in this postcritical regime throughout its flight.

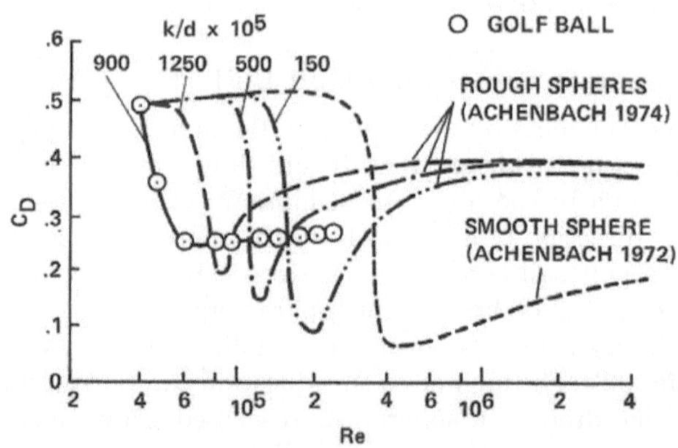

Figure 42. Variation of golf ball and sphere drag, where k is the sand-grain roughness height and d is the ball diameter (Bearman and Harvey 1976, Mehta 1985).

Given the proper spin, a golf ball can produce lift. Originally, golfers thought that all spin was detrimental. However, in 1877, British scientist P.G. Tait (1890, 1891) learned that a ball, driven with backspin about a horizontal axis produces a lifting force. As discussed above, the backspin results in a delayed separation over the top of the ball while enhancing that over the bottom part. This results in a lower pressure over the top compared to the bottom, a downwards deflected wake and hence an upward lift force or Magnus force. The dimples also help in the generation of lift. By keeping the ball in a postcritical regime, the dimples help promote an asymmetry of the flow in the wake. This asymmetry can be seen in Fig. 43. In this figure, the dye shows the flow pattern about a spinning golf ball. The flow is moving from left to right and the ball is spinning in a clockwise direction. The wake is being deflected downwards. This downward deflection of the wake implies that a lifting force is being applied to the golf ball. A hook or a slice can be explained in the same way. If the golf ball is given a spin about its vertical axis, the ball will be deflected to the right for a clockwise rotation and to the left for a counter-clockwise rotation. These deflections are also produced due to the Magnus effect.

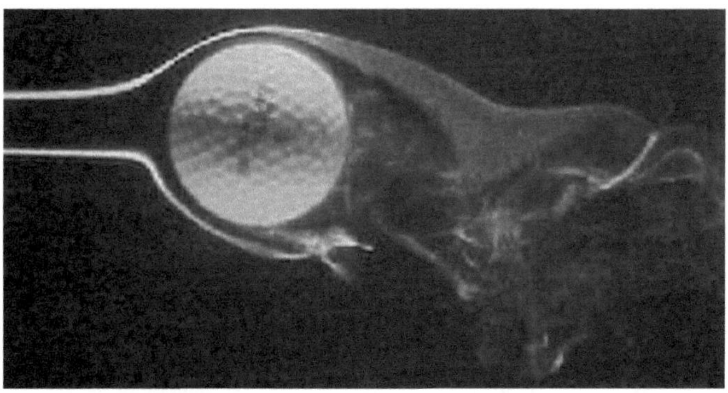

Figure 43. Flow visualization of a spinning golf ball at $Re = 1950$. Flow is from left to right and the ball is spinning in a clockwise direction at 0.5 revs/sec (Pallis and Mehta 2003).

Let us consider the forces acting on a golf ball in flight (Fig. 44) for ascending and descending conditions (Smits and Ogg 2004a,b). The lift (L) and drag (D) forces act in the direction normal to and along the direction of motion, respectively. During the ascent, there is a horizontal component of the lift force ($L2$) that opposes the forward motion of the ball and a vertical component ($L1$) that opposes the gravitational force (W). The drag force has a horizontal component ($D2$) that causes the ball to decelerate and also a vertical component ($D1$) that opposes the lift. Since with creating additional lift comes increased drag (due to induced drag contributions), it becomes clear that for ascending flight, it is critical to minimize the lift force. For a descending ball, the lift force not only opposes gravity

($L1$), but also has a component ($L2$) pulling the ball forward. The lift force therefore contributes to increased carry. In addition, by reducing the incoming angle to the ground, the roll distance is also increased. Of course, maximizing lift towards the end of the flight also results in additional (induced) drag. However, note that the drag force, apart from having a decelerating horizontal component ($D2$), also has a vertical component ($D1$) that helps to keep the ball in the air. So the dimples on a golf ball must be designed to affect the lift and drag forces in these desirable ways rather than just trying to maximize the lift-to-drag ratio, as was believed in the earlier days. A typical golf ball trajectory is shown in Fig. 45. The ball follows an almost linear flight path to start with. However, towards the end of the flight, as the velocity decreases, the aerodynamic forces lose importance to the gravitational force.

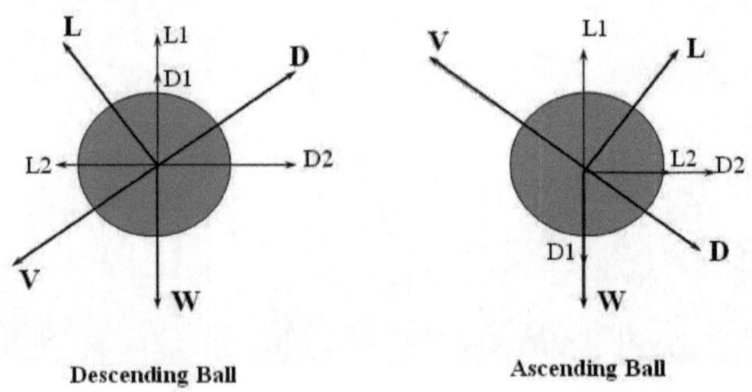

Figure 44. Schematic showing forces acting on a golf ball in flight for descending and ascending conditions (based on Smits and Ogg 2004a,b).

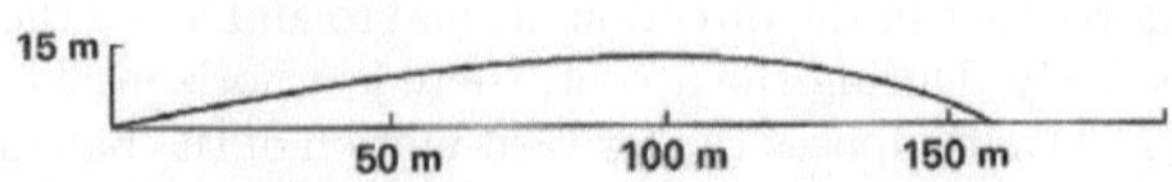

Figure 45. Typical golf ball trajectory. Initial conditions: velocity = 57.9 m/s, elevation = 10, spin = 3500 rpm (Bearman and Harvey 1976, Mehta 1985).

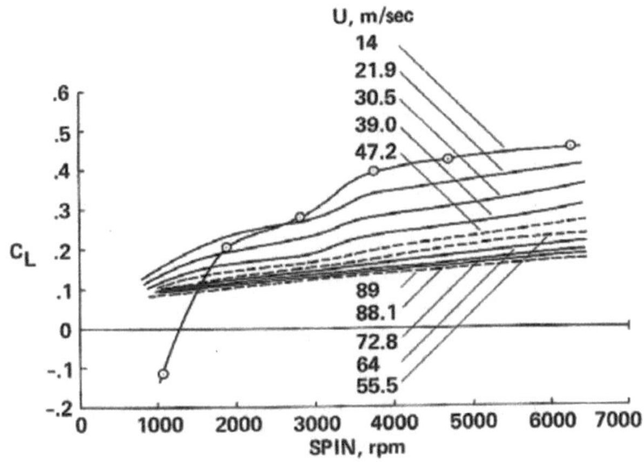

Figure 46. Lift coefficient of a conventional golf ball (Bearman and Harvey 1976, Mehta 1985).

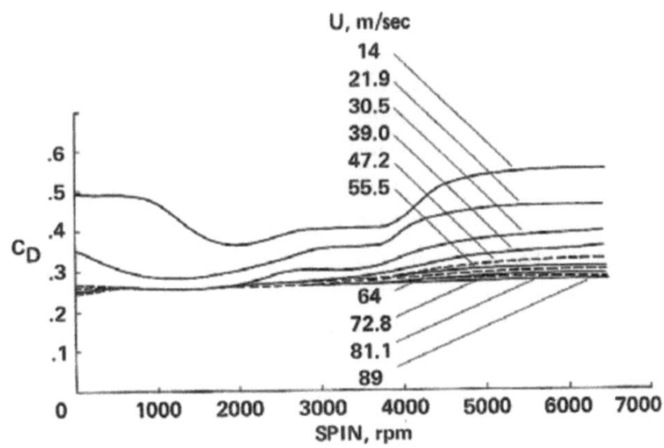

Figure 47. Drag coefficient of a conventional golf ball (Bearman and Harvey 1976, Mehta 1985).

5.3 Measurements of Lift and Drag Coefficients

The first comprehensive measurements of the aerodynamic forces on a golf ball were not made until after World War II by Davies (1949). Davies made the measurements by dropping spinning golf balls though a wind tunnel test section and measuring the point of impact on the floor. The ball was held between two shallow cups that were rotated with a variable speed motor about a horizontal axis perpendicular to the airstream. A trigger released the springs on the cups, and the ball was allowed to drop through the airstream. By spinning the ball in one direction and then the other, the lift and drag forces could be evaluated. Spin rates of up to 8000 rpm were investigated, but at only one speed of 32 m/s, giving a Re of about 9.4×10^4, somewhat lower than that typically attained by a ball leaving the tee ($\approx 2.1 \times 10^5$). Despite the limitations, the actual measured values of C_L and C_D were comparable to those measured later using more accurate methods.

Bearman and Harvey (1976) conducted a comprehensive study of golf ball aerodynamics using a large spinning model mounted in a wind tunnel over a wide range of Re (40,000 to 240,000) and S (0.02 to 0.3). Figures 46 and 47 show the variation of C_L and C_D, respectively, with varying spin. On the whole, C_L is found to increase with spin, as one would expect, and C_D also increases as a result of induced drag effects associated with lifting bodies. At a given spin rate, increasing U decreases the spin parameter, and hence C_L and C_D are also reduced. At postcritical Re, the relation between lift and spin rate is almost linear. However, this linear relationship does not hold as the spin rate approaches zero; and for $U \approx 14$ m/s and spin rate $<$ 1200 rpm, a negative lift is obtained at this precritical Re. Bearman and Harvey (1976) concluded that the lift in this regime cannot be explained by any simple attached flow circulation theory. This situation mainly results from the fact that transition is a very unstable phenomenon, which is easily influenced by parameters such as the details of the local surface roughness and free-stream turbulence. In summary, Bearman and Harvey (1976) found that C_L increased monotonically with S (from about 0.08 to 0.25), as one would expect, and that the C_D also started to increase for $S > 0.1$ (from about 0.27 to 0.32) due to induced drag effects. The trends were found to be independent of Reynolds number for $126{,}000 < Re < 238{,}000$. Bearman and Harvey (1976) also investigated the effects on range of the three main parameters (initial ball velocity, spin rate and elevation) and their results are presented in Figs. 48-50. In Fig. 48, the maximum range for the initial conditions given is obtained for a spin rate just over 4000 rpm. In Fig. 49, the range increases rapidly with velocity for initial velocities greater than about 30 m/s. The range seems to be a relatively weak function of the initial elevation (Fig. 50), although it is still increasing at an initial angle of 15°. In practice, as Bearman and Harvey (1976) point out, hitting the ball harder increases both the initial velocity and spin rate.

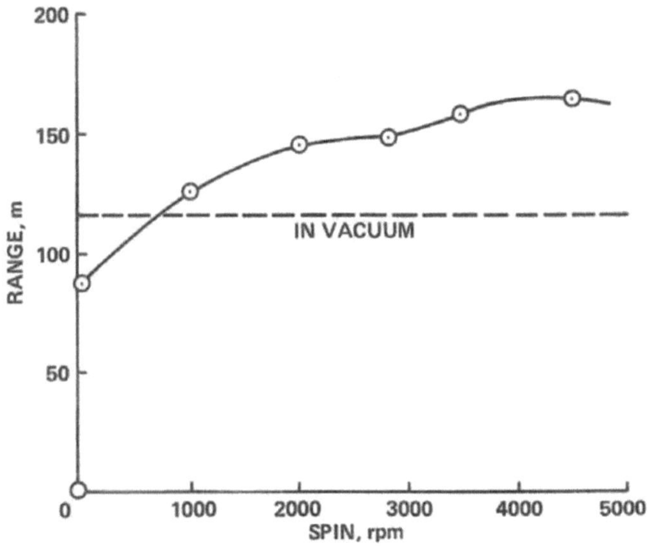

Figure 48. Effect of spin on range for a conventional golf ball. Initial conditions: velocity = 57.9 m/s, elevation = 10° (Bearman and Harvey 1976, Mehta 1985).

Bearman and Harvey (1976) also computed golf ball trajectories using the measured aerodynamic forces. The computation involved a step-by-step calculation procedure of the two components of the equation of motion:

$$\ddot{x} = -\tfrac{\rho A}{2m}(\dot{x}^2 + \dot{y}^2)(C_D \cos\beta + C_L \sin\beta)$$

$$\ddot{y} = \tfrac{\rho A}{2m}(\dot{x}^2 + \dot{y}^2)(C_L \cos\beta - C_D \sin\beta) - g$$

$$\beta = \tan^{-1}(\tfrac{\dot{y}}{\dot{x}})$$

where x and y are measured in the horizontal and vertical directions, respectively, and β is the inclination of the flight path to the horizontal. At each time step, the measured values (or interpolations) of C_L and C_D were used. Bearman and Harvey tried some realistic assumptions for the

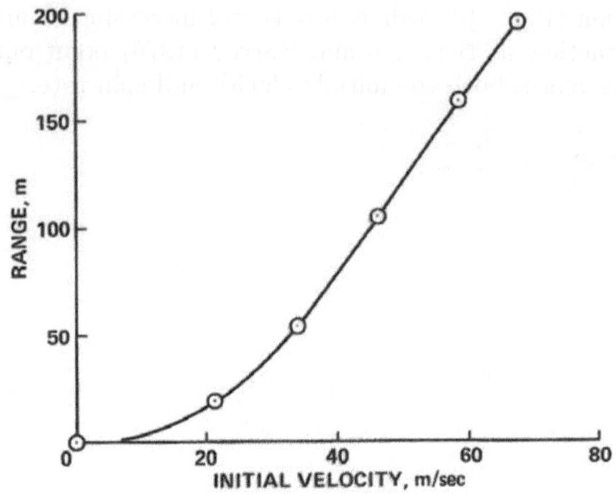

Figure 49. Effect of initial velocity on range for a conventional golf ball. Initial conditions: elevation = 10°, spin = 3500 rpm (Bearman and Harvey 1976, Mehta 1985).

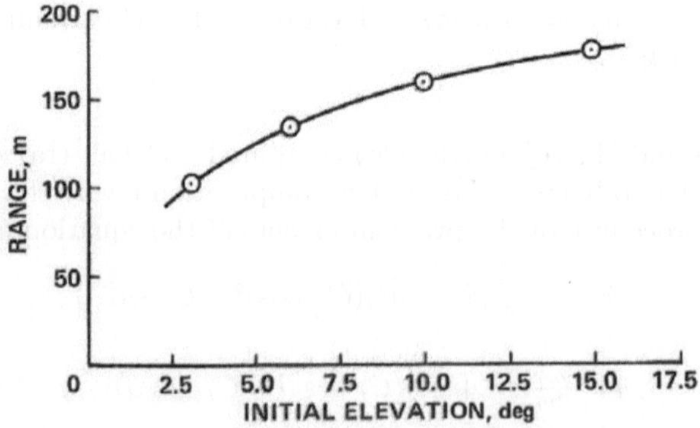

Figure 50. Effect of initial elevation on range for a conventional golf ball. Initial conditions: velocity = 57.9 m/s, spin = 3500 rpm (Bearman and Harvey 1976, Mehta 1985).

spin decay including one wherein the decay was assumed to be proportional to the square of the spin rate. On the whole, they concluded that the computed trajectories were not affected significantly by including the spin decay model. However, this is not necessarily true in general and newer models for the spin decay have been developed as discussed below.

Clearly, the aerodynamic performance of a golf ball is critically dependent on the flow induced by the dimples. The actual geometry of the dimples must therefore be expected to affect the flow regime and hence the aerodynamic forces on the golf ball. While this particular effect is difficult to understand and quantify accurately, some experiments have been performed to establish its importance (Beasley and Camp 2002). In general, for a given geometry and Re, the dimples would have to be deep enough to cause a sufficient disturbance in the laminar boundary layer so that transition is triggered. However, if the dimples are too deep, this may contribute to the skin friction drag. So there must be an optimum depth for the dimples at a given Re. This is indeed what has been observed when a ball with square dimples of varying depth was investigated (Fig. 51). The golf ball was launched using a driving machine and it is clear from the data in Fig. 51 that there is an optimum depth of about 0.25 mm at which the range is maximized. Sajima et al. (2006) also showed that the drag coefficient increased and the lift coefficient decreased when the dimple depth was increased from about 0.17 mm to 0.19 mm. Aoki et al. (2004) specifically investigated the effects of the number of dimples on the aerodynamic performance. The number of dimples was varied while keeping the ratio of depth to width constant at 0.096 (Fig. 52). As seen in Fig. 53, while the critical Re decreases as the number of dimples is increased, the C_D levels in the postcritical region increases. Also, the critical Re for the case of 328 dimples is the same as that for 504 dimples. This is a clear demonstration that simply maximizing the number of dimples on a golf ball does not necessarily give the optimum aerodynamic performance.

Bearman and Harvey (1976) also investigated the effects of changing the dimple shape. Apart from the model ball of conventional design (round dimples), they also measured forces on a "hex-dimpled" ball, which had 240 hexagonal dimples and 12 pentagonal dimples arranged in a triangular pattern. The results for the postcritical regime are compared in Fig. 54. In general, the hex-dimpled ball is superior to the conventional ball: it exhibits higher lift and lower drag properties. Tests using a driving machine showed that under normal driving conditions the hexdimpled ball traveled approximately 6 m farther than a conventional ball. Bearman and Harvey concluded that "hexagonal shaped dimples act as even more efficient (boundary layer) trips than round dimples, perhaps by shedding into

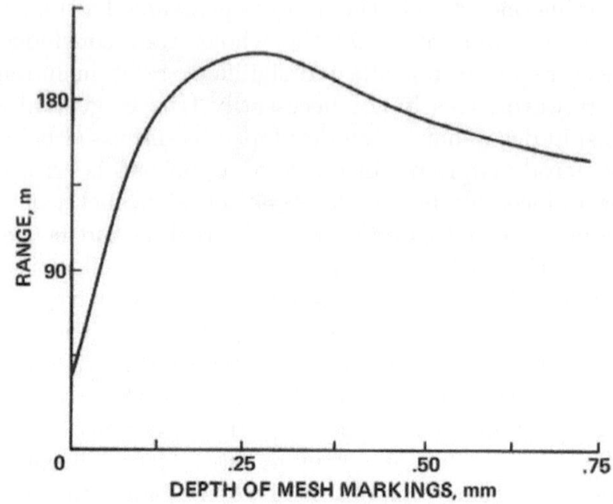

Figure 51. Effect of square dimple depth on range (Cochran and Stobbs 1968, Mehta 1985).

Table 1 Specifications of spherical surface.

N_D	b [mm]	c [mm]	k [mm]
smooth	-	-	-
104	3.897		
184	2.043	3.528	0.338
328	0.650		
504	0.297	3.046	0.292

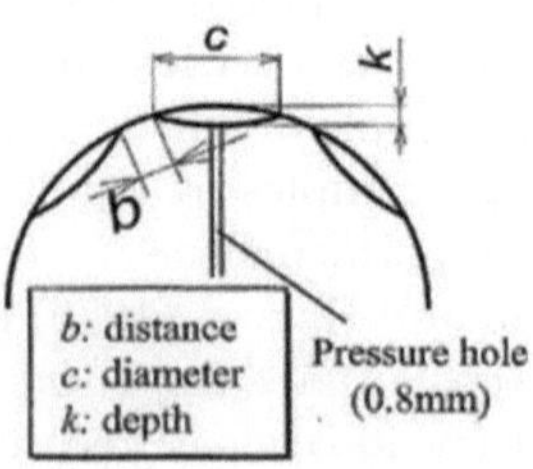

Figure 52. Details of dimple variations tested (Aoki et al. 2004).

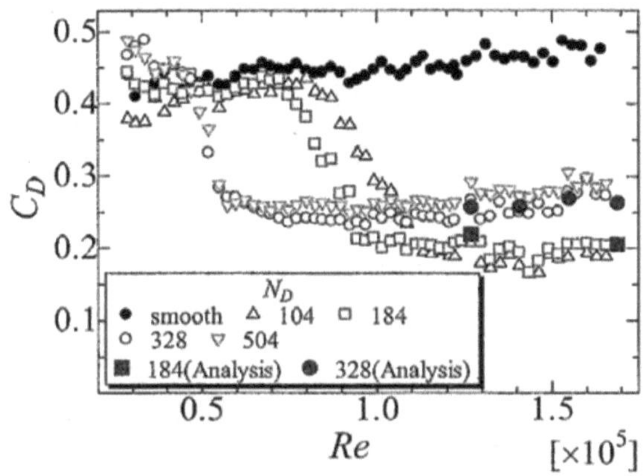

Figure 53. Variation of drag coefficient with Reynolds number: effects of number of dimples (Aoki et al. 2004).

the boundary layer more discrete (horse-shoe) vortices from their straight edges." It is interesting that after all these years, Callaway recently introduced the "Hex-ball," also with hexagonal dimples. In order to try and minimize the amount of sideways deflection on an inadvertently sliced drive, a ball was designed (marketed under the name: "Polara") with regular dimples along a "seam" around the ball and shallower dimples on the sides. The ball is placed on the tee with the seam pointing down the fairway. If the ball is hit optimally such that only backspin about the horizontal axis is imparted to it, the Polara ball will generate roughly the same amount of lift as a conventional ball. However, if the ball is heavily sliced, so that it rotates about a near-vertical axis, the reduced overall roughness increases the critical Re, and hence the sideways (undesirable) deflection is reduced. Basically, the ball crosses the critical Re and so the sideways force is reduced while the drag shoots up. In an extreme version of this design, where the sides are completely bald, a reverse Magnus effect can occur towards the end of the flight which makes a sliced shot, which initially goes towards the right, end up to the *left* of the fairway.

Aoyama (1990) adapted Davies' method for measuring the lift and drag forces on a spinning golf ball. However, he recorded the trajectory of the ball as it fell though the test section using short exposure video techniques. This approach should give more accurate results since there are no simplifying assumptions regarding the trajectory shape, the equations of motion, or the relative differences between the ball and the airstream. Lift and drag

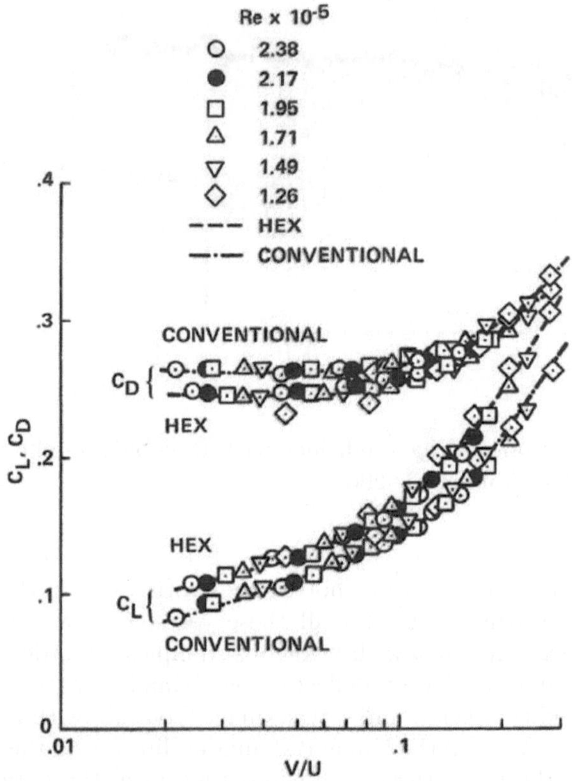

Figure 54. Comparison of conventional and hex-dimpled golf balls (Bearman and Harvey 1976, Mehta 1985).

coefficient measurements were obtained for ball velocities between 30 to 75 m/s and spin rates of 1000 to 3500 rpm. He showed that the lift and drag coefficients for the ball using the 384 icosahedron dimple pattern were consistently lower than those for a 336 "Atti" dimpled ball. The Atti design is an octahedron pattern, split into eight concentric straight-line rows, named after the producer of molds for golf balls (Smits and Ogg 2004a). Chikaraishi et al. (1990) estimated the aerodynamic forces on golf balls by hitting balls outdoors using a mechanical driving machine. They accurately measured the initial conditions of the ball as it left the tee and then measured the trajectory of the ball by means of video techniques. The lift and drag coefficients were then estimated for different balls and conditions. However, with the inherent difficulties of outdoor testing, such as the varying effects of atmospheric conditions, this form of golf ball testing did not catch on.

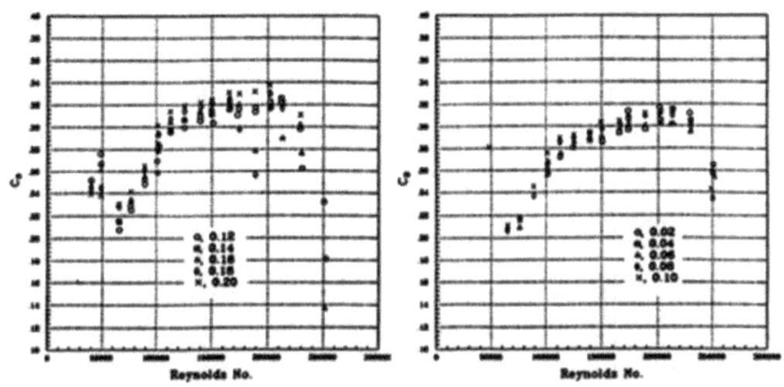

Figure 55. Drag coefficient as a function of Reynolds number for different values of the spin parameter (Smits and Smith 1994).

Smits and Smith (1994) made some wind tunnel measurements on spinning golf balls over the range, 40,000 $< Re <$ 250,000 and 0.04 $< S <$ 1.4, covering the range of conditions experienced by the ball when using the full set of clubs. Based on their detailed measurements, which included measurements of the spin decay rate, they proposed a new aerodynamic model of a golf ball in flight. Their measurements were in broad agreement with the observations of Bearman and Harvey (1976), although the new C_L measurements were slightly higher ($\approx$ 0.04) and a stronger dependence of C_D on the spin parameter was exhibited over the entire S range. A new observation was that for $Re >$ 200,000, a second decrease in C_D was observed, the first being that due to transition of the boundary layer (Fig. 55). Smits and Smith proposed that this could be due to compressibility effects since the local Mach number over the ball reached values of up to 0.5, although the exact fluid mechanic phenomenon has yet to be identified. Note that Bearman and Harvey (1976) used a 2.5 times larger model and so their Mach number was correspondingly lower. Smits and Smith (1994) proposed the following model for driver shots in the operating range, 70,000 $< Re <$ 210,000 and 0.08 $< S <$ 0.2:

$$C_D = C_{D1} + C_{D2}S + C_{D3}\sin(\pi(Re - A_1)/A_2)$$

$$C_L = C_{L1}S^{0.4}$$

$$\textit{Spin Rate Decay} = \frac{\partial\omega}{\partial t}\left(\frac{d^2}{4U^2}\right) = R_1 S$$

Suggested values for the constants are: $C_{D1} = 0.24, C_{D2} = 0.18, C_{D3} = 0.06, C_{L1} = 0.54, R_1 = 0.00002, A_1 = 90,000$ and $A_2 = 200,000$.

One of the more interesting features to note in the data presented in Fig. 55 is that the critical Reynolds number (Re_{cr}) for a spinning golf ball is also about 60,000, the same as that observed for a nonspinning golf ball (Fig. 42). This is due to the fact that the effective Re on the advancing side of a spinning ball is higher than that based on the ball translation velocity, and it is lower on the retreating side. As a result, as Re_{cr} is approached, boundary layer transition will occur first on the advancing side and later on the retreating side, compared to that on a nonspinning ball. So the two effects end up canceling each other and the overall Re_{cr} for a spinning ball ends up being comparable to that for a nonspinning ball. Therefore, the Re_{cr} for a golf ball, regardless of whether it is spinning or not, is around 60,000, the exact value is obviously determined by the detailed roughness created by the dimples.

In terms of empirical testing, the preferred method to obtain accurate aerodynamic data nowadays is to use indoor ranges (Smits and Ogg 2004a). The first successful Indoor Testing Range (ITR) was designed and constructed at the USGA Research and Test Center (Zagarola et al. 1994). The range uses a calibrated launching machine to provide precisely known initial velocity, launch angle and spin rate. The launching machine consists of four large wheels which drive a pair of belts. The ball is inserted into the gap between the belts and projected down the range. The ball velocity and spin rate are controlled by varying the relative belt speeds. The velocity of the ball is then measured at three down range stations, along with the vertical and horizontal position, all with a high degree of precision. This facility is used to obtain fundamental aerodynamic data which are used for conformance tests.

An Indoor Testing Range has also been developed by Callaway Golf in Carlsbad, California (Smits and Ogg 2004a). The launching machine is very similar to that used by the USGA, but the ball position as a function of time is found from image data acquired by pairs of stereoscopic cameras arranged at a large number of stations along its length. The sensors are calibrated with particular care so that the accuracy of the spatial data is maximized. The spin rate of the golf ball is measured by marking three orthogonal stripes on the ball and using image processing to determine the rotational orientation. The rotation rate is measured at several locations, including the one where the force coefficients are evaluated. This analysis also gives information regarding the spin decay characteristics of the ball. The lift and drag coefficients, and spin decay are measured at specific values of spin parameter and Reynolds number. Typical results for the lift and

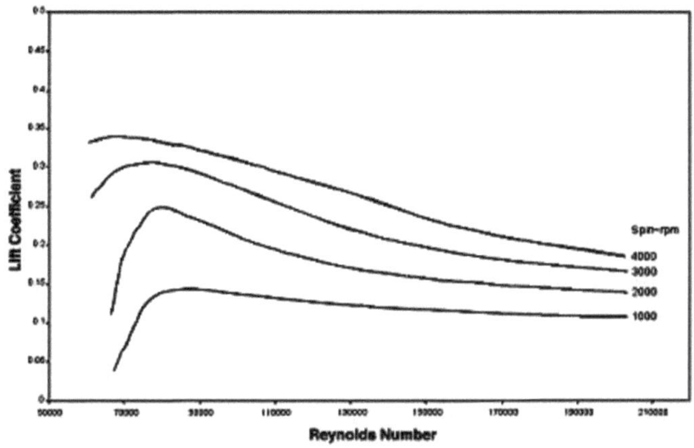

Figure 56. Lift coefficient as a function of Reynolds number for different values of the spin rate measured in an Internal Test Range (Smits and Ogg 2004a).

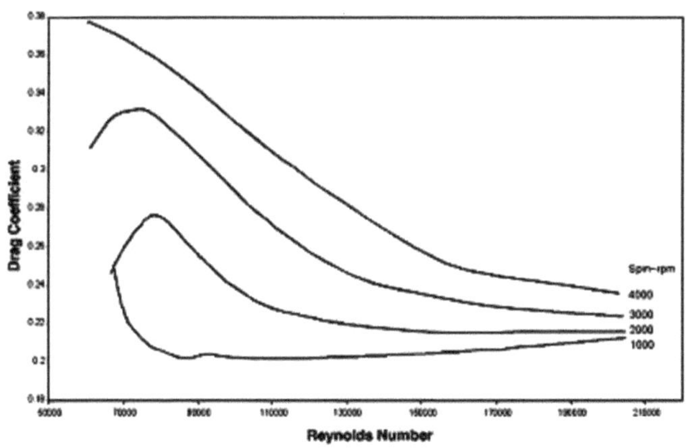

Figure 57. Drag coefficient as a function of Reynolds number for different values of the spin rate measured in an Internal Test Range (Smits and Ogg 2004a).

drag coefficients are shown in Figs. 56 and 57, respectively. In contrast to Bearman and Harvey's (1976) data shown in Figs. 46 and 47, the force coefficients here are plotted against *Re* for given spin rates. In both cases, a convergence in the data is noted as *Re* increases.

5.4 Conclusions

There has been more research done on golf balls than any other sports ball. This is obviously due to the fact that the aerodynamic performance of a golf ball is a critical element of the game. The fundamental design challenge is in designing a golf ball that exhibits the lowest drag at the highest *Re* (when it is driven) and high lift at the lowest *Re* (towards the end of the flight). In golf, as opposed to most other sports, the ball velocity and spin rate vary significantly during its flight. Therefore, in order to model the trajectory of the ball accurately, the values of the aerodynamic forces must be known over the whole range of Reynolds numbers and spin rates. Despite what is often said and written, golf ball aerodynamics is far from a mature subject. The precise nature of boundary layer transition and the effects on it of the intrinsic dimple details are still not fully understood, even for the relatively simple case of nonspinning balls. This in turn means that the detailed effects of a given dimple design on the lift and drag forces at defined values of *Re* and S cannot be easily predicted or determined. As a result, golf ball design is almost totally dependent on empirical testing. It is highly unlikely that Computational Fluid Dynamics (CFD) tools will be become useable for golf ball design in the foreseeable future.

6 SOCCER/VOLLEYBALL AERODYNAMICS

6.1 Introduction

The earliest evidence of soccer (or football as it is widely known) dates back to about 200 BC in China, where a form of the game was played that emphasized the ability of players to dribble a leather ball. The Greeks and Romans also participated in a variation of soccer that permitted ball carrying. The modern day outgrowth of soccer is known to have started in England and the first ball reportedly was the head of a dead Danish brigand. The earliest organized games were massive confrontations between teams consisting of two or three parishes each and goals three to four miles apart. By 1801, the game had been refined, requiring a limited and equal number of players on each side and confining the field length to about 80 to 100 yards, with a goal at each end. The goal was made of two sticks a few feet apart and the crossbars were lengths of tape stretched across the top. The first soccer club was formed in Sheffield, England in 1857. The London Football Association issued its first set of rules in 1863 and some order was brought to the sport. The word 'soccer' was derived from 'association.' The current 11-player teams were formally established in 1870 and the goalkeeper was established in the 1880s.

Figure 58. Adidas Fevernova soccer ball used in the 2002 World Cup.

The Soccer Laws mandate that the ball must be spherical, made of leather (or another suitable material), size 5 have a circumference of 68.6 to 71.1 cm (27 to 28 ins) and size 4 have a circumference of 63.5-66.0 cm (25 to 26 ins). The ball cannot be more than 450 g (16 oz) in weight and not less than 410 g (14 oz) at the start of the match. The earlier balls were made out of leather with a rubber bladder, the main problem with which was that the ball absorbed water in wet conditions, which made it a lot heavier and harder to control. In the last few decades, various synthetic (waterproof) materials have been introduced. A typical modern day soccer ball features a natural latex bladder with a polyurethane cover and a multi-layer, reinforced backing. Some of the materials and designs have led players to complain, as evidenced at the 2002 World Cup where the "Fevernova" ball (Fig. 58) was apparently 'lighter' and harder to control. The "Teamgeist" ball (Fig. 59) specifically designed for the 2006 World Cup drew complaints from the goalkeepers who claimed that the ball knuckled more than the traditional ball. Some novel patch designs have also emerged with the latest version displaying a surface covered with 'dimples' (Fig. 60).

Volleyball was invented by two Massachusetts natives in 1892. After

Figure 59. Adidas Teamgeist soccer ball used in the 2006 World Cup.

Figure 60. Puma dimpled soccer ball.

reaching Japan and Asia by 1900, the rules of the game were set in place over the next 20 years. The 'set' and 'spike' were created in the Philippines in 1916 and six-a-side play was introduced two years later. By 1920, the rules mandating three hits per side and back-row attacks were instituted. Japan, Russia and the United States each started national volleyball associations during the 1920s. The volleyball rules dictate that the ball shall be spherical, made of a flexible leather or leather-like case with an interior bladder made of rubber or a similar material. It shall be uniform and light in color or a combination of colors one of which must be light. The ball circumference must be between 65 to 67 cm (25.5 to 27 inches) and weigh between 260 to 280 grams (9 to 10 ounces) with an internal pressure of 0.30 to 0.325 kg/sq. cm (4.3 to 4.6 lbs./sq. inch).

6.2 Soccer Ball and Volleyball Aerodynamics

In recent years, there has been an increased interest in the aerodynamics of soccer balls and volleyballs. In soccer, the ball is almost always kicked with spin imparted to it, generally backspin or spin about a near-vertical axis, which makes the ball curve sideways (Carré et al., 2002). The latter effect is often employed during free kicks from around the penalty box. The defending team puts up a 'human wall' to try and protect a part of the goal, the rest being covered by the goalkeeper. However, the goalkeeper is often left helpless if the ball can be curved around the wall. A spectacular example of this type of kick was in a game between Brazil and France in 1997 (Asai et al., 1998). The ball initially appeared to be heading far right of the goal, but soon started to curve due to the Magnus effect and wound up in the back of the net. A 'toe-kick' is also sometimes used in the free kick situations to try and get the 'knuckling' effect.

There was a lot of talk regarding the flight (aerodynamics) of the "Teamgeist" ball being used in the 2006 World Cup. One of the main complaints, not surprisingly from the goalkeepers, was that it tended to knuckle more than the traditional ball. The knuckling effect makes the ball fly in an unpredictable manner with changes in direction or even "zigzagging." The notable difference in the construction of this ball was that it only had 14 panels and the panels were bonded together (the traditional ball has 32 panels which are stitched internally). These two new features made the Teamgeist ball smoother than the traditional ball and this affected its aerodynamic properties and hence flight in a significant way. A soccer ball will tend to knuckle when it is kicked without much spin at a critical speed (critical Re). In simple terms, at this critical speed, the flow very close to the ball surface around the apex region can be affected by the seams between the panels and this can produce a lateral force which makes the ball knuckle. This is a similar effect to that obtained on a nonspinning pitched baseball. The critical speed for a traditional soccer ball has been reported to be between 9 m/s (20 mph) Carré et al. (2004) and about 15 m/s (34 mph) Asai et al. (2006). At the time of the 2006 World Cup, it was hypothesized (Mehta 2006b) that with the smoother surface, the Teamgeist ball would have a much higher critical speed, perhaps closer to 20 to 25 m/s (50 to 56 mph). In match situations, free kicks are more likely to be executed at around 22 m/s than at 15 m/s (ball speeds of up to 36 m/s or 80 mph are encountered in World Cup matches). So perhaps what had inadvertently happened with this smoother ball was that the probability of the ball knuckling had suddenly increased. Another important point to note is that the magnitude of the lateral force is proportional to the square of the ball speed and so the knuckling effect will be exacerbated at the higher critical speed. So both the probability and magnitude of the knuckling effect are increased on this Teamgeist ball.

Some of the players also claim that the Teamgeist ball is lighter. If this were true then both the knuckling effect and the swerve of a spinning ball would be increased. But alas, in looking at the ball measurements, it turns out that the weight of the Teamgeist ball (441 to 444 grams) was at the upper end of the official FIFA specifications (420 to 445 grams) so the goalkeepers need to cross that off their list of excuses. However, Adidas also stated that the Teamgeist ball was virtually waterproof and this will have an effect on the ball aerodynamics. A ball that absorbs water and becomes heavier will knuckle and swerve less than a dry ball. So it is feasible that the Teamgeist ball swerves and knuckles more in wet conditions compared to the traditional ball. Adidas also claims that the Teamgeist ball has a truer round shape and that it will remain so for longer periods. This will tend to help the rolling properties and reduce the chances of the ball knuckling due to an odd shape. So while there was something to all the complaints from the goalkeepers about the weird flight of the Teamgeist ball, it seems that not all their complaints were justified.

In volleyball, two main types of serves are employed: a relatively fast spinning serve (generally with topspin), which results in a downward Magnus force adding to the gravitational force or the so-called 'floater' which is served at a slower pace, but with the palm of the hand so that very little or no spin is imparted to it. An example of a serve with topspin is shown in Fig. 61. The measured flight path implies that the downward force (gravity plus Magnus) probably does not change significantly during the ball's flight, thus resulting in a near-parabolic flight path. The floater (with little or no spin) has an unpredictable flight path, which makes it harder for the

Figure 61. Measured trajectory of a volleyball serve with topspin (Pallis and Mehta 2003).

returning team to set up effectively.

For a volleyball, the surface is relatively smooth with small indentations where the patches come together, so the critical Re would be expected to be less than that for a smooth sphere, but higher than that for a golf ball, perhaps similar to that of a soccer ball. The typical serving speeds in volleyball range from about 10 m/s to 30 m/s (22.4 to 67.1 mph) and Re_{cr} turns out to be about 200,000 which corresponds to U = 14.5 m/s (32.5 mph) as shown in Fig. 14. So it is quite possible to serve at a speed just above the critical (with turbulent boundary layer separation) and as the ball slows through the critical range, get side forces generated as non-uniform transition starts to occur depending on the locations of the patch-seams. Thus, a serve that starts off on a straight flight path (in the vertical plane), may suddenly develop a sideways motion towards the end of the flight. Even in the supercritical regime, wind tunnel measurements have shown that side force fluctuations of the same order of magnitude as the mean drag force can be developed on non-spinning volleyballs, which can cause the 'knuckling' effect (Wei et al. 1988).

6.3 Measurements of Aerodynamic Forces

Carré et al. (2002) made measurements of the aerodynamic forces acting on a soccer ball by projecting the ball using a machine and then tracking its flight in space and time using video cameras. The testing was conducted for both, nonspinning and spinning soccer balls. The trajectory was then simulated using the measured initial conditions and assumed values of the lift (C_L) and drag coefficients (C_D). An iterative solving algorithm was then used to determine C_D and C_L by minimizing the difference between the simulated trajectory and the measured values. The evaluated C_D for nonspinning soccer balls is compared in Fig. 62 with the smooth and rough sphere data of Achenbach (1972) and data for a nonspinning volleyball from Mehta and Pallis (2001). The soccer ball C_D shows a variation of between

0.05 and 0.35 over a very small Re range with a weak trend of increasing C_D with Re. It is unlikely that such strong Re effects can occur on a soccer ball and the critical Re also appears to be rather high. However, since all the measured C_D values lie below 0.5 (the C_D value for laminar boundary layer separation), the ball is probably in the postcritical range with turbulent boundary layer separation at these Re (2.5×10^5 to 4.5×10^5). The lift coefficient (C_L) measured in these studies for spinning soccer balls is shown in Fig. 63. The C_L increases with the spin parameter initially, but then tends to level off at a value of about 0.3. This is in complete contrast to the data of Maccoll (1928) for a smooth sphere and those due to Watts and Ferrer (1987) for a spinning baseball also plotted in this figure. On the whole, these series of experiments exemplify the difficulty of measuring aerodynamic forces on spinning balls accurately.

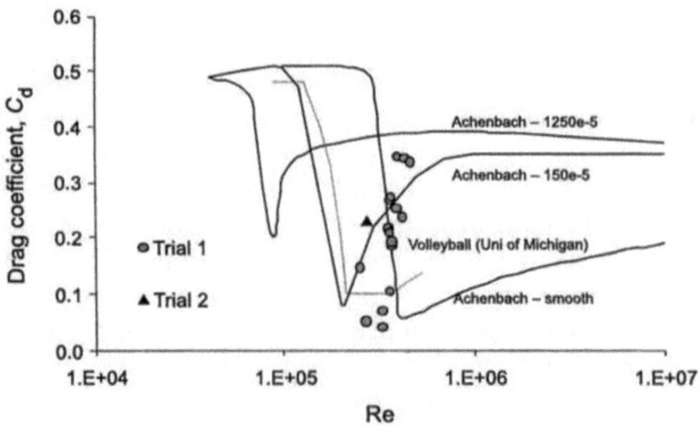

Figure 62. Drag coefficient versus Reynolds number for soccer balls and spheres with varying roughness (Carré et al. 2002).

In a later study, Carré et al. (2004) conducted more controlled wind tunnel studies to measure the aerodynamic forces on soccer balls. Because of the limiting size of the wind tunnel test section (305 × 355 mm), a scale model of a soccer ball was created which had a diameter of 66 mm (compared to 218 mm for a full-size ball). The model had a generic seam pattern, consisting of 20 hexagonal and 12 pentagonal patches. The model was mounted on an 'L'-shaped sting and attached to a two-component force balance. For the spinning ball tests, the ball was mounted on two thin axle supports from the sides and spun with an external motor. A mini soccer ball, a 140 mm diameter Adidas Fevernova (Fig. 58), was also tested. The flow speed in the wind tunnel was varied between 20 m/s to 70 m/s. In a

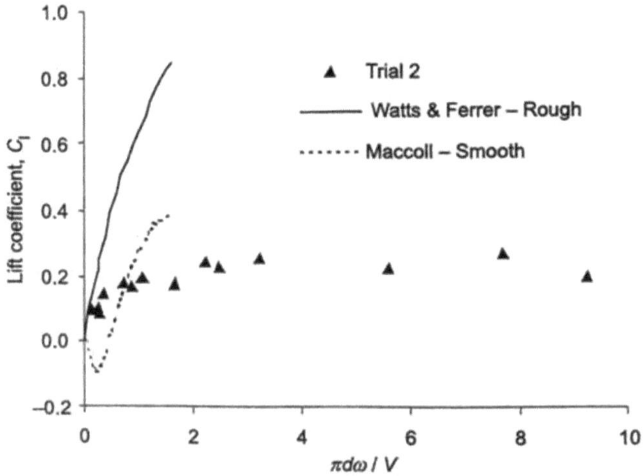

Figure 63. Lift coefficient versus spin parameter for soccer balls and spheres with varying roughness (Carré et al. 2002).

real match situation, a soccer ball has a velocity range of between 4.6 m/s (10 mph) to 32 m/s (70 mph), equating to a *Re* range of about 70,000 to 500,000. The measured C_D data for nonspinning soccer balls are shown in Fig. 64. Also shown are curves for a smooth sphere and a golf ball. The soccer ball data show a sharp drop in C_D from about 0.5 to 0.2 from a *Re* of 90,000 to 130,000 and then a slight increase in the postcritical region. So at a critical *Re* of 130,000, the critical ball velocity is about 9 m/s. The lift (or Magnus) coefficient data are plotted as a function of the spin parameter in Fig. 65. Once again, Maccoll's (1928) data for a smooth sphere and Watts and Ferrer's (1987) data for a baseball are also shown for comparison. In general, the C_L increases with the spin parameter as expected, and the data at the higher *Re* agree quite well with those of Watts and Ferrer (1987), in contrast to the earlier testing (Carré et al. 2002). At the lowest *Re* (90,000), a negative Magnus force was measured for $S < 0.25$. This occurs when boundary layer transition occurs on the advancing side, but a laminar separation is maintained on the retreating side. The separation is hence delayed on the advancing side (as opposed to the retreating side for positive Magnus force) and the direction of the Magnus force is reversed.

Spampinato et al. (2004) measured the aerodynamic forces on a full-scale nonspinning and spinning soccer ball in a wind tunnel. For the nonspinning case, they obtained a critical *Re* of about 200,000, comparable to that for

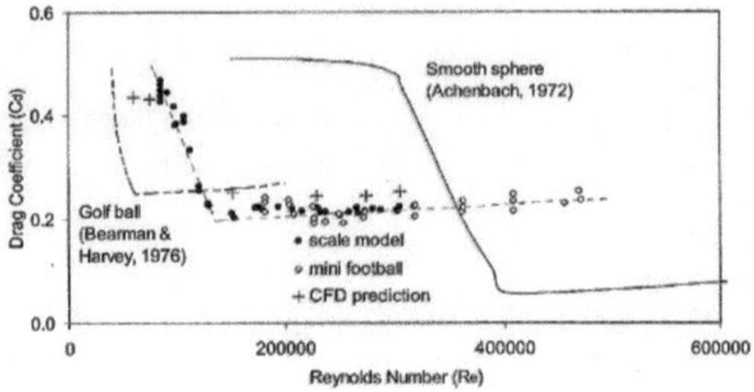

Figure 64. Drag coefficient versus Reynolds number for soccer balls and spheres with varying roughness. CFD predictions are also shown (Carré et al. 2004).

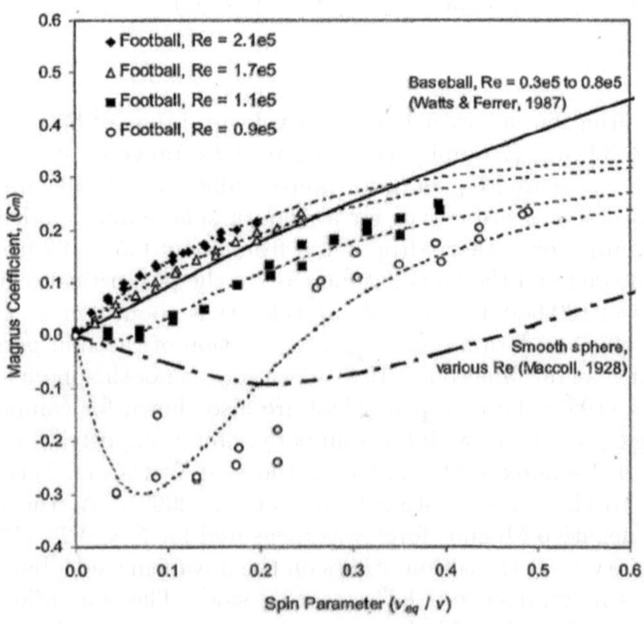

Figure 65. Lift (Magnus) coefficient versus spin parameter for spinning soccer balls, smooth spheres and spinning baseballs (Carré et al. 2004).

a volleyball. For the spinning cases, they measured lift (or side force) coefficients of just under 0.4 over a *Re* range of 200,000 to 500,000. In addition to the flight of a kicked soccer ball, aerodynamics also plays a role when the ball is thrown in after it goes out of bounds. Bray and Kerwin (2004) modeled the long throw using C_L and C_D values which best matched the carefully measured trajectories. From these data, a correlation of the force coefficients with spin rate was also developed.

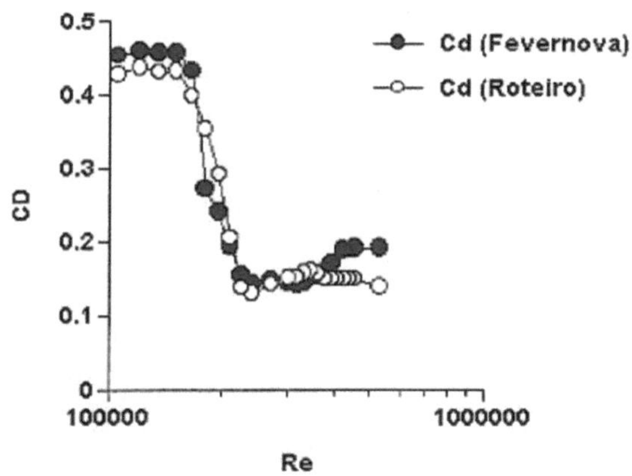

Figure 66. Drag coefficient versus Reynolds Number for non-spinning soccer balls (Asai et al. 2006).

More recently, Asai et al. (2006) tested full-scale nonspinning soccer balls in a wind tunnel and measured the drag coefficient as a function of *Re*. The results for two balls (Fevernova and Roteiro) presented in Fig. 66 show that the critical *Re* is about 220,000 which corresponds to a ball velocity of about 15 m/s. Some flow visualization studies were also conducted which clearly showed the effects of boundary layer transition in delaying separation and the effects of spin in generating an asymmetry in the separation locations. In a later study, the drag coefficient on the Teamgeist ball used in the 2006 World Cup was also measured (Asai et al. 2007). The results presented in Fig. 67 show that the critical *Re* for this ball is somewhat higher, as predicted. It is about 300,000, which corresponds to a ball velocity of about 20 m/s, compared to about 15 m/s for the traditional balls. The measured drag and lift coefficients for the spinning cases are presented in Figs. 68 and 69 as a function of the spin parameter. As expected, the lift coefficient increases with the spin parameter, reaching a

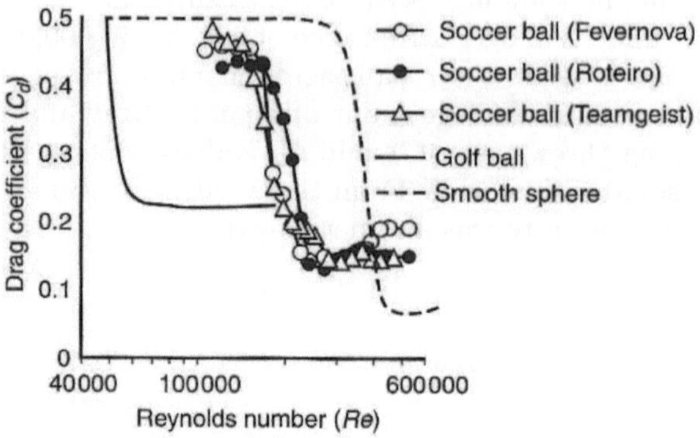

Figure 67. Relationship between drag coefficient and Reynolds number for non-spinning sphere, golf ball and soccer balls (Asai et al. 2007).

maximum of about 0.35 at $S = 0.3$. The drag coefficient also increases with S, mainly due to the contribution from induced drag effects. As far as aerodynamic force measurements on soccer balls are concerned, this data set is perhaps the most representative since full-scale real soccer balls were used in a controlled test environment.

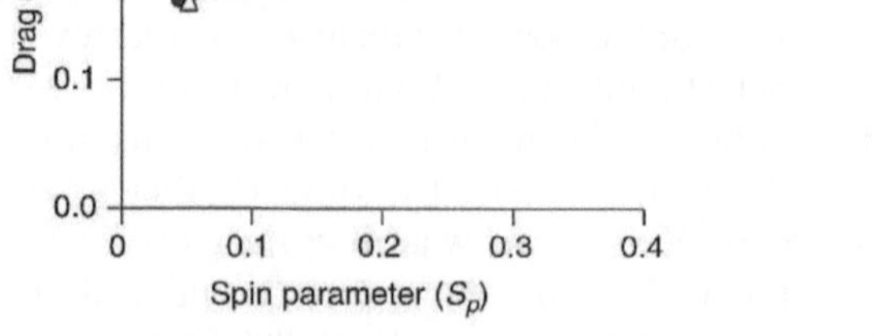

Figure 68. Relationship between drag coefficient and spin parameter for spinning soccer balls (Asai et al. 2007).

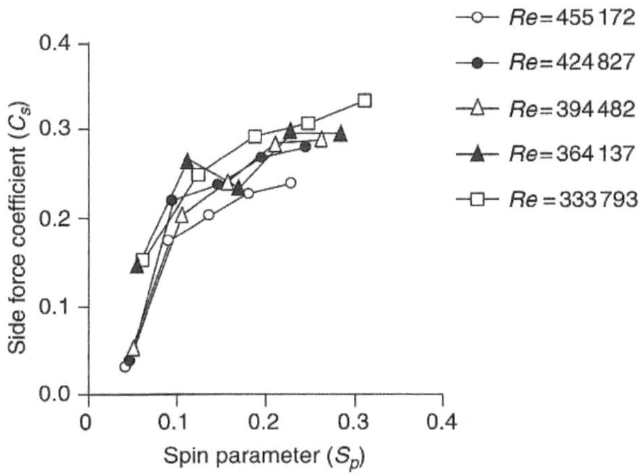

Figure 69. Relationship between side force coefficient and spin parameter for spinning soccer balls (Asai et al. 2007).

Barber et al. (2006) conducted some Computational Fluid Dynamics (CFD) studies of the postcritical flow over nonspinning soccer balls and compared the results with wind tunnel data. The effects of seam geometry were also investigated in this study and it was concluded that the seam width had a larger effect on the drag coefficient than the seam depth. The sharpness of the seam edges was also found to have an effect.

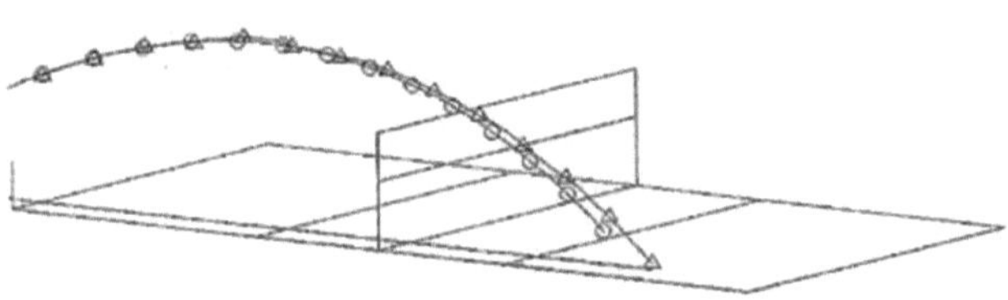

Figure 70. An observed volleyball serve (triangles) with *topspin* of 4.5 revs/sec travels farther than its computed counterpart with zero spin suggesting the negative Magnus effect. The initial velocity was 13.8 m/s (Cairns 2004).

Cairns (2004) modeled the lift and drag forces on spinning volleyballs.

Volleyballs were launched from a machine, videotaped and their spatial paths estimated. Cubic approximations to the lift coefficient in the equation of motion were computed using optimization routines. The drag force was studied using serves with zero spin. For the modeling, the drag measurements shown in Fig. 14 were used. Amongst the more important observations was an illustration that the negative Magnus effect can occur on a spinning volleyball in practice (Fig. 70). Wei et al. (1988) conducted a wind tunnel experiment on the dynamic loads on a nonspinning volleyball. The flow past the volleyball was visualized and the aerodynamic loads were measured using a strain gauge balance. The boundary layer separation locations were measured using hot films. The measurements made at flow speeds of between 7 and 25 m/s (Re = 100,000 to 360,000) showed large dynamic loads in the drag and lift forces throughout the speed range. The lift and drag force fluctuations were of the same order and comparable in magnitude to the mean drag. The fluctuations were found to be random in nature and occurred at a frequency of around 10 Hz. Perhaps the most important finding was that the large dynamic loads were measured on a stationary volleyball even when it was held in a symmetric orientation relative to the air stream. Wei et al. (1988) attributed the dynamic loads to the action of an unstable trailing vortex system which was observed and measured. However, in reality the vortex system is a result of unsteady and spatially nonuniform boundary layer transition and separation.

6.4 Conclusions

For both, soccer balls and volleyballs, there are two main flight modes of interest. For the most part, the ball is released with it spinning about an axis perpendicular to the line of flight. This results in the generation of a Magnus force. In volleyball, topspin is usually imparted and so the Magnus force acts downwards adding to the gravitational force. In soccer, spin is often imparted about a near-vertical axis so as to make the ball swerve trough the air, around the human wall during free kick situations, for example. With the relatively low critical Re for both balls, which is often encountered in practice, a negative Magnus effect can also occur.

With the nonspinning (toe) kick in soccer and (palm) serve in volleyball, a "knuckling" effect can be generated whereby the ball follows an unpredictable and time varying flight path. The "Teamgeist" soccer ball used in last year's World Cup apparently knuckled a lot more than the traditional balls. With fewer panels and a smoother surface, the critical Re was raised, thus increasing the probability and magnitude of the knuckling effect under typical match conditions.

7 BASEBALL AERODYNAMICS

7.1 Introduction

About the time of the American Civil War, a New England youngster named Arthur 'Candy' Cummings became fascinated by a new game called 'base ball.' He endlessly pitched clamshells on the beaches near his home. He found he could make the shells curve by holding and throwing them in a specific way. In 1867, 18 year old Candy Cummings, pitcher for the Brooklyn Excelsiors baseball team, tried out the pitch he had been perfecting in secret for years. Over and over throughout the game the ball made an arch and swept past the lunging batter into the catcher's glove. Today, in the Baseball Hall of Fame in Cooperstown, New York, there is a plaque that reads: 'Candy Cummings, inventor of the curveball.'

However, for more than 100 years after Cummings introduced this new pitch, people still questioned its flight: does a baseball really curve, or is it an optical illusion? Several attempts were made throughout the years to settle the question. In 1941 both, Life and Look magazines used stop-action photography to determine if curveballs really curved. Life concluded that they do not; Look claimed the opposite. As late as 1982, Science magazine commissioned an imaging study to measure the flight of a curveball and confirm the extent of the ball movement.

Although all sports ball aerodynamics are a balancing act between competing forces, the stitches on the baseball and the orientation of the stitches as it meets the oncoming air is key to the vast variety of pitches in baseball. The construction of a baseball starts with the 'pill,' a small sphere of cork and rubber enclosed in a rubber shell. The pill is tightly wound with three layers of wool yarn and finished with a winding of cotton/polyester yarn. This core is coated with a latex adhesive over which a leather cover is sewn. The cover consists of two 'figure-eight' pieces of alum-tanned leather which are hand-stitched using a red cotton thread in a 104-stitch pattern. The rules of baseball require that the ball shall weigh not less than 0.14 kg (5 oz) nor more than 0.15 kg (5.25 oz) and measure not less than 22.9 cm (9 inches) nor more than 23.5 cm (9.25 inches) in circumference.

7.2 Aerodynamics Principles as Applied to Baseballs

When a spinning baseball is pitched, combined with the effect of air pressure resulting in drag and lift (or side force), is gravity. Since gravity forces objects to move faster over time, its effect on the ball is more pronounced as it reaches home plate. A pitch that drops half a foot in the first half of its flight falls another two feet in the second half. On a classic curveball, there is an additional downward (Magnus) force due to the applied topspin and this together with gravity makes the ball drop even faster (Watts & Bahill, 2000).

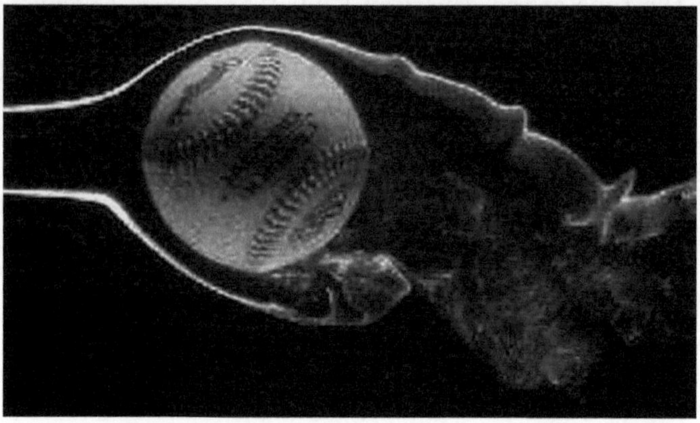

Figure 71. Flow visualization of a spinning baseball at $Re = 3400$; flow is from left to right and the ball is spinning in a clockwise direction at 0.5 revs/sec (Pallis and Mehta 2003).

Let us first consider the aerodynamics of a pitch, such as the curveball, where spin is imparted to the baseball in an attempt to alter its flight just enough to fool the batter. The baseball for this particular pitch is released such that it acquires topspin about the horizontal axis. As discussed above, under the right conditions, this results in a (downward) Magnus force that makes the ball curve faster towards the ground than it would under the action of gravity alone. In Fig. 71, the flow over a clockwise spinning baseball is shown in a water channel using luminescent dyes at a relatively low Re (3400) and a spin rate parameter (S) of 2.5. At such a low Re, the flow over the baseball is expected to be subcritical, but the asymmetric separation and deflected wake flow are clearly evident, thus implying an upward Magnus force, simulating a fastball. Note the indentation in the dye filament over the upper surface due to the seam. At higher Re, the rotating seam would produce an effective roughness capable of causing transition of the laminar boundary layer. The flow over a counter-clockwise spinning baseball at higher Re (1×10^5) is shown in Fig. 72. Again the asymmetric boundary layer separation and deflected wake are clearly apparent. In this case, however, the baseball will experience a downward force, thus simulating a curveball. Spin rates of up to 35 revs/sec and speeds of up to 45 m/s (100 mph) are achieved by pitchers in baseball.

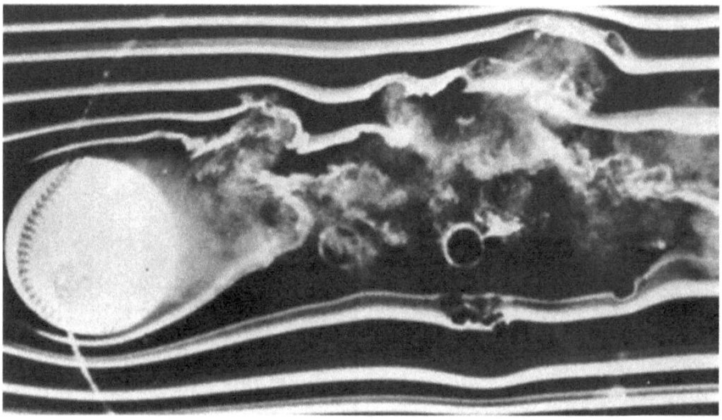

Figure 72. Smoke photograph of flow over a spinning baseball. Flow is from left to right and the flowspeed is 21 m/s (Re = 99,235). The baseball is spinning in a counterclockwise direction at 15 revs/sec. Photograph by F.N.M. Brown, University of Notre Dame (Brown 1971, Mehta 1985).

A good pitcher develops a variety of curving pitches and each with a descriptive name. The 'round house' curve ball is thrown by a right handed pitcher to a right handed batter with both topspin and a twist of the wrist as if turning a door handle. This creates a high pressure area not only above but to the right of the ball, causing movement both down and to the left. The ball arrives lower and farther away than it would if it had been thrown in a straight path. Turn in the opposite direction and the ball curves toward the batter rather than away - the 'screwball.' A baseball thrown with more force and spin about a near vertical axis, drops more slowly and seems to slide to the right or left on the same plane as it was thrown; hence the name, 'slider.' A ball thrown with topspin but with less force than the slider, and without the turning motion, doesn't move left or right but simply sinks as it arrives - the 'sinker.' A fastball is thrown with considerable initial velocity (almost 44.4 m/s or 100 mph) and backspin which generates a lift force opposing gravity. Recently, there has been a lot of discussion and controversy regarding a new pitch, the 'gyroball.' According to the Japanese originators, the ideal gyroball is one that is released with it spinning along the axis of motion, like a bullet or an American football. Well, it turns out that for this type of pitch, no additional aerodynamic forces are generated on the ball; there is obviously a drag force and a gravitational force, regardless of the orientation of the ball during its flight. However, if the axis of spin is tilted somewhat from the line of flight, then a lateral force can be generated on the ball. Even if the baseball is released in the true gyroball orientation, there is some possibility of deceiving the batter. For one thing, it will drop faster than a fastball (with backspin), but slower than a curveball (with topspin). Another source of deception is the relatively straight flight path; with the rotation readily visible, the batter will expect the ball to deviate sideways.

The 'knuckleball' is held and released with low velocity and zero or very little spin; the aerodynamics of this pitch are left entirely to random effects of air pressure. Some believe that a knuckleball thrown without any spin will be at the mercy of any passing breeze. Thus, the ball dances through the air in an unpredictable fashion. However, the real reason for the 'dance' of a knuckleball is the effect of the seam on boundary layer transition and separation. Researchers have learned that, depending on the ball velocity and seam orientation, the seam can induce boundary layer transition or separation over a part of the ball thus creating a side force. With a baseball rotating very slowly during flight, not only does the magnitude of the force change, but the direction also changes (Watts & Sawyer, 1975). This is why the ball appears to dance. It is important to note that even if the pitcher throws the ball with no rotation, the flow asymmetry will cause the ball to rotate. The flow asymmetry is developed by the unique stitch pattern on a baseball. Although banned over 60 years ago, spit or its modern counterpart, sunscreen cream is still sometimes used so that the ball may be squirted out of the fingers at high speed. In Fig. 73, the ball is not spinning, but it is oriented so that the two seams help in causing transition in the boundary layer on the upper side of the baseball. The boundary layer on the lower surface is seen to separate relatively early in a laminar state. Once again, the downward deflection of the wake confirms the presence of the asymmetric boundary layer separation, which would produce an upwards lift force on the baseball. Frohlich (1984) realized the importance or effectiveness of pitching at about the critical *Re* where the "drag crisis occurs." Although he erroneously attempted to attribute the "rising fastball" and the ball that drops "sharply" to this drag crisis, the fact that the flow field and hence aerodynamic forces undergo a major change at the critical *Re* was realized.

Apart from pitched baseballs, aerodynamics also play a significant role in the flight of the ball after it is hit by the batter. Sawicki et al. (2003) came up with an improved batted flight model that incorporated experimental lift and drag measurements (including the drag crisis at the critical *Re*). They concluded that the lift is enhanced by undercutting the ball during batting, since this produces backspin on the ball. As a result, an optimally hit curve ball will travel farther than an optimally hit fastball or knuckleball due to

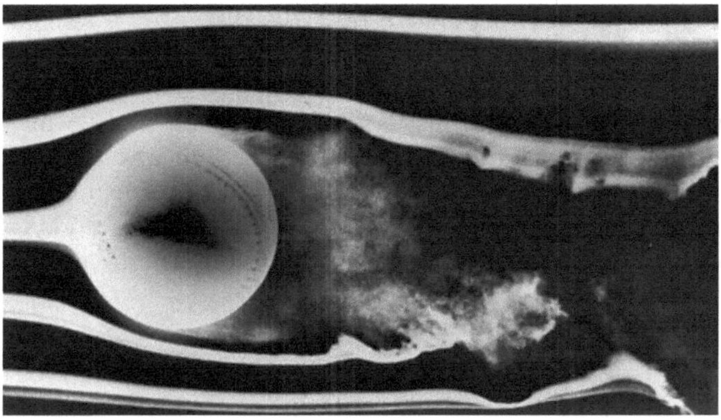

Figure 73. Smoke photograph of flow over a stationary (nonspinning) baseball. Flow is from left to right. Photograph by F.N.M. Brown, University of Notre Dame (Brown 1971, Mehta 1985).

increased lift during flight.

7.3 Measurements of the Aerodynamic Forces

The effects of spin and speed on the lateral deflection of a baseball were first investigated by Briggs (1959). In his first set of experiments, he fired spinning baseballs from an airgun and measured lateral deflections, but this technique did not prove very successful. In his second set of measurements, spinning baseballs were dropped through a 1.8 m wind tunnel. The spinning mechanism was mounted on top of the tunnel and consisted of a suction cup mounted on the shaft that supported the ball. The spinning baseball was released by a quick-acting valve that cut off the suction. The lateral deflection was taken as one half of the measured spread of the two points of impact, with the ball spinning first clockwise and then counterclockwise. The mechanism was aligned so that a baseball spinning about a vertical axis was dropped through a horizontal airstream. Hence, the lateral deflection directly represents the side force due to the spin. The measured deflections are shown in Fig. 74. The straight lines through the origin show that within the limits of experimental accuracy, the lateral deflection is proportional to spin rate. However, Briggs' (1959) extrapolation to zero deflection at zero spin is not accurate, since a nonspinning baseball can also develop a lateral force, as discussed above. Briggs' overall findings were that the lateral deflection is directly proportional to spin rate and it is independent of the speed between 20 and 40 m/s. Briggs also evaluated the final deflections that would be obtained in practice over the distance of 18.3 m (60.5 ft). He used a simple model, which assumed that the side force was constant and that the distance traveled was proportional to the square of the elapsed time, thus giving a parabolic flight path. For a speed of 23 m/s and spin rate 20 revs/sec, a lateral deflection of about 0.28 m was obtained, whereas at a spin rate of 30 revs/sec, it was about 0.43 m (corresponding to F/mg ≈ 0.2). In the case of a curveball, the gravitational force adds to the force due to spin, and so the final deflection would be much greater.

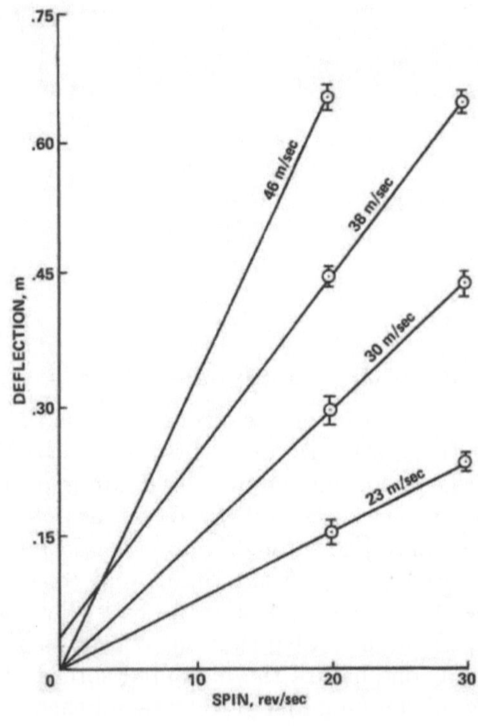

Figure 74. Lateral deflection of a baseball, spinning about a vertical axis, when dropped across a horizontal airstream. These values are for the same time interval (0.6 seconds), the time required for the ball to cross the airstream (Briggs 1959, Mehta 1985).

In experiments described by Allman (1982), the actual flight paths of curveballs pitched by a professional pitcher were photographed using the stroboscopic technique. Analysis of the flight paths confirmed that the baseball travels in a smooth parabolic arc from the pitcher's hand to the catcher, and deflections of over a meter were recorded. Therefore, the assumption of a constant side force for a spinning baseball seems to be valid. Since most of the deflection takes place in the second half of the flight, a batter often gets the impression that the ball "breaks" suddenly as it approaches home plate.

In some wind tunnel measurements of the lateral or lift force (L) on spinning baseballs, Watts and Ferrer (1987) concluded that the lift force coefficient, C_L, was a function of the spin parameter only, for S = 0.5 to 1.5, and at most only a weak function of Re, for Re = 30,000 to 80,000. Their trends agreed well with Bearman and Harvey's (1976) golf ball data obtained at higher Re (up to 240,000) and lower spin parameter range (S =0 to 0.3). Based on these correlations, Watts and Bahill (2000) suggested that for spin rates typically encountered in baseball ($S < 0.4$), a straight line relation between C_L and S with a slope of unity is a good approximation.

Alaways (1998) analyzed high-speed video data of pitched baseballs (by humans and a machine) and used a parameter estimation technique to determine the lift and drag forces on spinning baseballs (Always and Hubbard 2001). For a nominal pitching velocity range of 17 to 35 m/s (38 to 78 mph) and spin rates of 15 to 70 revs/sec, Alaways gave a C_D range of 0.3 to 0.4. This suggests that the flow over a spinning baseball in this parameter range is in the supercritical regime with turbulent boundary layer separation. As discussed above, an asymmetric separation and a positive Magnus force would therefore be obtained in this operating range. Alaways' lift measurements on spinning baseballs obtained for Re = 100,000 to 180,000 and S =0.1 to 0.5, were in general agreement with the extrapolated trends of the data due to Watts and Ferrer (1987). However, one main difference was that Alaways found a dependence of seam orientation (2-seam versus 4-seam) on the measured lift coefficient. The C_L was higher for the 4-seam case compared to the 2-seam for a given value of S, as shown in Fig. 75. Watts and Ferrer (1987) had also looked for seam orientation effects, but did not find any. Alaways concluded that the seam orientation effects were only significant for $S < 0.5$, and that at higher values of S, the data for the two orientations would collapse, as found by Watts and Ferrer (1987). The main difference between these seam orientations is the effective roughness that the flow sees for a given rotation rate. As discussed above, added effective roughness puts the ball deeper into the supercritical regime, thus helping to generate the Magnus force. It is possible that at the higher spin rates (higher values of S), the difference in apparent roughness between the two seam orientations becomes less important. Alaways (1998) also estimated the drag coefficients for spinning baseballs in his study (Fig. 76). Most of the data lie in the Reynolds number range, Re = 100,000 to 300,000

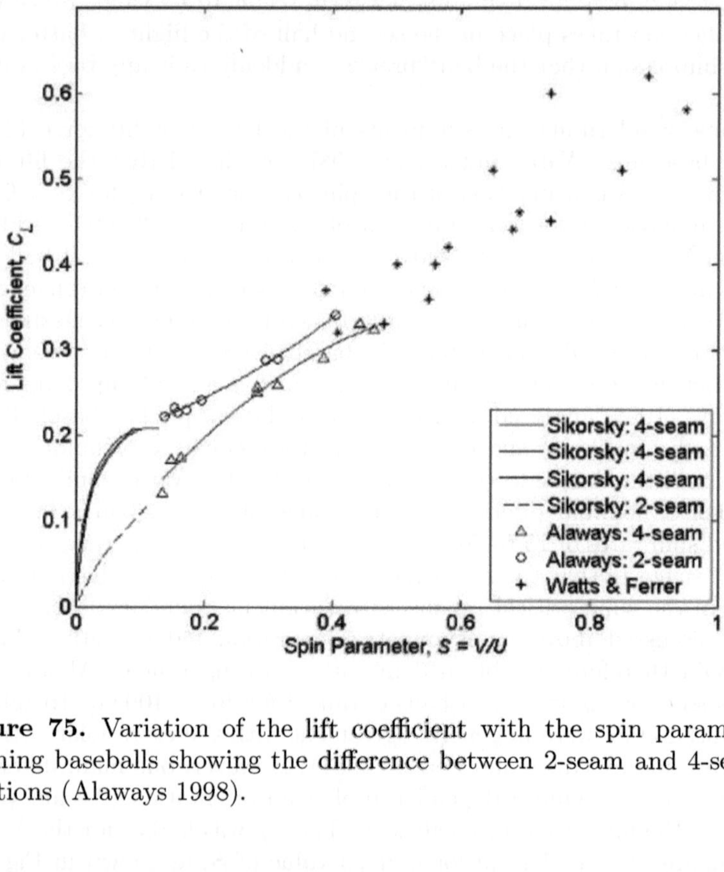

Figure 75. Variation of the lift coefficient with the spin parameter for spinning baseballs showing the difference between 2-seam and 4-seam orientations (Alaways 1998).

and since the C_D values are below 0.5, these results suggest that the critical *Re* for spinning baseballs is probably just below 100,000.

The main significance of the seam orientation is realized when pitching the fastball. Fastball pitchers wrap their fingers around the ball and release it with backspin so that there is an upward (lift) force on the ball opposing gravity. The fastball would thus drop slower than a ball without spin and since there is a difference between the 2-seam and 4-seam C_L, the 4-seam pitch will drop even slower. However, even the 4-seam fastball cannot generate enough lift to overcome the weight of the ball, which is what would be needed to create the so-called "rising fastball." The maximum measured lift in Alaways' (1998) study was equivalent to 48% of the ball's weight, so a truly rising fastball is not likely to occur in practice (Bahill and Baldwin

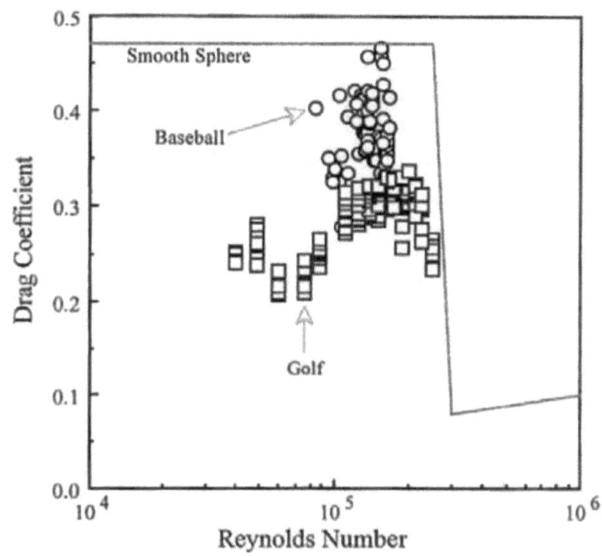

Figure 76. Estimated drag coefficients versus Re (Alaways 1998).

2004). A popular variation of the fastball is the split-finger and forkball. The split-finger, as the name implies, is held between the two pitching fingers and it is released with the same arm action and velocity as a regular fastball. Although some backspin is imparted to the ball, it is less than that for the fastball and it therefore drops a bit faster, thus giving it that sinker look. Apparently, and contrary to popular belief, the forkball is not quite the same pitch (Private Communication, LeRoy Alaways, 2000). For the forkball, the pitching fingers fork out and the ball is kept close to the palm with the thumb tucked under. The thumb is used to push the ball out and topspin is imparted to it, which produces an additional downward force and a flight trajectory below that of a ball with no spin.

More recently, Nathan et al. (2006, 2008) made some measurements of the Magnus force on spinning baseballs. A pitching machine was used to project the spinning balls and a high-speed motion analysis system was used to measure the initial conditions and to track the trajectory over the first 5m of the flight. Measurements were obtained over a speed range of 22 to 49 m/s (50- 110 mph) and spin rates in the range 25 to 75 revs/sec, corresponding to a $Re = 110{,}00$ to $240{,}000$ and $S = 0.09$ to 0.595. The measured values of C_L are compared with the data of Briggs (1959), Alaways (1998) and Watts and Ferrer (1987) in Fig. 77. The dashed line represents the fit to Alaways (1998) data proposed by Sawicki et al. (2003). In the range $S < 0.3$, Nathan et als data agree extremely well with the previous data and the Sawicki et al. fit and for higher values of $S(\approx 0.5)$ their data agreed well with those of Alaways (1998). They also found that the C_L was not a strong function of Re in the spin parameter range, S =0.15 to 0.25.

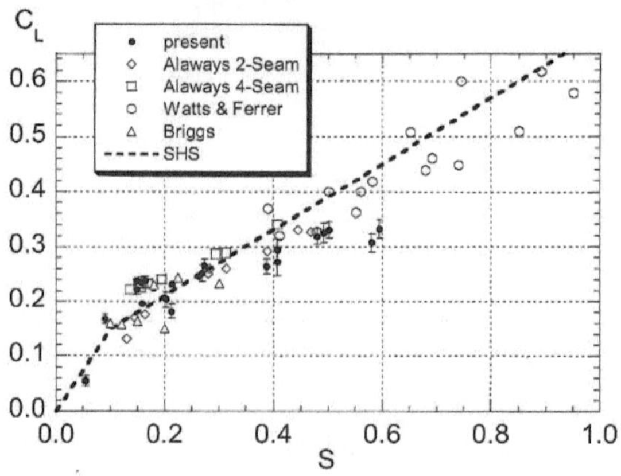

Figure 77. Lift coefficient measurements on spinning baseballs (Nathan et al. 2006) compared with previous data and the parameterization of SHS (Sawicki et al. 2003).

Watts and Sawyer (1975) investigated the nature of the forces causing the "erratic motions" of the so-called "knuckleball." The knuckleball, which is released with zero or very little spin, has some very interesting aerodynamic characteristics. Watts and Sawyer (1975) investigated the nature of the flow field by mounting a stationary baseball in a wind tunnel and measuring the drag and lateral forces for different seam orientations. The datum position of the baseball is defined in Fig. 78, and Fig. 79 shows the measured forces on the ball at U =21 m/s for the various orientations ($\phi = 0 - 360°$). At $\phi = 0°$, the normalized lateral force was zero, but as the ball orientation was changed, values of $F/mg = \pm 0.3$ were obtained with large fluctuations ($F/mg \approx 0.6$) at $\phi = 50°$. These large fluctuating forces were found to occur when the seam of the baseball coincided approximately with the location where boundary layer separation occurs. The separation point was then observed to jump from the front to the back of the stitches and vice versa, thereby producing an unsteady flow field, and hence side force. The frequency of the fluctuation was of the order of 1 Hz, a value low enough to cause a change of direction in the ball's flight. A discontinuous change in the lateral force was also observed at $\phi = 140$ and $220°$. Watts and Sawyer (1975) concluded that this was associated with the permanent movement of the separation point from the front to the rear of the seam (and vice versa). They claim that the data near all four of the "critical" positions ($\phi = 52, 140, 220$, and $310°$) were "quite repeatable."

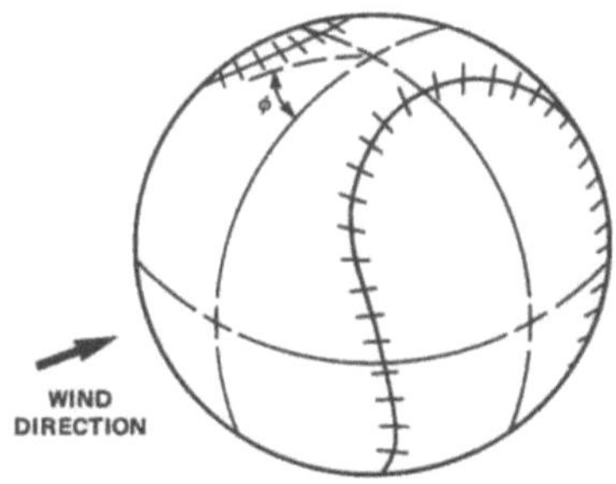

Figure 78. Datum position of baseball at $\phi = 0°$. The ball can be rotated in the direction ϕ to a new position (Watts and Sawyer 1975, Mehta 1985).

Watts and Sawyer (1975) computed trajectories using their measured forces and the simple assumption that the lateral force on the baseball acts in a direction perpendicular to the original direction of level flight. The lateral force was assumed to be periodic in time, which implied that the lateral deflection would decrease with increasing spin. They computed trajectories for the cases when the ball was initially oriented at $\phi = 90°$ and spin imparted such that the ball rotated a quarter- or a half-revolution during its flight to the home plate (Fig. 80). Clearly, the pitch with the lower spin rate has the maximum deflection and change of direction and would therefore be the more difficult one to hit. Those experiments were performed at $U = 21$ m/s, which corresponds to Re of about 100,000. In Fig. 14, the baseball data are for a non-spinning baseball held in a symmetric orientation (the seams seen in Fig. 71 were in the horizontal plane and facing the flow). The critical Re is about 155,000 ($U = 30$ m/s) and so the flow regime in Watts and Sawyer's experiments was probably subcritical with laminar boundary layer separation. However, the present data suggest that if the ball was released at about 30 m/s (67 mph), there exists a possibility of generating a turbulent boundary layer over parts of the ball and hence a strong separation asymmetry and side force. Of course, note that the C_D versus Re trends and details of the generated flow fields will depend strongly on the seam orientation. And indeed, in some wind tunnel measurements of a nonspinning baseball, Aoki et al. (2002) found the critical *Re* to be about 110,000 (the orientation of the baseball was not reported). However, Sawicki et al. (2003) reported that the critical *Re* for spinning baseballs (data collected at the 1996 Atlanta Olympics) was about 160,000.

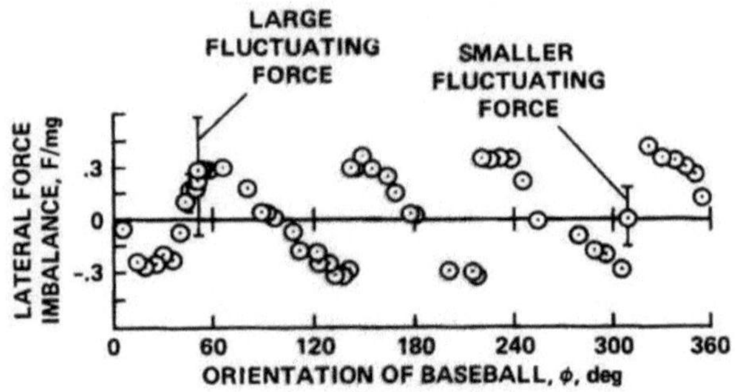

Figure 79. The variation of the lateral force imbalance with orientation of the baseball (Watts and Sawyer 1975, Mehta 1985).

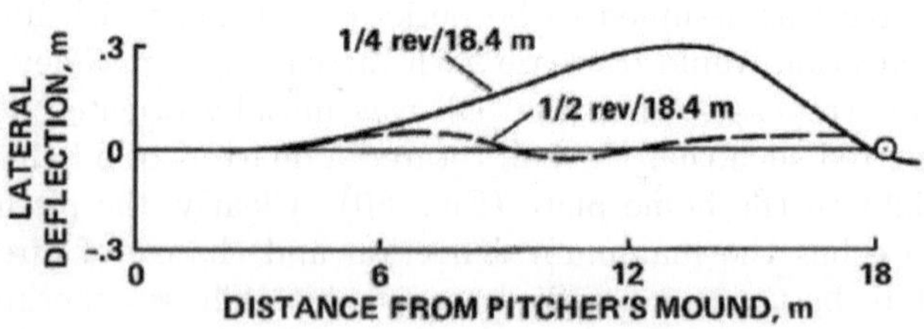

Figure 80. Typical computed trajectories for a slowly rotating baseball with $U = 21$ m/s (Watts and Sawyer 1975, Mehta 1985).

For the erratic flight of a knuckleball, Watts and Sawyer (1975) suggest that this could happen when the ball is so released that the seam lies close to the separation point. However, as Weaver (1976) rightly points out, a baseball thrown with zero or near-zero spin will experience a torque due to the flow asymmetry that will cause the ball to rotate. It therefore seems that it would be difficult to pitch a baseball in practice that maintains, for the whole flight, an attitude where the erratic lateral forces occur. Thus, sudden changes in flight path are not very likely to occur. And indeed, on studying the actual flight paths of professionally pitched knuckleballs (Allman 1983), it was found that while the direction of lateral deflection was unpredictable, there were no sudden, erratic changes in flight path.

7.4 Conclusions

Baseball aficionados were not convinced that a baseball could actually curve until the late 1940s when some visual evidence was obtained and published. Two basic aerodynamic principles are used to make a baseball curve in flight: spin about an axis perpendicular to the line of flight and asymmetric boundary-layer separation due to seam location on non-spinning baseballs. The magnitude of the aerodynamic forces (lift and drag) on spinning and non spinning baseballs have been measured in several investigations.

For the classic curveball, topspin is imparted about a near-horizontal axis and this produces a downward Magnus force which adds to the gravitational force and makes the ball drop faster than a ballistic trajectory. On the other hand, a fastball has backspin imparted to the ball which produces an upwards Magnus force. In baseball, this force is always less than about 50% of the ball's weight and so a rising fastball is purely a baseball myth. For a lot of the other pitches, such as a slider and screwball, spin is imparted about a near-vertical axis thus making the ball swerve sideways, The latest vogue regarding the "gyroball" is more hype than substance. If the baseball is thrown in the true advertised way, spinning along the axis of motion (like a bullet), then no additional aerodynamic lift or side force is generated. However, since the batter sees the ball spinning, there is a good chance of some deception.

The critical *Re* for a baseball is about 155,000 which corresponds to a ball velocity of about 30 m/s (67 mph). It is therefore no coincidence that knuckleball pitchers tend to release the ball at around this speed. Sudden erratic changes in the knuckleball direction are not likely to occur in practice. However, the ball can undergo changes in direction during flight. All other (spinning) pitches have higher velocities and they are therefore in the postcritical regime. This helps in generating more lift (or side) force due to the differential *Re* effects on the advancing and retreating sides.

ACKNOWLEDGEMENTS

I am indebted to the following publishers and establishments for allowing me to reproduce previously published and copyrighted material in this chapter: *Annual Review of Fluid Mechanics, International Sports Engineering Association (ISEA), New Scientist Magazine, ONERA, Routledge/Taylor & Francis Publishing Group, Springer Publishing Company, The Minerals, Metals and Materials Society [TMS], Woodhead Publishing Limited* and *University of Notre Dame.* I am also grateful to my good friend and colleague, Dr. Jani Macari Pallis for allowing me to use material from articles on which she was my coauthor.

Bibliography

E. Achenbach, *Experiments on the flow past spheres at very high Reynolds number*, *Journal of Fluid Mechanics*, 54, pages 565-575, 1972.

E. Achenbach, *Vortex shedding from spheres*, *Journal of Fluid Mechanics*, 62, pages 209-221, 1974a.

E. Achenbach, *The effects of surface roughness and tunnel blockage on the flow past spheres. Journal of Fluid Mechanics*, 65, pages 113-125, 1974b.

F. Alam, A. Subic & S. Watkins, *An experimental study on the aerodynamic drag of a series of tennis balls*, In *Proceedings of the International Congress on Sports Dynamics: Sports Dynamics-Discovery and Application*, pages 295 - 300, 1-3 September, Melbourne, Australia, 2003.

F. Alam, A. Subic & S. Watkins, *Effects of spin on aerodynamic properties of tennis balls*, In *The Engineering of Sport 5 (Hubbard M., Mehta, R.D. and Pallis, J.M., eds.)* Proceedings of the 5th International Sports Engineering Association Conference, Davis, California, USA. Vol 1, pages 83-89, 2004.

F. Alam, A. Subic & S. Watkins, *An experimental study of spin effects on tennis ball aerodynamic properties*, In *The Impact of Technology on Sport (Subic, A. and Ujihashi, S., eds.)*, pages 240-245, Tokyo, Japan, 2005a.

F. Alam, A. Subic & S. Watkins, *Measurement of aerodynamic drag forces of a rugby ball and Australian rules football*, In *The Impact of Technology on Sport (Subic, A. and Ujihashi, S., eds.)*, pages 276-279, Tokyo, Japan, 2005b.

F. Alam, W. Chee, A. Subic & S. Watkins, *A comparison of aerodynamic drag of a rugby ball using EFD and CFD*, In *The Engineering of Sport 6: Developments for Sports (Moritz, E. and Haake, S. eds.)* Proceedings of the 6th International Sports Engineering Association Conference, Munich, Germany. Vol 2, pages 145-150, 2006.

F. Alam, W. Tio, A. Subic & J. Naser, *Effects of spin on tennis ball aerodynamics: an experimental and computational study*, In *Proceedings of the 16th Australasian Fluid Mechanics Conference*, pages 324-327, 3-7 December, Gold Coast, Australia, 2007.

L.W. Alaways, *Aerodynamics of the curve-ball: an investigation of the effects of angular velocity on baseball trajectories.* Ph.D. dissertation, University of California, Davis, USA, 1998.

L.W. Alaways & M. Hubbard, *Experimental determination of baseball spin and lift*, *Journal of Sports Sciences*, 19, pages 349-358, 2001.

W.F. Allman, *Pitching rainbows*, *Science* 83, 3 (8), pages 32-39, 1982.

W.F. Allman, *Flight of the knuckler*, *Science* 83, 4 (5), pages 92-93, 1983.

K. Aoki, Y. Kinoshita, E. Hirota, J. Nagase, & Y. Nakayama, *The surface structure and aerodynamics of baseballs*, In *The Engineering of Sport 4 (Ujihashi, S. and Haake, S., eds.)* Proceedings of the 4th International Sports Engineering Association Conference, Kyoto, Japan, pages 283-289, 2002.

K. Aoki, M. Nonaka, & T. Goto, *Effect of dimple structure on the flying characteristics and flow patterns of a golf ball*, In *The Engineering of Sport 5 (Hubbard M., Mehta, R.D. and Pallis, J.M., eds.)* Proceedings of the 5th International Sports Engineering Association Conference, Davis, California, USA. Vol 1, pages 49-55, 2004.

S. Aoyama, *A modern method for the measurement of aerodynamic lift and drag on golf balls*, In *Science and Golf (Cochran, A.J., ed.)*, pages 199-204, E. & F.N. Spon, London, UK, 1990.

T. Asai, T. Akatsuka & S.J. Haake, *The physics of football*, *Physics World*, 11-6, pages 25-27, 1998.

T. Asai, K. Seo, O. Kobayashi, & R. Sakashita, *Flow visualization on a real flight non-spinning and spinning soccer ball*, In *The Engineering of Sport 6: Developments for Sports (Moritz, E. and Haake, S., eds.)* Proceedings of the 6th International Sports Engineering Association Conference, Munich, Germany. Vol 1, pages 327-332, 2006.

T. Asai, K. Seo, O. Kobayashi, & R. Sakashita, *Fundamental aerodynamics of soccer ball*, *Sports Engineering*, 10, No. 2, pages 101-109, 2007.

T. Bahill & D.G. Baldwin, *The rising fastball and other perceptual illusions of batters*, *Biomedical Engineering Principles in Sports (Hung, G.K. and Pallis, J.M., eds.)*, Kluwer Academic, New York, USA, pages 257-287, 2004.

S. Barber, S.J. Haake, & M.J. Carré, *Using CFD to understand the effects of seam geometry on soccer ball aerodynamics*, In *The Engineering of Sport 6: Developments for Sports (Moritz, E. and Haake, S., eds.)* Proceedings of the 6th International Sports Engineering Association Conference, Munich, Germany. Vol 2, pages 127-132, 2006.

N.G. Barton, *On the swing of a cricket ball in flight*, *Proceedings of the Royal Society London.* A, 379, pages 109-131, 1982.

P.W. Bearman, & J.K. Harvey, *Golf ball aerodynamics*, *Aeronautical Quarterly*, 27, pages 112-122, 1976.

D. Beasley, & T. Camp, *Effects of dimple design on the aerodynamic performance of a golf ball*, In *Science and Golf IV (Thain, E., ed.)*, pages 328-340, 2002, Routledge, London and NY.

J.H. Bell & R.D. Mehta, *Contraction design for small low-speed wind tunnels, NASA-CR 177488*, 1988.

K. Bentley, P. Varty, M. Proudlove, & R.D. Mehta, *An experimental study of cricket ball swing, Imperial College Aero Technical Note 82-106*, 1982.

L.J. Briggs, *Effect of spin and speed on the lateral deflection of a baseball, and the Magnus effect for smooth spheres, American Journal of Physics*, 27, pages 589-96, 1959.

A.M. Binnie, *The effect of humidity on the swing of cricket balls, International Journal of Mechanical Sciences*, 18, pages 497-9, 1976.

W. Bown, & R.D. Mehta, *The seamy side of swing bowling, New Scientist*, 139, No. 1887, pages 21-24, 1993.

L.O. Bowen, *Torque and force measurements on a cricket ball and the influence of atmospheric conditions, Transactions of Mechanical Engineering*, IE Australian, ME20, No. 1, pages 15-20, 1997.

K. Bray, & D.G. Kerwin, *Modeling the long throw in soccer using aerodynamic drag and lift*, In *The Engineering of Sport 5 (Hubbard M., Mehta, R.D. and Pallis, J.M., eds.)* Proceedings of the 5th International Sports Engineering Association Conference, Davis, California, USA. Vol 1, pages 56-62.

F.N.M. Brown, *See the wind blow, Aeronautical Engineering Department Report*, University of Notre Dame, South Bend, Indiana, USA, 1971.

T.M.C. Brown, & A.J. Cooke, *Aeromechanical and aerodynamic behavior of tennis balls*, In *Tennis Science and Technology (Haake, S.J. and Coe, A., eds.)* Proceedings of the 1st International Conference on Tennis Science and Technology. Blackwell Science, Oxford, UK, pages 145-153, 2000.

T.W. Cairns, *Modeling the Lift and Drag Forces on a Volleyball*, In *The Engineering of Sport 5 (Hubbard M., Mehta, R.D. and Pallis, J.M., eds.)* Proceedings of the 5th International Sports Engineering Association Conference, Davis, California, USA. Vol 1, pages 97-103, 2004.

M.J. Carré, T. Asai, T. Akatsuka, & S.J. Haake, *The curve kick of a football II: flight through the air, Sports Engineering*, 5, No. 4, pages 193-200, 2002.

M.J. Carré, S.R. Goodwill, S.J. Haake, R.K. Hanna, & J. Wilms, *Understanding the aerodynamics of a spinning soccer ball*, In *The Engineering of Sport 5 (Hubbard M., Mehta, R.D. and Pallis, J.M., eds.)* Proceedings of the 5th International Sports Engineering Association Conference, Davis, California, USA. Vol 1, pages 70-76, 2004.

M. Cavendish, *Balls in flight, Science Now*, 1, pages 10-13, 1982.

S.G. Chadwick, & S.J. Haake, *The drag coeffcient of tennis balls*, In *The Engineering of Sport. Research, Development and Innovation (Subic, A.J. and Haake, S.J., eds.)* Proceedings of the 3rd International Conference on the Engineering of Sport. Blackwell Science, Oxford, UK, pages 169-176, 2000a.

S.G. Chadwick, & S.J. Haake, *Methods to determine the aerodynamic forces acting on tennis balls in flight*, In *Tennis Science and Technology (Haake, S.J. and Coe, A., eds.)* Proceedings of the 1st International Conference on Tennis Science and Technology. Blackwell Science, Oxford, UK, pages 127-134, 2000b.

S.G. Chadwick, *The aerodynamics of tennis balls,* Ph.D. thesis, University of Sheffield, UK, 2003.

A. Chase, *A slice of golf*, *Science* 81, 2(6), pages 90-91, 1981.

T. Chikaraishi, Y. Alaki, K. Maehara, H. Shimosaka, & F. Fokazawa, *A new method on measurement of trajectories of a golf ball*, In *Science and Golf (Cochran, A.J. ed.)*, pages 193-198, E. & F.N. Spon, London, UK, 1990.

A. Cochran, & J. Stobbs, *Search for the Perfect Swing*, pages 161-162, Lippincott: Philadelphia/New York, USA, 1968.

A.J. Cooke, *An overview of tennis ball aerodynamics*, *Sports Engineering*, 3, No. 2, pages 123-129, 2000.

J.C. Cooke, *The boundary layer and seam bowling*, *The Mathematical Gazette*, 39, pages 196-199, 1955.

J.M. Davies, *The aerodynamics of golf balls*, *Journal of Applied Physics*, 20, pages 821-828, 1949.

C. Frohlich, *Aerodynamic drag crisis and its possible effect on the flight of baseballs*, *American Journal of Physics*, 52 (4), pages 325-334, 1984.

S.R. Goodwill, & S.J. Haake, *Aerodynamics of tennis balls - effect of wear*, In *The Engineering of Sport 5 (Hubbard M., Mehta, R.D. and Pallis, J.M., eds.)* Proceedings of the 5th International Sports Engineering Association Conference, Davis, California, USA. Vol 1, pages 35-41, 2004.

S.R. Goodwill, S.B. Chin, & S.J. Haake, *Aerodynamics of spinning and non-spinning tennis balls*, *Journal of Wind Engineering and Industrial Aerodynamics*, 92, pages 935-958, 2004.

C. Grant, A. Anderson, & J.M. Anderson, *Cricket ball swing - the Cooke-Lyttleton theory revisited*, In *The Engineering of Sport - Design and Development. (Haake, S.J., ed.)*. Blackwell Publishing, Oxford, UK, pages 371-378, 1998.

Guinness, *Guinness World Records, Millennium Edition*. Guinness World Records Ltd, Bantam, London, UK, 2000.

S.J. Haake, S.G. Chadwick, R.J. Dignall, S. Goodwill. & P. Rose, *Engineering tennis — slowing the game down*, *Sports Engineering*, 3, No. 2, pages 131 -143, 2000.

S.J. Haake, S.R. Goodwill, & M. J. Carré, *A new measure of roughness for defining the aerodynamic performance of sports balls*, *Journal of Mechanical Engineering Science*, Part C, 221, pages 789-806, 2007.

J.H. Horlock, *The swing of a cricket ball*, *ASME Symposium on the Mechanics of Sport*, 1973

A. Imbrosciano, *The swing of a cricket ball.*, *Project Report, Newcastle College of Advanced Education* Newcastle, Australia, 1981.

R.A. Lyttleton, *The swing of a cricket ball*, *Discovery*, 18, pages 186-191, 1957.

J. Maccoll, *Aerodynamics of a spinning sphere*, *Journal of the Royal Aeronautical Society*, 32, page 777, 1928.

R.D. Mehta, & D.H. Wood, *Aerodynamics of the cricket ball*, *New Scientist*, 87, No. 1213, pages 442-447, 1980.

R.D. Mehta, K. Bentley, M. Proudlove, & P. Varty, *Factors affecting cricket ball swing*, *Nature*, 303, pages 787-88, 1983.

R.D. Mehta, *Aerodynamics of sports balls*, *Annual Review of Fluid Mechanics*, 17, pages 151-189, 1985.

R.D. Mehta, *Cricket ball aerodynamics: myth versus science*, In *The Engineering of Sport. Research, Development and Innovation.* (*Subic, A.J. and Haake, S.J., eds.*) Blackwell Science, London, pages 153-167, 2000.

R.D. Mehta & J.M. Pallis, *Sports ball aerodynamics: effects of velocity, spin and surface roughness*, In *Materials and Science in Sports. (Froes, F.H. and Haake, S.J., eds.)*, pages 185-197, The Minerals, Metals and Materials Society [TMS], Warrendale, USA, 2001a.

R.D. Mehta & J.M. Pallis, *The aerodynamics of a tennis ball*, *Sports Engineering*, 4, No. 4, pages 1-13, 2001b.

R.D. Mehta & J.M. Pallis, *Tennis ball aerodynamics and dynamics*, In *Biomedical Engineering Principles in Sports (Hung, G.K. and Pallis, J.M., eds.)*, pages 99-124, Kluwer Academic/Plenum Publishers, Norwell, MA, USA, 2004.

R.D. Mehta, *An overview of cricket ball swing*, *Sports Engineering*, 8, No. 4, pages 181-192, 2005.

R.D. Mehta, *Swinging it three ways*, *The Wisden Cricketer*, 3, No. 7, pages 50-53, 2006a.

R.D. Mehta, *The unpredictable flight of the World Cup ball explained*, *Sports Traders Magazine*, Vol. 27, No. 4, page 38, August 2006b.

R.D. Mehta, *Cricket ball tampering*, *New Scientist*, Issue 2569, page 23, September 16, 2006c.

R.D. Mehta, *Swing is not Colour Deep*, *Cricinfo Magazine*, 1, 12, page 20, December, 2006d.

R.D. Mehta, *The art and science of ball tampering*, *Cricket International Quarterly Magazine*, 1, No. 1, pages 11-13, March, 2007a.

R.D. Mehta, *Does a white cricket ball swing more than a red one?*, *Cricket International Quarterly Magazine*, 1, No. 1, pages 42-43, March, 2007b.

A.M. Nathan, J. Hopkins, L. Chong, & H. Kaczmarski, *The effect of spin on the flight of a baseball*, In *The Engineering of Sport 6: Developments for Sports (Moritz, E. and Haake, S. eds.)* Proceedings of the 6th International Sports Engineering Association Conference, Munich, Germany. Vol 1, pages 23-28, 2006.

A.M. Nathan, *The effect of spin on the flight of a baseball*, *American Journal of Physics*, 76, Part 2, pages 119-124, 2008.

I. Newton, *New theory of light and colours*, *Philosophical Transactions of the Royal Society London*, 1, pages 678-688, 1672.

D. Oslear, & J. Bannister, *Tampering with Cricket,* Collins Willow (Harper Collins) Publishers, London, UK, 1996.

J.M. Pallis, & R.D. Mehta, *Tennis science collaboration between NASA and Cislunar Aerospace*, In *Tennis Science and Technology (Haake, S.J. and Coe, A., eds.)* Proceedings of the 1st International Conference on Tennis Science and Technology. Blackwell Science, Oxford, UK, pages 135-144, 2000.

J.M. Pallis, & R.D. Mehta, *Balls and ballistics*, In *Materials in Sports Equipment (Jenkins, M., ed.)*, pages 100-125, Woodhead Publishing Limited, Cambridge, England, UK, 2003.

J.M.T. Penrose, D.R. Hose, & E.A. Trowbridge, *Cricket ball swing: a preliminary analysis using computational fluid dynamics*, In *The Engineering of Sport. (Haake, S.J., ed.)* A.A. Balkema, Rotterdam, Holland, pages 11-19, 1996.

Lord Rayleigh, *On the irregular flight of a tennis ball*, *Messenger of Mathematics*, 7, pages 14-16, 1877.

W.J. Rae & R.J. Streit, *Wind-tunnel measurements of the aerodynamic loads on an American football*, *Sports Engineering*, 5, No. 3, pages 165-172, 2002.

T. Sajima, T. Yamaguchi, M. Yabu, & M. Tsunoda, *The aerodynamic influence of dimple design on flying golf ball*, In *The Engineering of Sport 6: Developments for Sports (Moritz, E. and Haake, S., eds.)* Proceedings of the 6th International Sports Engineering Association Conference, Munich, Germany. Vol 1, pages 143-148, 2006.

G.S. Sawicki, M. Hubbard, & W.J. Stronge, *How to hit home runs: optimum baseball bat swing parameters for maximum range trajectories*, *American Journal of Physics*, 71 (11), pages 1152-1162, 2003.

A.T. Sayers, & A. Hill, *Aerodynamics of a cricket ball*, *Journal of Wind Engineering and Industrial Aerodynamics*, 79, pages 169-182, 1999.

K. Sherwin, & J.L. Sproston, *Aerodynamics of a cricket ball*, *International Journal of Mechanical Engineering Education*, 10, pages 71-79, 1982.

A.J. Smits, & D.R. Smith, *A new aerodynamic model of a golf ball in flight*, In *Science and Golf II, (Cochran, A.J., ed.)*, pages 341-347, E. & F.N. Spon, London, UK, 1994.

A.J. Smits, *A Physical Introduction to Fluid Mechanics*, John Wiley & Sons, New York, NY, USA, 2000.

A.J. Smits, & S. Ogg, *Golf ball aerodynamics*, In *The Engineering of Sport 5 (Hubbard M., Mehta, R.D. and Pallis, J.M., eds.)* Proceedings of the 5th International Sports Engineering Association Conference, Davis, California, USA. Vol 1, pages 3-12, 2004a.

A.J. Smits, & S. Ogg, *Aerodynamics of the golf ball*, In *Biomedical Engineering Principles in Sports (Hung, G.K. and Pallis, J.M., eds.)*, pages 1-27, Kluwer Academic/Plenum Publishers, Norwell, MA, USA, 2004b.

J.P. Spampinato, N. Felten, P. Ostafichuk, & L. Brownlie, *A test method for measuring forces on a full-scale spinning soccer ball in a wind tunnel.* , In *The Engineering of Sport 5 (Hubbard M., Mehta, R.D. and Pallis, J.M., eds.)* Proceedings of the 5th International Sports Engineering Association Conference, Davis, California, USA. Vol 1, pages 111-117, 2004.

A. Stepanek, *The aerodynamics of tennis balls — the topspin lob*, *American Journal of Physics*, 56, pages 138-141, 1988.

P.G. Tait, *Some points in the physics of golf. Part I*, *Nature*, 42, pages 420-423, 1890.

P.G. Tait, *Some points in the physics of golf. Part II*, *Nature*, 44, pages 497-498, 1891.

P.G. Tait, *Some points in the physics of golf, Part III*, *Nature*, 48, pages 202-205, 1890b.

S. Taneda, *Visual observations of the flow past a sphere at Reynolds numbers between* 10^4 *and* 10^6, *Journal of Fluid Mechanics*, 85, pages 187-192, 1978.

M. Van Dyke, *An Album of Fluid Motion*, The Parabolic Press, Stanford, California, USA, 1982.

R.G. Watts & C. E. Sawyer, *Aerodynamics of a knuckleball*, *American Journal of Physics*, 43, pages 960-963, 1975.

R.G. Watts, & E. Sawyer, *The lateral force on a spinning sphere: aerodynamics of a curveball*, *American Journal of Physics*, 55, pages 40-44, 1987.

R.G. Watts, & A.T. Bahill, *Keep your eye on the ball: Curve balls, Knuckleballs, and Fallacies of Baseball*, W. H. Freeman: New York, NY, USA, 2000.

Q. Wei, R. Lin, & Z. Liu, *Vortex-induced dynamic loads on a non-spinning volleyball*, *Fluid Dynamics Research*, 3, pages 231-237, 1988.

H. Werlé, *Transition and separation - visualizations in the ONERA water tunnel*, In *Recherche Aérospace*, 1980-5, pages 35-49, 1980.

B. Wilkins, *The Bowlers Art*, A&C Black Publishers Ltd., London, UK, 1991.

M.V. Zagarola, B. Lieberman, & A.J. Smits, *An indoor testing range to measure the aerodynamic performance of golf balls*, In *Science and Golf II (Cochran, A.J. and Farally, M.R., eds.)*, pages 348-354, E&F.N. Spon, London, UK, 1994.

MIX
Papier aus verantwortungsvollen Quellen
Paper from responsible sources
FSC® C105338

If you have any concerns about our products,
you can contact us on
ProductSafety@springernature.com

In case Publisher is established outside the EU,
the EU authorized representative is:
Springer Nature Customer Service Center GmbH
Europaplatz 3, 69115 Heidelberg, Germany

Printed by Libri Plureos GmbH
in Hamburg, Germany